Statistics Translated

STATISTICS TRANSLATED

A Step-by-Step Guide to Analyzing and Interpreting Data

Steven R. Terrell

THE GUILFORD PRESS
New York   London

© 2012 The Guilford Press
A Division of Guilford Publications, Inc.
72 Spring Street, New York, NY 10012
www.guilford.com

Printed in the United States of America

This book is printed on acid-free paper.

Last digit is print number: 9 8 7 6 5 4 3 2 1

Library of Congress Cataloging-in-Publication Data

Terrell, Steven R.
  Statistics translated : a step-by-step guide to analyzing and interpreting data /
    by Steven R. Terrell. — 1st ed.
      p. cm.
    Includes index.
    ISBN 978-1-4625-0301-8 (pbk. : alk. paper)
    ISBN 978-1-4625-0321-6 (cloth : alk. paper)
  1. Statistics—Study and teaching. I. Title.
HA35.T47 2012
519.5—dc23

                                                          2011043366

For Dalia, my beloved wife

Acknowledgments

First, I would like to thank C. Deborah Laughton, Publisher, Methodology and Statistics, at The Guilford Press for her untiring support, ideas, critique, and insight during this project; there's nothing to say except she's the editor that every author dreams of. My job was made far easier by her professional and caring staff. I offer a special thanks to Mary Beth Wood, Louise Farkas, and others at Guilford for their tremendous help along the way. William Meyer—all I can say is "thank you so much." The book looks great and I truly appreciate what you've done! I am also most grateful to the following reviewers: Brian Withrow, Department of Criminal Justice, Texas State University; Tammy Kolb, Center for Education Policy Analysis, Neag School of Education, University of Connecticut; Cyndi Garvan, College of Education, University of Florida; Chris Ohana, Mathematics and Science Education Groups, Western Washington University; Robert Griffore, Human Development and Family Studies Department, Michigan State University; Christine McDonald, Department of Communication Disorders and Counseling, School, and Educational Psychology, Indiana State University; Val Larsen, James Madison University; Adam Thrasher, Department of Health and Human Performance, University of Houston; Melissa Gruys, Department of Management, Wright State University; Lynne Schrum, Elementary Education Program, George Mason University; Dave Edyburn, School of Education, University of Wisconsin–Milwaukee; and M. D. Roblyer, Instructional Technology and Distance Education, Nova Southeastern University.

Many of my friends played a role in this book without knowing it. When I was stuck for an idea for a case study, all I had to do was think of what they do in their professional lives, and the ideas flowed from there. Thank you; you guys are the best.

Finally, thank you to my students at Nova Southeastern University. By using earlier drafts of this book, you provided me the feedback that helped me write what you needed to know—statistics can be simplified.

Contents

Statistics Translated

Introduction: You Do Not Need to Be a Statistician to Understand Statistics!

One thing I have noticed about teaching statistics is that it is unusual to find a student taking my class on purpose; most of them are there only to fulfill the requirements of their chosen major. It is even rarer to find a student pleasantly anticipating what the semester will bring. Instead of beginning with students anticipating a great class, I am faced with people looking as if they are walking those last few steps to the gallows. They have heard the horror stories; they have no clue how statistics affects their lives, and many are not really interested in finding out!

Despite this, I look forward to each new class because I know I can make statistics easy, fun, and informative. I know this because, when I first studied statistics, I had exactly the same feelings as some of my students today. Many years later, I love this stuff! How did I get this way? I am glad you asked.

A Little Background

I took my first statistics class as a prerequisite to a graduate program. I am the first to admit that I didn't understand a thing. The formulas were mysterious, the professor even more so, and, for the life of me, I couldn't tell the difference between a *t*-test and a teacup. I struggled through, made a passing grade, and went on to graduate school. As part of my curriculum, I had to take another statistics class. Much like the first, I couldn't make heads or tails out of what the teacher was saying, and the C grade I earned was proof of that. "At least I passed," I thought. "I will never have to take statistics again!"

Never say never, I soon learned, for I had barely finished my master's degree when I decided to continue on for the doctorate. Faced with four required research and statistics classes, I resolved to do my best and try to make it through. I figured I had already passed the prior two classes and could prevail again. Besides, I thought, if I couldn't do the work, I didn't deserve the degree.

Although I entered my first doctoral class looking and feeling much like the students I described earlier, things soon took a change for the better. During the class I discovered the teacher was a "real person," and this "real person" was explaining statistics in a way I could understand. Not only could I understand what she was saying, I could understand the largely unintelligible textbook! I came out of the class feeling like a million dollars. I knew then I could, and would, earn my degree and nothing could stand in my way. While this might sound melodramatic, it is really ironic what that class did for me. Not only did I learn what I needed to learn, but I became so interested that I wound up taking quite a few extra statistics courses.

When I finished my doctorate, I liked what I was doing so much I convinced the dean where I was working to allow me to teach statistics to our students. I was really excited about teaching this subject for the first time, and, although that was over 20 years and many classes ago, I still love what I am doing. People look at me funny when I say this, but most of them are like the people I have described earlier. What they do not know is I have learned a few things that make it far easier to become a consumer of statistics; I will tell you what they are.

Many Students Don't Know What They're Getting Into

The first thing I have discovered is most people do not really know the definition of *statistics*. To me, it seems odd they would be sitting in a class where they do not know what they are about to study, but it occurs all the time. Although we will cover this topic in great detail later in the book, let me give you a simple definition for now:

> *The science of statistics is nothing more than the use of arithmetic tools to help us examine numeric data we have collected and make decisions based on our examination. The first of these tools,* descriptive statistics, *helps us understand our data by giving us the ability to organize and summarize it.* Inferential statistics, *the second tool, helps us make decisions or make inferences based on the data.*

Second, I have found that students greatly overestimate the proficiency in math they need in order to calculate or comprehend statistics. Whenever I find people worrying about their math skills, I always ask them, "Can you add, subtract, multiply, and divide? Can you use a calculator to determine the square root of a number?" More often than not, their answer is "yes" so I tell them, "That's great news! You have all the math background you need to succeed!" I have found that many people, when I tell them this, breathe a big sigh of relief!

Next, I have discovered that a class in statistics isn't nearly as bad as people anticipate if they approach it with an open mind and learn a few basic rules. Since most people want to be able to use statistics to plan research, analyze its results, and make sound statistically based decisions, they want to become *consumers of statistics* rather than statisticians per se. As consumers, they want to be able to make decisions based on data, as well as understand decisions others have made. With that in mind, this book will help you; all you have to do is approach it with an open mind and learn a few, easily understood steps.

Fourth, I have found students get confused by all of the terminology thrown around in a statistics course. For instance, for a given statistical problem you could find the correct answer by using procedures such as the Scheffé test, Tukey's HSD test, or the Bonferroni procedure. The truth is, you could use any one of these techniques and get very similar answers. I was recently at a meeting where a very prominent statistician jokingly said we do this "to confuse graduate students"; there certainly seems to be a lot of truth in what he said! These types of terms, as well as the ambiguity of having many terms that mean basically the same thing, seem to intimidate students sometimes. Knowing that, I like to stick to what students need to know most of the time. We can deal with the exceptions on an "as needed" basis.

Along these same lines, I have discovered many instructors try to cover far too much in an introductory statistics class, especially when they are dealing with people who want to be consumers of statistics. Research shows that 30 to 50% of studies published in prestigious research journals use a small set of common descriptive and inferential statistics. My experience, however, is that the average student tends to use the same basic statistical techniques about 90% of the time. Knowing that, we are left with a fairly manageable problem: we need to be able to identify when to use a given statistical test, use it to analyze data we have collected, and interpret the results. To help you do that, I have developed a set of six steps to guide you. Using these steps, you'll be able to work with the statistics you need the vast majority of the time.

A Few Simple Steps

You may be asking yourself, "Why are these steps so important? How could something so important be easy? Exactly what are these steps?" If so, we can begin to answer your questions with a short story:

Do you remember back in high school when you took beginning algebra? Can you remember what you did the first few days of class? If you can't, let me refresh your memory.

You spent the first few class sessions learning about the basic rules of algebra. These included topics such as the commutative property, the associative property, the distributive property, and other things that seemed really boring at the time. If you were like me, you probably questioned spending valuable time worrying about rules that did not seem to apply to algebra. If you were really like me, you didn't pay attention since you were saving your energy and attention until it could be used for "real" math.

Unfortunately, for many of us, our plan backfired. We soon discovered that learning algebra was a linear process and we had tried to start in the middle of the line; learning those boring rules was important if we were going to succeed. At that point, in order to understand what was going on, we found it necessary to go back and learn what we should have already learned. Some of us went back and learned; some of us didn't.

We can apply this analogy to my steps for statistics. In the beginning we will spend a lot of time discussing these six steps and the fundamentals of statistical decision making. After we get through the basics, we will move into the "real statistics." For now, just read through these steps and the brief description of what each step entails.

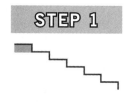

Identify the Problem

At the outset, researchers need to understand both the problem they are investigating and why investigating it is significant. They may begin by asking questions such as "How do I know this is a problem?", "Why is investigating this problem important?", "What causes this type of problem?", and "What can possibly be done to address this problem?" Once they have answered those questions, they then need to develop their *problem statement*. In essence, in this step, they are identifying "what" we are going to investigate; the remaining five steps will help us understand "how" to investigate it.

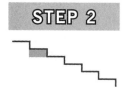

State a Hypothesis

A *hypothesis* is a researcher's beliefs about what will be found or what will occur when he or she investigates a problem. For example, let's suppose that our school is faced with lower than average scores on a statewide math test. We've heard that it is possible that the problem may be based on gender; apparently males do not perform as well in math as their female classmates, and this may cause the classes' overall average to be lower. Since we have considerably more males than females in our school, we then have to ask ourselves, "How can I find out if this is true?" Or, if we plan on using some type of intervention to increase the achievement of our male students, we could ask, "What do I think will occur based on what I do?" These questions lead us to make a statement regarding our beliefs about what will occur. As I said earlier, stating the hypothesis is the first step in understanding "how" we can investigate the problem.

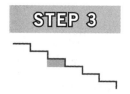

Identify the Independent Variable

When statisticians state a hypothesis, they have to identify what they believe causes an event to occur. The first major part of the hypothesis is called the *independent variable* and is the "cause" we want to investigate. In most instances, the independent variable will contain two or more *levels*. In our preceding example, our independent variable was gender, and it had two levels, male and female. In other cases, we may be investigating a problem that has more than one independent variable; the levels of each independent variable will still be identified in the manner we've just described.

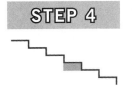

Identify and Describe the Dependent Variable

Once researchers identify what they believe will cause an event to occur, they then have to determine what they will use to measure its effect. This second major part of the hypothesis is called the *dependent variable*. The researcher will collect data about the dependent variable and then use *descriptive statistics* to begin to understand what effect, if any, the independent variable had on it. In our example, we're interested in student achievement—our dependent variable. Again, as we said, a hypothesis may have more than one dependent variable.

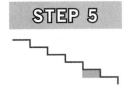

Choose the Right Statistical Test

Although there are many different inferential statistical tests, information about the independent variable, the dependent variable, and the descriptive statistics will, in most instances, tell the researcher exactly which test to use. We will learn to compute both the descriptive and inferential statistics manually and we will also learn to use computer software to assist us in becoming good consumers of statistics.

Use Data Analysis Software to Test the Hypothesis

Our computations or results from the statistical software will give the researcher information that can be used to test their hypothesis. The researcher can then determine if the independent variable actually had any effect on the dependent variable.

As you can see, these rules are very straightforward; you'll be surprised how comfortable and confident you will become as we discuss them.

Summary

As I have said, if you understand these six steps and what we are about to discuss, you can become a proficient consumer of statistics. Let me remind you that this isn't going to be an exhaustive overview of statistics. Again, about 90% of the time, students use a given set of statistics over and over. That's what this text is about— the common set of statistical tools that you, as a consumer of statistics, need to be familiar with.

In the next few chapters, we are going to go through each of these steps, one by one and explain, in detail, what is necessary to accomplish the step. After we have looked at each step in detail, we will then apply the knowledge we have gained by analyzing data and making decisions based on our analysis. Believe me, it is going to be far easier than you think, so let's get started!

Do You Understand These Key Words and Phrases?

consumer of statistics

descriptive statistics

independent variable

levels of the independent variable

dependent variable

hypothesis

inferential statistics

problem statement

CHAPTER 1

Identifying a Research Problem and Stating Hypotheses

Introduction

The first two steps in the six-step model should help good consumers of statistics understand exactly what they intend to investigate. As I said earlier, Step 1 focuses on "what" will be investigated by clearly defining the problem; Step 2 starts the process of describing "how" it will be investigated. Unfortunately, far too many people try to get to Step 2 without understanding how it relates to the problem at hand. Because of that, let's see what constitutes a good problem statement.

STEP 1 Identify the Problem

I always caution my students not to get caught up in worrying about hypotheses, variables, and statistical tests before they know and can accurately state the problem they want to investigate. In fact, my very words are "Stop running around with a solution looking for a problem!" Admittedly, I have also said, "Don't find a hammer and then look for a nail," and "Don't buy a plane ticket before you know where you want to go." The point is, I want them to ask, "Exactly what is the problem I want to investigate?" prior to ever thinking of "How will I collect and analyze data?"

At this point, many would argue that I'm crossing the line between a research methods book and one dedicated to statistical decision making. Their reasoning is that the problem statement leads to the development of a hypothesis only after the researcher has read the relevant literature, has learned as much as possible about the problem to be investigated, and can develop a sound basis for his or her hypothesis. My argument is, if you're a consumer of statistics, most of the time the problem statement will be clearly stated for you; if not, in most instances it's clearly implied.

I'll admit that this point is arguable, but the bottom line is this: whether we're consumers of statistics or conducting actual research, we still need to be able to recognize a good problem statement or develop one when necessary. Either way, keep in mind that, in Step 1, we are identifying "what" we are going to investigate; the remaining five steps will help us understand "how" to investigate it. Knowing that, we're going to move forward by discussing the characteristics of a good problem statement; we'll then learn how to correctly write a problem statement.

Characteristics of a Good Problem Statement

The problem statement represents the specific issue, concern, or controversy that we want to investigate. A good research problem for inferential decision making must meet six criteria:

1. The problem is interesting to the researcher.
2. The scope of the problem is manageable by the researcher.
3. The researcher has the knowledge, time, and resources needed to investigate the problem.
4. The problem can be researched through the collection and analysis of numeric data.
5. Investigating the problem has theoretical or practical significance.
6. It is ethical to investigate the problem.

Obviously, before we can determine if a problem meets these six requirements, we first need to learn how to find a problem to investigate; how do we do this? We can explain the entire process with an example.

Finding a Good Research Problem

Problems that we can investigate are all around us: they come from personal experience, issues at your workplace or institution, or readings about a topic you are interested in. Other potential areas for investigation can be found by attending professional conferences, speaking with experts in the field you are interested in, or replicating the work of others in an attempt to better understand or apply the results of research they have conducted.

For example, suppose I am working in a graduate school of psychology and our students' passing rate on the state licensure test is below average; that's clearly a problem I could investigate:

> ▧ *Our school has a lower than average passing rate on the state licensure test.*

This type of problem is generally called a *practical* research problem in that it focuses on an issue within our own school or organization. We may also investigate *research-based* problems: those that come from conflicts or contradictions in previous findings or a desire to extend the knowledge about a particular problem area. For example, let's say that, in our readings, we have found conflicting research on the effectiveness of using age-appropriate popular press material such as magazines to teach reading in elementary schools; we could easily write a problem statement such as

> There is conflicting research regarding the use of popular press materials to teach reading in our elementary schools.

Despite the source of the problem we're investigating or the type of research it represents, we must keep the six characteristics of a good problem in mind. As you're reading these characteristics, keep in mind that we're not going to look at them in any particular order; they all must be met before we can move forward.

The Problem Is Interesting to the Researcher

Many times, I run into students who simply want to finish a project, thesis, or dissertation; they don't care what they work on, as long as they're working on something. Unfortunately, this approach often fails simply because they are not interested enough to see it through to the end. Far too often this leads to students staying in school longer than anticipated or dropping out of school before they earn their degree.

For example, I once had a student who was near her wits' end trying to find a topic for her senior project. She was a very bright student, so I asked her to work with me on a project investigating achievement in elementary school science. She jumped at the chance to work on the project—not necessarily because she was interested but because she saw it as a way to graduate. Unfortunately, she did not finish the project with me; she simply didn't have the energy, effort, and enthusiasm she needed to work on something that didn't really interest her. Two good things did come from it, however. She did graduate after she found a suitable project, and I ultimately finished what I was working on; both of us met our goals because we were working on something we were interested in.

On the other hand, several years ago one of my favorite students approached me with a practical research problem. His school's passing rate in algebra was abysmal, and he wanted to investigate the effect of using laptop computers on achievement. I warned him against choosing this topic because I felt that using technology in a classroom had already been investigated time and time again. He assured me, however, that he was approaching the problem from a new perspective, so I agreed to let him continue. He was so excited and motivated that he finished the best dissertation I have ever chaired. He was interested in his problem, and his personal goal was to finish. He did an excellent job and his students' achievement went up; it was a "win-win" situation.

The Scope of the Problem Is Manageable by the Researcher

When my students tell me they cannot find a problem to work on for their project, thesis, or dissertation I tell them, "It's not that you can't find a problem, there are literally thousands to choose from. Your problem is that you cannot find one with a scope you can manage." For example, look at the following "problem statement" presented to me by one of my students:

■ *One of the most dynamic affordances of the Internet is the ability to offer educational programs to students throughout the globe. This has led to nearly 90% of all universities and colleges in the United States offering Internet-based courses. While convenient for students, research has shown that attrition from Internet-based classes is significantly higher than that from classes taught in a traditional environment.*

The problem here is certainly clear; students in Internet-based courses drop out more often than students in traditional classrooms. This certainly seems to be a serious problem and one that bears investigating. While important, however, is this something my student could investigate as written? Of course not! The scope is much too broad. In this case, it seemed that my student was suggesting an investigation of Internet-based education throughout the United States. In such cases, I tell my students not to attempt to "boil the ocean"; if the issue is so large that it cannot be easily investigated, focus on narrowing it down into something more specific and manageable. This idea is shown in Figure 1.1.

For example, the problem statement above could be easily reworded as a practical problem statement by simply stating

■ *The attrition rate in our school's Internet-based classes is higher than the attrition rate from traditional classes.*

By narrowing down the first statement, the student was able to create a problem statement that could be easily investigated.

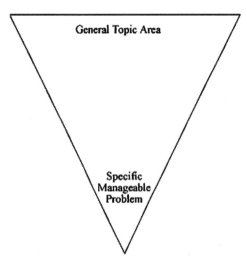

FIGURE 1.1. Narrowing down and focusing the problem statement.

The Researcher Has the Knowledge, Time, and Resources Needed to Investigate the Problem

This requirement is pretty much a "no brainer," isn't it? It only makes sense that the researcher meets all of these criteria; lacking any of them will ultimately either cause the research to fail or, at best, add significantly to the amount of time it would take to complete. I'll give you an example.

I teach a learning theory class in which one of the things we talk about is brain hemispheric preference; left-brainers tend to be more linear thinkers, whereas right-brainers tend to approach learning from a more holistic perspective. This really caught the attention of one of my students who wanted to investigate achievement in the introductory computer programming courses she taught. She told me that since programming is so linear and logical she felt that "lefties" would have significantly higher achievement than "righties." This viewpoint seemed interesting to me, but when I asked her how much she knew about cognitive psychology, she said the only thing she knew about hemispheric preference was what I talked about in class. Although she did successfully complete her dissertation, it took quite some time for her to get up to speed in the skills and knowledge needed. What did she ultimately determine? Interestingly, it didn't make any difference which side of the brain a student preferred; the only thing that mattered was if the student used his or her brain at all.

The Problem Can Be Researched through the Collection and Analysis of Numeric Data

This requirement is also quite obvious. Since our definition of statistics included "the use of arithmetic tools to help us examine numeric data," it only stands to reason that we cannot collect data from such things as interviews, transcripts, or recordings (these are commonly referred to as *qualitative* data). Later in the book we will talk about different types of numeric data, but for now let's just accept the fact that numeric data are required and move forward.

Investigating the Problem Has Theoretical or Practical Significance

Simply put, whether or not a problem has *theoretical* or *practical* significance means that it can pass the "who cares?" test. In looking back at the problems dealing with using the Internet and algebra, we can readily see that investigating each problem is worthwhile. Both studies were practically significant because they dealt with problems in the classroom; fortunately, both investigations led to positive results.

In other cases, we might investigate a problem that has theoretical significance. In my dissertation, for example, I investigated the following problem:

■ *Research has shown a causal relationship between high levels of intrinsic motivation and achievement. This study will investigate the use of alternate feedback to increase intrinsic motivation.*

I based my study on a theory that suggested different types of feedback would lead to higher student motivation. In doing that, I looked at differences between elementary school students using graphical "cause-and-effect" report cards versus traditional report cards. Unfortunately, my intervention did not work as planned. While that may sound like bad news, my dissertation chair pointed out that at least I knew that what I tried didn't work; I should try something different next time!

Let's look at a few more examples and determine if their investigation is either theoretically or practically significant:

■ *Patients being treated for depression use different-colored drinking glasses.*

I can picture it now; a student rushes up to me and says that he learned in a psychology class that rooms painted in brighter colors have been shown to lead to lower levels of depression in hospitalized patients. The student would then proclaim that an investigation of the effect of different colored drinking glasses would address both a practical and a theoretical problem by measuring depression between patients using brightly colored drinking glasses versus those who drink from clear or white glasses. After I picked myself up off the floor, I would explain to the student that there is a rather large difference in the size of a room and the size of a drinking glass; that means there is a much larger effect of color in the first! There would be absolutely no reason to investigate this "problem."

How about this?

■ *Since youth who receive their driver's license at 16 have a high number of accidents, the legal age for obtaining a license should be 18 or older.*

Again, this problem would seem to have some practical significance, but there are obviously two reasons we might not want to investigate it. First, is it really the age that affects the number of accidents, or is it the amount of experience the driver has? Doesn't it stand to reason that youngsters who receive their license at 18 would have, percentage-wise, the same number of accidents as those who started earlier? As beginning drivers, they would have the same level of experience as their 16-year-old peers. Second, this problem doesn't warrant a full-scale investigation; it would be easy to simply look at the percentage of drivers in each age group who have an accident to determine whether there is a difference.

Let's look at one last problem:

■ *Young females consistently outperform their male classmates when computers are used as part of their class work.*

At first glance, this problem seems like one that does not bear investigation; again, couldn't we simply compare grades between the two genders to see if it's true? We certainly could, but perhaps there is more here than meets the eye. This was an actual problem statement presented to me by one of my doctoral students. My first reaction was not to allow her to work on it, but she explained to me how it could have both practical and theoretical experience. She pointed out that the literature indicates, and her study subsequently showed, that one of the reasons this statement might be true is the way technology is used in the classroom. Young females tend to use computers as part of their assignments (e.g., spreadsheets, word processors). Young males, on the other hand, tend to be more interested in understanding how the computer actually works and playing games; they care less than their female peers about how they can actually use them as part of a given assignment. Based on her argument, I agreed that the problem had both practical and theoretical significance and so I allowed her to move forward.

It Is Ethical to Investigate the Problem

The history of research is filled with studies that were clearly unethical—from injecting viruses into terminally ill patients (the researchers' excuse was that they were going to die anyway) to simulating electric shock to investigate a person's reaction to authority. It's obvious to most researchers when a study is not ethical, but, at most institutions, standards have been established to approve research studies using human subjects before they are conducted. In most cases, the school, institution, or business where the research is to be conducted has personnel who will review the proposed research to ensure whether it should be approved. As a researcher, if you're not sure, always ask.

Writing the Problem Statement

As you will recall, in Step 1 of our six steps, we're interested in "what" the researcher plans to investigate; the remaining steps will describe how it will be investigated. We have just learned what a problem statement is and have discussed the characteristics of a good problem statement. The final piece of the puzzle is learning how to write a good problem statement. Again, some would argue that this crosses the line between a research methodology book and a statistics book. As I have said, as consumers of statistics, the problem statement will usually be clearly stated or implied, but we still need to be able to recognize one when we see it. Knowing that, let's look closer at the characteristics of a well-written problem statement.

Problem Statements Must Be Clear and Concise

The first thing to keep in mind as you write a problem statement is that the reader must understand what the problem is; the problem must be stated as clearly and concisely as possible. We alluded to this point in an earlier section, but let's look at another example to ensure we understand exactly what this means. It's common for me to receive a problem statement such as this:

> *Reduced or free lunches and self-esteem.*

What does this mean? Is what the author is trying to say clear and concise? In this case, it is not. In order to state the problem more clearly, the author needs to establish the relationship between the reduced or free lunches and student self-esteem. It could be presented in this manner:

> *This study will investigate the effect of free or reduced lunches on the self-esteem of children at Main Avenue Elementary.*

Thinking back, does it meet our six criteria? First, we can only hope that it's interesting to the researchers and that they have the necessary time, skills, and resources; if not, why would they want to investigate the problem? Next, the problem can certainly be analyzed through the collection and analysis of numeric data, and investigating it has practical significance. It appears that it would be ethical to investigate the problem, but, again, an institutional review panel could confirm that for us. Finally, the scope of the problem is manageable since it is limited to one elementary school.

How about this?

- *Epilepsy and diet.*

In this case, we don't know much about what the researchers are proposing to investigate, but it could be reworded to make it clearer:

- *Is there a relationship between the number of seizures experienced between children who follow a high-fat diet and those who do not?*

Here again, we're assuming the researcher is interested in the problem and has the necessary knowledge, skills, and time. Investigating this problem has both theoretical and practical significance, and numeric data can be collected. There is a bit of an issue with the ethics of investigating this problem in that (1) the subjects are children and (2) they are being treated for a medical condition; both of these raise a red flag with review boards and would be closely examined to ensure the health and well-being of the children while in the study. The scope of the problem, however, seems to be our biggest concern. Who are the children the author wants to work with? All of those in the United States? The particular state they are in? Who? In short, this problem statement should reflect a much narrower scope—one the researcher has access to and has the ability to work with.

The Problem Statement Must Include All Variables to Be Considered

The second criterion for writing a good problem statement is that each of the variables to be investigated must be included. For example, in the first problem statement above, we wanted to investigate the relationship between free or reduced lunch (the cause) and self-esteem (the effect). No other variables were included in the problem statement. In the second statement, we included diet type (the cause) and the number of seizures (the effect); again, no other variables were mentioned.

If we wanted to investigate the effect of different types of incentives such as higher pay, more vacation, or better office conditions on employee morale, would there be anything wrong with this problem statement?

- *This study proposes to investigate the relationship between different types of incentives given to employees and levels of motivation.*

In this case, we have included a variable that describes the cause—different types of incentives—but instead of a variable measuring morale, we have included one that measures motivation—these are two completely different things. Employees could be highly motivated for fear of losing their jobs, but morale could be very low. While this problem statement looked good at first glance, remember the first criterion: a problem statement must be clear and concise.

Finally, is this a good problem statement if we are interested in investigating the ability of a dog to be trained and whether the dog was purchased from a breeder or from a pet store?

■ *This study will investigate whether there is a relationship between where a puppy was purchased and its ability to be effectively trained.*

This statement seems to be OK; the location where the dog was purchased is the cause we want to investigate and the dog's ability to be trained is the effect. Thinking back in this chapter, however, are there other issues with this problem statement that should concern us? I would argue that there is no practical or theoretical significance in investigating the problem; after all, what difference does it make? When it comes to purchasing a dog, aren't there a lot more factors involved in selecting one than their potential to be trained? Besides, if the dogs are well trained, who cares where they came from?

The Problem Statement Should Not Interject the Researcher's Bias

Here again lies one of the biggest pitfalls that underlie research: far too many people want to "prove" things or conduct research that supports their personal beliefs or goals. For example, if I was arguing for the need for more technology in our public schools, I might write:

■ *This study will prove that purchasing computers to use in the elementary school curriculum will increase achievement.*

That's a fairly large assumption on my part, isn't it? Suppose I do conduct research and show that there is increased achievement after obtaining new technology? Have I proven anything? Absolutely not! There are far too many things that affect achievement: the students themselves, a change in the curriculum or new teachers who work in a manner different from their predecessors, and so on. At the same time, what would happen if achievement went down? Would we go back to Apple, Dell, or Gateway and ask for our money back? Of course not; we do not know if the technology had any effect on student achievement. There are far too many factors that influence achievement to let us assume we can "prove" anything. Instead of the problem statement above, what about the following?

■ *This study will investigate the effect of technology on achievement at Pinecrest Elementary School.*

In this case, we are very clear and concise. We have included all variables to be considered, in this case technology and achievement, and we have not interjected our personal bias. This seems to be a very good problem statement.

Summary of Step 1: Identify the Problem

Understanding and clearly stating the problem you are investigating is the first step in becoming a good consumer of statistics. As we discussed, the problem statement in Step 1 tells you "what" you are going to investigate, while the remaining five steps will describe "how" you will investigate the problem. Keep in mind, while formulating a good problem statement, you must ensure you meet the following criteria:

1. You must be interested in the problem you are investigating.
2. The scope of the problem you want to investigate must be manageable.
3. You must be comfortable in terms of the knowledge, time, and resources necessary for you to investigate the problem.
4. You must be able to collect and analyze numeric data.
5. There must be some practical or theoretical reason for you to research the problem.
6. It must be ethical for you to investigate the problem.

Once you are sure you have met these criteria and begin writing the actual problem statement, you must ensure that you are clear and concise, that all variables to be investigated are included, and that you do not interject your personal bias. Following all of these rules ensures that we have a problem statement we can use to state a hypothesis in Step 2 of our six-step model.

STEP 2 — State a Hypothesis

In this section, we want to accomplish a few things. First, we need to better define a hypothesis and then understand why and how hypotheses are stated for the problem we are investigating. After that, we will discuss *directional* and *nondirectional* research hypotheses and learn why statisticians always test hypotheses stated in the *null* format. After covering these basics, we are going to take a brief look at hypothesis testing. We will cover hypothesis testing in greater detail later in the book, but this overview will give us an idea of where we are going.

An Example of Stating Our Hypothesis

Let's suppose I am the principal of a high school and I have decided to spend the day reviewing scores from the most recent administration of a standardized science test called the ABC. Interestingly, as I am reviewing the science scores, a pattern seems to emerge. Students taking science in the morning appear to consistently score higher on the ABC than students taking science in the afternoon. Based on my observations, I surmise there is some type of meaningful relationship between the time of day students take science and how well they do on the ABC test.

Before moving forward, what have I just identified? The problem of course! Students taking science in the morning appear to score higher than students taking science in the afternoon. It meets all the criteria for being a problem I can investigate. I am interested in the achievement of my students, I am certainly comfortable with the problem area and the scope of the problem is manageable. The problem has practical significance, it is ethical to investigate, and I can collect numeric data for analysis. It sounds good so far . . .

What about the manner in which the problem is actually written? While it seems obvious, let's write it out to make sure:

> ▧ *Science achievement scores of the morning students are higher than science achievement scores of the afternoon students.*

Yes, this is very clear, and all of the variables to be considered are included; we're comparing the achievement of students in the morning to that of students in the afternoon. I have not interjected my bias; I have simply stated what I have observed. Throughout the rest of this chapter, we're going to focus on writing hypotheses. While we won't state the problem each time, as I said in the last chapter, it will be clearly inferred.

Given that, I decide to follow up on my observations by going to the library and try to determine if there have been studies investigating the effect of time of day on achievement. Much to my surprise, I discover a lot of information suggesting that students studying science in the morning seem to outperform students taking science in the afternoon. Based on this finding, my observations are supported; the time of day a student takes science might have an effect on his or her class achievement.

Now, based on what I have observed and read, I can make a statement about what I believe is occurring in my school. In other words, I can state a *hypothesis* since, by definition, a hypothesis is nothing more than a statement expressing the researcher's beliefs about an event that has occurred or will occur. A well-stated hypothesis has four requirements:

1. It must provide a reasonable explanation for the event that has occurred or will occur.
2. It must be consistent with prior research or observations.
3. It must be stated clearly and concisely.
4. It must be testable via the collection and analysis of data.

Knowing these four things, I can state a formal hypothesis based on what I have seen:

> ▧ *Students taking science in the morning have higher levels of achievement than students taking science in the afternoon.*

By reading this hypothesis, we can see it meets our four criteria. First, it is stated clearly and is consistent with both our prior observations and reading. It is a reasonable explanation for what has occurred, and we are able to collect data for statistical analysis. Given these things, it appears we have met the criteria—we have a well-stated hypothesis.

A Little More Detail

Before we move forward, it is a good idea to go into a little more detail about how we state hypotheses and what we do with them once we state them. We have seen that there are four basic rules, and we have looked at a good example that follows those rules. Some of these rules are very straightforward, but there are nuances we should clear up to avoid possible misunderstandings later on.

The Direction of Hypotheses

As we just saw, both my observations and the research suggested that students studying science in the morning have higher achievement than those taking science in the afternoon. Since it would contradict both my observations and readings, it would not make sense to hypothesize that the time of day during which a student takes science has no effect on achievement. Let's start by looking at how we can use directional hypotheses to test either a "greater than" or "less than" scenario.

Since hypotheses have to be stated to reflect our readings or observations, it is necessary to state either *directional* or *nondirectional* hypotheses. The use of a directional hypothesis shows we expect a "greater than" or "less than" relationship in our results. Nondirectional hypotheses indicate that we expect a difference in our results, but we do not know if it will be a "greater than" or "less than" relationship. We can use both types of hypotheses to make comparisons between data we collect and an exact value (e.g., the average body temperature of a group of patients to 98.6°), or we can use them to compare two or more groups of data we have collected (e.g., the scores on an achievement test between the students from two separate schools).

Using Directional Hypotheses to Test a "Greater Than" Relationship

In many instances, we might be interested in hypothesizing that one variable about which we have collected data is greater than an exact value. For example, if we administered an IQ test to a group of students, we might hypothesize that the average IQ of our students is greater than the national average of 100. This is a direct comparison of our collected data to an exact value, so we would use the following hypothesis:

■ *The average IQ of our students is greater than 100.*

When we are not comparing something against an exact value, we can state hypotheses that compare data collected from two or more groups. For example, with the hypothesis we stated concerning the difference in achievement between science students in the morning and afternoon, we expect the morning group to have higher scores than the afternoon group. We are not stating an exact value; we are just saying that one group will do better than the other:

■ *Students taking science in the morning have higher levels of achievement than students taking science in the afternoon.*

Using Directional Hypotheses to Test a "Less Than" Relationship

We can do the same thing when we are looking at a "less than" scenario. For example, suppose we are concerned with the rising number of disciplinary problems in our school and have read studies showing that students who are required to wear uniforms have fewer disciplinary problems than students in schools where uniforms are not required. Before requiring our students to wear uniforms, we decide to check this assertion for ourselves. It would be easy to identify schools in our district where uniforms are required and schools where uniforms are not required.

After identifying the schools, it would be a simple matter to collect the necessary data to investigate the following hypothesis:

■ *Children attending schools where uniforms are required have fewer disciplinary problems than children in schools where uniforms are not required.*

We can also make a comparison to an exact value. For example, imagine we are investigating the high number of dropouts from our doctoral program and want to know if our graduation rate is less than the national average of 50%. In order to conduct our study, we would state the following hypothesis:

■ *The graduation rate in our doctoral program is less than 50%.*

Nondirectional Hypotheses

In the preceding examples, the hypotheses were consistent with the literature or our experience; it would not make sense to state them in any other manner. What happens, though, when the prior research or our observations are contradictory?

To answer this question, let's use the debate currently raging over block scheduling as an example. For anyone who hasn't heard of block scheduling, it involves students in middle and high schools attending classes every other day for extended periods. For example, a student might be enrolled in six classes, each of which meets in a traditional class for 1 hour each day. In a school with block scheduling, the student meets for three 2-hour classes one day and meets for 2 hours with each of the other three classes the next day. Proponents feel that longer periods allow teachers to get into more detail and make for more meaningful class sessions. Opponents of block scheduling criticize it because they believe many teachers do not know how to effectively use a longer class period or, as is the case with many math teachers, because they believe students need to be exposed to their subject matter every day.

Instead of both sides of the issue giving us their personal beliefs or feelings, wouldn't it be a good idea to see what the literature says? Interestingly enough, the reviews of block scheduling are just as mixed in the academic research journals. Some articles suggest that block scheduling negatively affects student achievement, and others seem to show that it increases student achievement. Knowing this, how do we write our hypothesis? The answer is simple: we have to state a *nondirectional* hypothesis:

> ▪ *There will be a difference in achievement between students taking algebra in a block schedule and students taking algebra in a traditional schedule.*

In this example, the hypothesis implies no direction; instead, we are stating we believe a difference in algebra achievement will exist between the two groups. No "greater than" or "less than" direction is implied.

Just as was the case with the directional hypotheses, we can use a nondirectional hypothesis to make a comparison to an exact value. Suppose, for example, someone told me the average age for doctoral students in the United States is 42, but I did not feel that accurately describes my students. In order to investigate my belief, I ask each student their age and then test the following hypothesis:

> ▪ *The average age of my students does not equal 42.*

Notice here, again, I have not said "less than" and I have not said "greater than." All I care about is if a difference exists. Later in the book we will see that different statistical tests are used when we compare data we have collected against a specific value or compare data we have collected from two or more groups.

Hypotheses Must Be Testable via the Collection and Analysis of Data

Like a good problem statement, a good hypothesis is one about which we can collect numeric data. In order to analyze it, we will use inferential statistics; specialized, mathematical tools developed to help us make decisions about our hypotheses. Knowing that, it is imperative that we word our hypotheses in a way that reflects or infers the collection of some type of data.

For example, in our hypotheses involving the teaching of algebra, it would be easy to collect final grades and use them to make comparisons between block and traditional schedules. In our example about IQ, we could test our students and then compare their average score to the average national score. While these examples are very straightforward, we need to "throw in a monkey wrench" at this point.

Research versus Null Hypotheses

In each of the preceding examples, we have stated what is called a *research hypothesis* or, as it is sometimes called, an *alternate hypothesis*. The logic behind the name is very straightforward. First, it is stated based on prior research or observations, and, second, it is what the researcher wants to investigate. In order to address issues we will cover later, it is imperative that we state a *null hypothesis* for every research hypothesis we state. The null hypothesis is nothing more than the exact opposite of the research hypothesis and, like everything else, can best be explained with an example.

Stating Null Hypotheses for Directional Hypotheses

If we look back at our example in which we wanted to give an intelligence test to our students, we hoped to show that our students have higher than average levels of intelligence. In order to do that, we stated the following directional research hypothesis:

■ *The average IQ of our students is greater than 100.*

Obviously if something is not greater than 100, it could be less than 100 or exactly equal to 100. Since, in our research hypothesis, we are only interested in the "greater than" scenario, we can state our null hypothesis in this manner:

■ *The average IQ of our students is not greater than 100.*

It is just as easy to develop a null hypothesis for comparing the achievement of students taking science in the morning to students taking science in the afternoon. Our directional research hypothesis was

■ *Students taking science in the morning have higher levels of achievement than students taking science in the afternoon.*

In this case, the opposite of "higher levels" could either be less than or equal to:

■ *Students taking science in the morning do not have higher levels of achievement than students taking science in the afternoon.*

Using our example concerning school uniforms, our research hypothesis was

■ *Children attending schools where uniforms are required have fewer disciplinary problems than children in schools where uniforms are not required.*

Here we are doing exactly the opposite of the prior two research hypotheses; now we are using a "less than" research hypothesis. Given that, we will state our null hypothesis in the following manner:

■ *Children attending school where uniforms are required do not have fewer disciplinary problems than children attending school where uniforms are not required.*

Finally, in our example concerning doctoral graduation rates, we stated:

■ *The graduation rate in our doctoral program is less than 50%.*

Because we are only worried about the greater than scenario, our null hypothesis would read:

■ *The graduation rate in our doctoral program is not less than 50%.*

Issues Underlying the Null Hypothesis for Directional Research Hypotheses

Stating the null hypothesis for a directional hypothesis is a thorny issue. Although the manner we just used is technically correct, it is common practice to write the null hypothesis for a directional hypothesis to simply say that no difference exists. For example, we just stated the following research hypothesis:

■ *Children attending schools where uniforms are required have fewer disciplinary problems than children in schools where uniforms are not required.*

Its null counterpart was

■ *Children attending schools where uniforms are required do not have fewer disciplinary problems than children in schools where uniforms are not required.*

In thinking about this null hypothesis, it is actually hypothesizing that one of two things will occur:

1. Children attending schools where uniforms are required have *exactly* the same number of disciplinary problems as children in schools where uniforms are not required.
2. Children attending schools where uniforms are not required will have a *greater* number of disciplinary problems than students in schools where uniforms are required.

Given this, as well as other issues we will discuss shortly, it is better to state the null hypothesis for the directional hypothesis in the following manner:

■ *There will be no difference in the number of disciplinary problems between children attending schools where uniforms are required and children attending schools where uniforms are not required.*

Stating the null hypothesis in this manner ignores the "greater than" or "less than" condition stated in the research hypothesis by saying that no difference exists; doing so better reflects what we are actually trying to hypothesize. No difference would mean we have an equal number of disciplinary problems in the two groups. If we subtracted the average of one group from the average of the other group, the answer would be zero or "null." This may not seem logical right now, but it is the correct form for statistical decision making, and we will be stating it in this manner for the rest of the book.

Stating Null Hypotheses for Nondirectional Hypotheses

In our example where we compared achievement between students in a block schedule to students in a regular schedule, we had the following research hypothesis:

- *There will be a difference in achievement between students taking algebra in a block schedule and students taking algebra in a traditional schedule.*

Stating the null hypothesis for a nondirectional hypothesis is very logical; the exact opposite of "there will be a difference" is "there will be no difference." Putting that into a null format, you would write

- *There will be no difference in achievement between students taking algebra in a block schedule and students taking algebra in a traditional schedule.*

We can do the same thing with the nondirectional research hypothesis where we wanted to compare the age of our students to a known value:

- *The average age of my students does not equal 42.*

This one is not as straightforward, but think about it this way. Since we have the words "is different" in the hypothesis, the opposite of that is "is not different." We can put our hypothesis into the correct form by stating:

- *The average age of my students equals 42.*

Again, in this case, if the average age of my students is 42 and the average age of the population I am comparing it to is 42, when I subtract one from the other, the answer is zero. Zero obviously corresponds to null, again the name of the type of hypothesis.

A Preview of Testing the Null Hypothesis

Later in the book, we are going to discuss, in detail, the concepts underlying testing a null hypothesis. There are, however, a few basics we need to cover before we get there. In order to do so, let's suppose we are investigating math achievement between boys and girls and want to test the following null hypothesis:

- *There will be no difference in math ability between boys and girls.*

This would be easy to test; we could administer tests to the students and then look at the difference in scores between the boys and the girls, then determine if there is a difference between the two groups. That would be the correct thing to do, right? Unfortunately, it is not that simple and leads us to briefly talk about *sampling error*, one of the basic building blocks of inferential statistics.

Sampling Error

Any time we are comparing data we have collected to an exact value or to another set of data we have collected, there is only a very small chance they would be *exactly* the same. This, of course, could wreak havoc with our decision making unless we recognize and control for it. Let's look at a couple of examples to help get a better understanding of this problem.

Suppose, before we collect any data, we already know that the boys and girls have *exactly* the same ability level in math. This means, theoretically, that if we gave them a math exam, the average score for the boys should be *exactly* equal to the average score for the girls. While this could occur, there is a very high probability that the average scores of the two groups will *not be exactly* the same. This is because, although math knowledge would be the highest contributor to an individual score, many, many other factors might affect individual scores. Factors such as how well students slept the night before an exam, their mental attitude on the day of the exam, whether they had breakfast, or even how well they guessed on some answers might affect their score. Because of this, we could give the students the same exam repeatedly and two things would occur. First, the average score for both groups would change slightly each time, and second, in only a very few instances would the scores for the two groups be exactly equal. The fact that the overall average changes daily is caused by *sampling error*, as is illustrated in Table 1.1.

TABLE 1.1. Ability Level for Boys and Girls in Math

Day	Boys average	Girls average	Overall average
1	88	90	89
2	82	88	85
3	84	90	87
4	85	85	85

The same type of problem exists when we take samples of data from a large group. For example, suppose we have a group of 900 students and we are interested in finding the average reading ability for the entire group (we call all possible participants in a group the *population* and any value we know about the population is called a population *parameter*). Giving a reading examination to all 900 of them is possible, but testing a representative sample would be far easier (a *sample* is a subset of the population, and any value we know about it is called a sample *statistic*). Theoretically, if you randomly pick a sample of adequate size from a population, any measure you take of the sample should be about the same as the measure of the entire population. In this case, if we randomly chose a group of 30 students and tested them, their average reading ability should reflect that of the entire population. Unfortunately, again, our results will be affected by sampling error; let's use an example to demonstrate this.

Let's use a traditional range of zero to 100 for our reading scores. Suppose, however, that I already know the average reading ability is 70 for the 900 students I want to survey. I also know that the average score of 70 represents 300 students who scored 50 on the exam (we will refer to them as Group A), 300 who scored 70 on the exam

(Group B), and 300 who scored 90 on the exam (Group C). All of our data are shown in Table 1.2.

TABLE 1.2. Average Reading Scores for Three Groups

Group	N	Average
A	300	50
B	300	70
C	300	90
Population total	900	70

This means that, when I select 30 students as my random sample, I hope to get a set of scores with an average of 70, exactly that of the population. Because of the random nature by which I choose the students, however, the sample average will usually not be exactly equal to the population average. Most of the time, the average score of my sample will be about 70, but at other times I might randomly select more students with lower scores, resulting in an average less than 70. In other samples, a higher number of students with scores greater than 70 would result in a sample average larger than the population average. In short, the luck of the draw would affect the average of my sample; sometimes it would be exactly equal to the population average, but, in many instances, it would be higher or lower. Interestingly, however, the more samples we collect the closer the sample mean will be to the actual population mean. In fact, given an infinite number of samples, they would be identical. We will discuss that idea later in the book when we're introduced to the idea of a normal distribution, but we'll put that aside and move forward for now.

This idea of sampling error is further demonstrated in Table 1.3. In this table, I have created five samples by randomly selecting 30 scores from the population. In the first sample, I have 8 scores of 50 from Group A, 11 scores of 70 from Group B, and 11 scores of 90 from Group C. This gives me a sample average of 72, slightly higher than the population average of 70. In the second sample, however, you can see that I have more scores from the A and B groups than I did in the first sample; this causes my sample average to fall below 70. The number of values in each group are about the same in the third and fourth sample, but the average goes even lower in the fifth sample; a large number of scores from the A group have pulled our sample average down to 64. Again, this will even out over time, with most of the sample averages clustering around 70; with enough samples it would equal the population mean of 70 exactly. Despite that, it's clear that randomly selecting samples can cause misleading results.

TABLE 1.3. Five Samples of Average Reading Scores for Three Groups

Sample	A (50)	B (70)	C (90)	Average
1	8	11	11	72
2	12	12	6	66
3	9	9	12	72
4	10	10	10	70
5	14	11	5	64

In order to make sure we're getting the hang of this, look at Table 1.4. In the first column we have a population of scores where the average is 10. If I take random samples of five of the scores, shown in the remaining four columns, I compute sample averages ranging from 7 to 13. Again, this shows that random sampling can dramatically affect our calculations.

TABLE 1.4. The Effect of Random Sampling on Average Scores

Population scores	Sample 1	Sample 2	Sample 3	Sample 4
5, 6, 7, 8	5	11	5	8
9, 10, 11, 12	6	12	7	9
13, 14, 15	7	13	9	10
	8	14	11	11
	9	15	13	12
Population average = 10	Sample average = 7	Sample average = 13	Sample average = 9	Sample average = 10

Where Does That Leave Us?

Given these examples, whether we're comparing two samples to one another or one sample to an exact value, one of two things is true. First, any difference we find could be due to chance; we call this sampling error. Second, differences could be representative of a true difference between the groups and not be caused by sampling error. In the following examples, we will see the problem this creates for a statistician.

First, using our example of science students before and after lunch, let's imagine we have a population of 1,000 students and randomly select 50 students from both the morning and afternoon groups. After our calculations, we might find that the average score for the morning students was 89% and the average score for the afternoon students was 88%. Based on these scores, would we be comfortable in saying the two groups are really different, or are they different due to chance? It seems that although the morning students' scores are higher than the afternoon students' scores, the difference is so small it could be due to sampling error.

Right now, all we have are the average scores from the two groups, so we cannot make a sound decision; we will need to use inferential statistics to help us do that. We will then find that either the research hypothesis appears to be true and the difference is not due to sampling error, or that the research hypothesis appears to be false and the difference is due to sampling error. Because statisticians cringe when they hear words such as "true" and "false," let's look at a better way of describing our results.

Rejecting and Failing to Reject Our Hypothesis

The key phrase in the preceding paragraph was "the difference is due to sampling error" since it reflects the uncertainty of making decisions when sampling is involved. Because we are never 100% sure of our decision, we cannot use words such as "true," "false," or "prove" as they indicate certainty. Since sampling error is involved, we instead say we *reject* or *fail to reject* our null hypothesis. This means we believe the groups are different and we reject the null hypothesis, or they are not different and we fail to reject the null hypothesis. When we reject our null hypothesis, we are supporting the research hypothesis that corresponds to it. Similarly, when we fail to reject our null hypothesis, we are not supporting the research hypothesis. Table 1.5 shows the relationship between the null and research hypotheses.

TABLE 1.5. Actions Regarding the Null and Research Hypotheses

Action regarding the null hypothesis	Action regarding the research hypothesis
When we reject the null hypothesis, the groups appear to be different. Any differences appear to be due to reasons other than sampling error.	Support the research hypothesis; what we have hypothesized appears to be accurate. Any differences appear to be due to reasons other than sampling error (i.e., the differences are real).
When we fail to reject the null hypothesis, the groups appear not to be different. Any differences appear to be due to sampling error.	Fail to support the research hypothesis; what we have hypothesized does not appear to be accurate. Any differences appear to be due to sampling error.

Going back to our example, suppose we now found the average grade for students taking science in the morning was 89% and the average score for students taking science in the afternoon was 83%. Although we would need to verify these findings using the appropriate statistical test, we would feel fairly safe in saying the morning students had higher achievement than the afternoon students and it wasn't due to sampling error. Knowing this, we would reject the null hypothesis; the difference between the groups does not seem to be due to chance. By saying we reject the null hypothesis, we are saying we support our research hypothesis. We have to take it a step further though.

All Hypotheses Must Include the Word "Significant"!

In order to remember that we are looking for differences that are "real" and not due to sampling error, we have to reword our hypotheses in a way that reflects that. Changing our hypotheses is easy; all we do is include the word "significant." For our research hypothesis dealing with the difference between morning and afternoon students, this means that we would write the following:

▦ *Students taking science in the morning have significantly higher levels of achievement than students taking science in the afternoon.*

Our null hypothesis would read:

▦ *There will be no significant difference in levels of achievement between students who take science in the morning and students who take science in the afternoon.*

For our example of children wearing uniforms versus those children not wearing uniforms, our research hypothesis would be

■ *Children attending schools where uniforms are required have significantly fewer disciplinary problems than children in schools where uniforms are not required.*

Our null hypothesis would read

■ *There will be no significant difference in the number of disciplinary problems between children attending schools where uniforms are required and children attending schools where uniforms are not required.*

By adding the word "significant" to our hypotheses, we are supporting our desire to ensure that any difference we find when we are analyzing data is "real" and not due to chance. This idea is shown in Table 1.6.

TABLE 1.6. Actions Regarding the Null and Research Hypotheses When Differences Are Significant

Action regarding the null hypothesis	Action regarding the research hypothesis
When we reject the null hypothesis, the groups appear to be significantly different. Any differences appear to be due to reasons other than sampling error.	When we reject the null hypothesis, we support the research hypothesis; what we have hypothesized appears to be accurate. Any differences appear be "real" and due to reasons other than sampling error.
When we fail to reject the null hypothesis, the groups appear not to be significantly different. Any differences appear to be due to sampling error.	When we fail to reject the null hypothesis, we fail to support the research hypothesis; what we have hypothesized does not appear to be accurate. Any differences appear to be due to sampling error.

Statistical Words of Wisdom

Everything we just said can be summed up in one paragraph:

When you are calculating inferential statistics, always test the null hypothesis. When you test the null hypothesis, you are not trying to "prove" anything. You are simply trying to determine if your results cause you to reject your null hypothesis, thereby supporting your research hypothesis, or if your results cause you to fail to reject the null hypothesis, subsequently failing to support your research hypothesis.

These are statistical "words of wisdom." The task of learning and understanding statistics will be much easier if you do not forget them.

Summary of Step 2: State a Hypothesis

Stating the research or alternate hypothesis for a study you want to conduct is a straightforward process. After identifying the problem you want to investigate, you must state the relationship you predict will emerge. The prediction is based on either your observation of past performance or on other sources such as literature you have read. When you have determined the prediction you want to make, you need to state it as both a null hypothesis and a research hypothesis. The research hypothesis will be either directional or nondirectional, depending on what the literature or your experience tells you.

Testing your hypothesis means you will do one of two things. You will either reject the null hypothesis and support the research hypothesis, or you will fail to reject the null hypothesis and not support the research hypothesis. As you will see in the next two chapters, using statistics to test the hypothesis will require you to identify and describe the independent and dependent variables stated in the hypothesis. This will help you determine the exact statistical test you will need to use to test the hypothesis.

Do You Understand These Key Words and Phrases?

alternate hypothesis	directional hypothesis
ethical research	fail to reject the null hypothesis
hypothesis	nondirectional hypothesis
null hypothesis	parameter
population	practical significance
problem statement	qualitative data
quantitative data	rejecting the null hypothesis
research hypothesis	sample
sampling error	significantly different
statistic	theoretical significance

Quiz Time!

Before we move forward, let's make sure we've got a good handle on what we've covered by working on the problem statements and case studies below. Once you are finished with both sections, you can check your work at the end of the book.

Problem Statements

Evaluate each of the problem statements below; are all six criteria for a good problem statement met? Are they clear, do they include all variables, and do they avoid interjecting personal bias? If the problem statement does not meet the criteria, what is wrong?

1. Student absenteeism at East Highland High School.

2. Teachers in many older public schools are concerned with the toxic effects of mold and asbestos they believe may be present in their buildings. Teachers in this study will locate, inspect, and report on types of potential health hazards in their schools.

3. Due to ever-increasing concerns regarding toxins and other pollutants in public water supplies, many at-home vegetable growers are becoming leery of watering their plants with other than distilled water. This research paper will investigate the effect of distilled versus tap water on plant growth.

4. The tremendous growth of the beef industry in Central Iowa has left many smaller, family-owned ranches facing the possibility of bankruptcy. This study will investigate the attitudes of small ranchers toward three options: going into bankruptcy, merging with larger ranching conglomerates, or attempting to stay lucrative by developing specialty beef products aimed at niche markets.

5. Is there a difference in efficacy between different treatment therapies on levels of clinical depression?

6. Many elementary school teachers find it difficult to instill an interest in science and math into younger students. Research has shown that student engagement can be positively affected when students are interested in a given topic and understand the relevance of it. This study will focus on the effects of lesson plans based on different types of radiation and the ability to use them to measure interplanetary space.

7. This study will investigate average miles per gallon between three grades of gasoline in imported cars.

8. Medical journals have repeatedly discussed the relationship between prenatal lifestyle and the health of newborn infants. This study will investigate the effect of different diets on weight gain during pregnancy.

9. A comparison of crime rates in urban, suburban, and rural areas.

Case Studies

State both the research and null hypothesis for each of the following scenarios:

The Case of Distance Counseling

When a person feels he needs some type of counseling, he typically enlists the services of a local counselor. By working directly with the client, the therapist is able to address the problems and concerns with which the client is trying to cope. What happens when a person has time or location constraints and cannot find a professional to work with?

Some people feel the answer to this dilemma is "distance therapy." Much like technology-based distance education, the therapist is able to work with a client using Internet-based chat sessions, phone consultations, and e-mail. While many people debate the efficacy of this approach, very little research has been conducted on it; no one really knows if either counseling method works better or worse than the other.

Given what we know, suppose we wanted to determine if it takes therapists the same number of sessions to deal effectively with depression cases in a traditional versus a distance approach. How would we state our research and null hypotheses?

The Case of the New Teacher

A new teacher, straight out of college, was complaining to one of her friends. "I have the hardest time with my classes. I send home notes asking for parent–teacher conferences, and very few of the parents ever respond. I do not know what to do." Her friend, a teacher at a school in a very prominent area of town, told her "Didn't you know that's to be expected? Your school is in an area of town with a lower socioeconomic base. There are many reasons why parents there do not have the time to come meet with you. Since we are in a higher socioeconomic area, we experience exactly the opposite. When we send out a note asking to meet with them; most of our parents contact us within a day or two." The young teacher, not having much experience, then told her friend: "I can't believe that; parents are parents, and they should all want to hear what the teacher says about their child. I would like to collect data to help determine if there really are fewer parent–teacher conferences in my school than in yours." In order to do that, how would she state her null and research hypotheses?

The Case of Being Exactly Right

A local chemical processing plant has been built to handle production waste from local industries. The purpose of the plant is to effectively clean the waste to ensure nothing hazardous is emitted into the atmosphere. Their equipment must run at exactly 155° for the process to work correctly; anything significantly higher or lower will result in the inability to guarantee effective cleanliness. If engineers working at the plant were developing software to control the process, what research and null hypotheses would they work with?

The Case of "Does It Really Work?"

With the rise of distance education, more courses are being offered in an Internet-based format. In addition to the traditional general studies courses such as English and history, some colleges are starting to offer courses that have been traditionally lab-based. Knowing that, the biological sciences faculty at one small college approached the dean about developing an online anatomy and physiology course. The dean was receptive but cautioned the teachers that they would have to ensure the distance education students would have the same experiences and learn as much as their on-campus peers.

After developing and using the system for one semester, the faculty wanted to demonstrate to the dean that their efforts were successful. Using final exam scores as a measure of achievement, how would they state their null and research hypotheses?

The Case of Advertising

The superintendent of a small school district was worried about the low number of teachers applying for jobs within the system. "I can't understand it," she told a close friend. "We are spending several thousand dollars a month in advertising, and we are getting very few applicants." Her friend, the owner of a marketing consulting firm, asked several questions about the content of her advertisements and then asked where they were published. The superintendent quickly answered, "They are in the newspaper, of course; isn't that where all job announcements are placed?" The consultant explained to the superintendent that many job applicants look for positions advertised on the Internet. Many young people, she explained, do not bother to read the newspaper, and they certainly do not look for jobs there.

The superintendent decided it couldn't hurt to try advertising on the Internet and was interested in seeing if the number of applicants really increased. If she wanted to test the effectiveness of the advertisement by comparing the number of applicants in the month prior to the advertisement to the number of applicants in the month after the advertisement was placed on the Internet, how would she state her research and null hypotheses?

The Case of Learning to Speak

Teachers of foreign languages seem to fall into two groups. The first group believes that a full understanding of the formal grammar of a language is important before one even tries to speak it. The other group believes in "language immersion"; students naturally learn the grammar of a language when they are forced into a setting where they have no choice. If we wanted to compare the two methods to see if there was a significant difference in language skills after one year of instruction, how would we state the null and research hypotheses?

CHAPTER 2

Identifying the Independent and Dependent Variables in a Hypothesis

Introduction

Now that we understand how to state our hypothesis, let's take a closer look at it. By doing so, we will find much of the information that will ultimately help us select the appropriate statistical tool we will use to test it. Let's refer to one of the research hypotheses we stated in Chapter 1 to get a better feel for what we are trying to do:

> Students taking science in the morning have significantly higher achievement than students taking science in the afternoon.

In this hypothesis, we can see that we are looking at the relationship between the time of day students take science and their performance in the course. Although we ultimately have to rely on statistical analysis to confirm any relationship, for now we can think of our *independent variable* as the "cause" we are interested in investigating and our *dependent variable* as the "effect" we want to measure. In this chapter we'll do several things; first we'll talk about identifying the independent and dependent variables, and from there we'll move to a discussion of different data types and how we can statistically describe them.

STEP 3

Identify the Independent Variable

In the preceding example, identifying the independent variable is easy; we are interested in the time of day a student takes science. It's important to understand, however, that independent variables can be developed in two different ways. First, if we are interested in determining the effect of something that occurs naturally and doesn't require intervention by the researcher or statistician, they are called *nonmanipulated independent variables* (these are sometimes called *quasi-independent variables*). In cases where the researcher has to actively manipulate a variable by randomly setting up groups or assigning objects to a class, the independent variable is described as being *manipulated* or *experimental*.

Nonmanipulated Independent Variables

In our hypothesis where we stated that students taking science in the morning have significantly higher achievement than students taking science in the afternoon, we didn't assign students to a particular science class; we looked at the existing classes. This is a perfect example of a nonmanipulated independent variable since the researcher played no part in determining which subjects were in either group.

What is the independent variable in the next hypothesis?

■ *Male students skip significantly more days of class than female students.*

The independent variable is gender and, since we do not assign students to one gender or the other, it isn't manipulated. We are interested in seeing the relationship between gender (the independent variable or cause) and the number of skipped classes.

How about this hypothesis?

■ *College freshmen have significantly lower grade point averages than students in other classes.*

We are not assigning individual students to the various classes; we are using their current status. In this case, "year in college" is the nonmanipulated independent variable, and we are trying to determine its effect on grade point average.

What is the independent variable in this hypothesis?

■ *Teachers with a master's degree earn significantly more than teachers with only a bachelor's degree.*

In this case, the type of degree earned is the independent variable and we are not assigning teachers to either group; either they have a master's degree or they do not. The teachers' salaries are the dependent variable.

Another Way of Thinking about Nonmanipulated Independent Variables

There is another way of developing a nonmanipulated independent variable that is somewhat tricky because it sounds like the researcher is intervening. Suppose, for instance, we had a researcher who was interested in investigating the following hypothesis:

■ *Low-achieving math students spend significantly less time on homework than high-achieving math students.*

This is pretty much a "no-brainer" hypothesis, but it serves to illustrate the point. Because we are going to compare two groups, low achievers and high achievers, our independent variable has to be achievement. Our only problem is, how do we determine which students we are going to label as low achieving and which we are going to label as high achieving?

This can be easily done. Suppose, for instance, we have the results of a recent administration of a standardized math test. We can label those as falling into the lower 50% as low achievers and those falling into the upper 50% as high achievers. While this may sound like we are manipulating the independent variable, we really are not. Although we have decided the parameters of group membership, the students naturally fell into either the low-achieving or high-achieving group by their own performance.

It is important to understand we can divide the students by whichever standard we choose, as long as it logically makes sense. For example, we could have also divided the students into four ability groups or arranged them so that anyone in the top 30% was labeled as a high achiever and everyone else was labeled as low achievers. Regardless of how we do it, setting up the independent variable in this manner is justified.

Manipulated or Experimental Independent Variables

In other cases, the researcher has to actively assign participants to one group or another in order to test the stated hypothesis. For example, if we wanted to test the effect of computer-based instruction versus lecture-based instruction on accounting students, we might state the following hypothesis:

■ *Students taking accounting in a computer-based format will have significantly higher achievement than students taking accounting in a lecture-based format.*

Notice, in this case, that we have stated a directional research hypothesis and, if the classes haven't already been formed, we could randomly assign part of the students to the computer group and the others to the lecture group. Since we have actively intervened by creating the groups, this is a manipulated, or as it is sometimes called, an experimental independent variable. Remember, though, if the classes already existed, it would be a nonmanipulated independent variable.

Now, suppose we are interested in looking at the effect of taking an after-lunch nap on the behavior of children in kindergarten. Further, suppose that we have six classes and decide to randomly assign three classes to the "nap" group and three classes to the "no nap" group. In this case, we are dealing with existing classrooms, but we are still randomly assigning them to either the "nap" or "no nap" group. This meets the criteria for a manipulated independent variable, and our progress so far is shown in Figure 2.1.

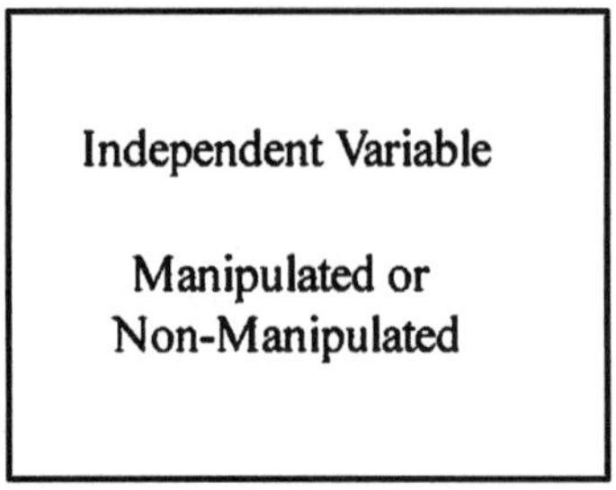

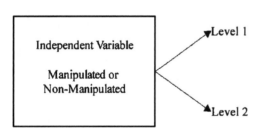

FIGURE 2.1. Types of independent variables.

Levels of the Independent Variable

Although we have already alluded to it, we need to develop an understanding of the relationship between the *levels* of the independent variable. The levels are nothing more than the values that the independent variable can represent. Using our example of science students taking class in the morning and in the afternoon, we will be able to understand this better.

If someone were to ask us what our independent variable in the study is, we would tell them it is the time of day instruction is delivered. In this case, we have identified two levels: "morning" and "afternoon." Since we are using two preexisting groups, they are nonmanipulated.

If we were to look at the study where we divided students into a computer group and a lecture group, we would be actively intervening in the process. The independent variable could be called "type of instruction." Since we actively placed students in groups, it is a manipulated or an experimental variable and has two levels: students in the lecture group and students in the computer group. This means, of course, that we need to expand our prior picture, as shown in Figure 2.2. When looking at Figure 2.2, we should not assume that all independent variables have only two levels. Theoretically, an independent variable can have an infinite number of levels but, in reality, seeing an independent variable with more than four or five levels is unusual. Although we won't discuss it too much in this book, we can also have more than one independent variable. As we will see later, the number of independent variables helps us to select the statistical test we will employ to test our hypothesis. For now, let it suffice to say, if we have more than one independent variable, we will be using *multivariate* statistics. Hypotheses having only one independent variable may qualify as *univariate* statistics, but that will depend on the number of dependent variables involved. That will be discussed in Step 4 of our six-step model.

FIGURE 2.2. Levels of the independent variable.

Summary of Step 3: Identify the Independent Variable

The independent variable is the "cause" we want to investigate when we state our hypotheses. As we saw, variables are nonmanipulated (quasi-independent) if we are investigating a cause that occurs naturally; they are manipulated (experimental) if we are actively involved in assigning group or treatment membership. Whether they are manipulated or not manipulated, independent variables must have at least two levels (we will see one slight exception to that rule later in the book), and we are interested in determining the "effect" that each of these levels has on our dependent variable. Hypotheses with only one independent variable generally use univariate inferential statistics, while those with more than one independent variable use multivariate statistics. All of this information, along with the ability to identify and describe dependent variables, will help us pick the correct statistical tool to use in testing our hypotheses.

STEP 4 — Identify and Describe the Dependent Variable

Unlike Steps 1–3, it will take us several chapters to cover everything we need to in order to accomplish Step 4. The remainder of this chapter focuses on identifying the dependent variable, determining the type of data it represents, and computing measures of central tendency. The next two chapters discuss measures of dispersion and relative standing as well as ways by which we can graphically display our data. Using all of that information, we'll be able to move to Step 5, choosing the right statistical test for our hypothesis. For now, however, let's look at the dependent variable.

Identifying Your Dependent Variable

As we've already said, the dependent variables represent the "effect" we want to measure. This might include such things as academic achievement, sales, grade point average, weight, height, income, or any of the other myriad things about which we like to collect data. The cause leading to this effect, of course, is the independent variable and its levels. We can add this at Figure 2.3 to make this idea even clearer.

We can better understand this concept by looking at the following research hypothesis:

> ▪ *Businesses that advertise only on television will have significantly higher sales than businesses that advertise only in print or only on the radio.*

What is the independent variable in this hypothesis? Remember, when we are identifying the independent variable we are looking for the "cause" we want to inves-

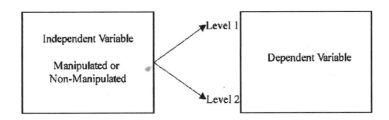

FIGURE 2.3. The relationship between independent and dependent variables.

tigate. In this case, we could label it "advertising." What are the levels of the independent variable? In this case we have three: (1) television, (2) radio, and (3) print. The dependent variable, or effect, is what we are measuring. In this case, it is the total amount of sales for the businesses in each category represented by the three levels of the independent variable. In short, we are trying to investigate the effect of the advertising method (the cause) on the amount of sales (the effect).

Using that same process, identify the independent variable, its levels, and the dependent variable in the following hypothesis:

■ *Customer satisfaction for airlines that charge customers for checking their bags will be significantly lower than customer satisfaction of airlines that do not charge for bags that are checked in.*

In this case, we're trying to investigate cause and effect by looking at the difference in customer satisfaction toward airlines that charge for baggage and those that don't. We could label our independent variable "airline," and there are two levels: those airlines that charge customers to check their baggage and those that do not. The dependent variable is customer satisfaction.

Again, in some cases we may have more than two levels of the independent variable:

■ *There will be a significant difference in the amount of rainfall in each of the four seasons.*

Here, our independent variable is "season," and there are four levels; fall, winter, spring, and summer. Our dependent variable would require that we measure the amount of rainfall during each of the four seasons.

Let's look at one more hypothesis before moving forward:

■ *There will be a significant difference in the number of absences and dropouts between students in a rural school district, students in a suburban school district, and students in an urban school district.*

In this case, the independent variable is "school district" and there are three levels: rural, suburban, and urban. What about the dependent variable? Wait a minute: this looks odd, doesn't it? We seem to have two dependent variables: absences and dropouts. How do we handle this? Actually, this is a very common problem. As we'll see later in the book, just as is the case of multiple independent variables, any time we're dealing with more than one dependent variable we will use multivariate inferential statistics to test our hypotheses. We won't concern ourselves with that idea just yet, so let's move forward; our next task involves understanding the type of data our dependent variable represents.

What Type of Data Are We Collecting?

If you are like me, I was surprised to find that there are four distinct types of data: *nominal data*, *ordinal data*, *interval data*, and *ratio data*. These data types are hierarchical from the lowest (nominal) to the highest (ratio) and each level has a unique set of properties.

Nominal Data

Nominal data, sometimes called *categorical* or *discrete*, are data that are measured in categories. For example, if we are interested in determining the number of males and females in our class, the data value "gender" is nominal in nature. In doing this, we only need to count the number of persons falling into each category. The important thing to remember is that the categories are mutually exclusive because the items being counted can only fall into one category or another. For example, a person cannot be counted as both a male and a female. Other examples of nominal data include ethnicity, college class, or just about any other construct used to group persons or things. In Table 2.1, we can see where we have asked 100 people the political party to which they belong. Each of the parties represents a distinct category; a person cannot belong to more than one.

TABLE 2.1. Nominal (Categorical) Data Representing Political Party

Republican	Democrat	Independent	Other
40	35	20	5

In Table 2.2, we have asked 500 students to tell us the primary method they use to get to school. There are five possible choices and, since we have asked for only the primary method of transportation, a student could only choose one. Since all we want to know is how many students fall into each group, it meets the definition of nominal data.

TABLE 2.2. Nominal (Categorical) Data Representing Primary Mode of Transportation

Drive	Carpool	Public transportation	Walk	Bike
210	105	85	62	38

Ordinal Data

The second type of data are called ordinal data or, as it is sometimes called, *rank* data. An example we are all familiar with is the use of grade point average to assign positions within a class. The student with the highest grade point average is ranked first, with the remainder of the students ranked in order down to the student with the lowest grade point being ranked last. There are a couple of things about ordinal data that are worthwhile to remember.

First, there can be ties in rankings. Suppose, for instance, we were asked to pick the class president based on scores on the SAT. We might find that the top three students all have scores of 800 while the fourth-place person has a SAT score of 400. If we were to rank them (shown in the following table), we would see that three people are tied for first place and the fourth-place person lags a considerable distance behind. Knowing that, it is always wise to remember that rank tells you relative position and nothing else.

Second, you can see that Student D is ranked fourth and you might ask yourself, "Why fourth? The last ranking we have is first." You are right, but we had three students tied for first place; that means they actually cover the first, second, and third places in the rankings. In order to find the fourth ranking, we have to skip past those three values and look at the value in the fourth position. This is confusing sometimes, but look at Table 2.3 closely and it will become apparent very quickly.

TABLE 2.3. Ordinal (Rank) Data for SAT Scores

Student	SAT	Rank
A	800	1
B	800	1
C	800	1
D	400	4

It is also important to remember that, while students are in rank order, the position between their ranks is not proportionate. In Table 2.4, you cannot judge one student's ability based solely on her position in the list. The first three students have SAT scores of 800, 799, and 798, respectively. Although it is highly unlikely, the fourth student in the list might have an SAT score of only 300. While the student could surely claim, "I have the fourth highest SAT score in my class," a college admissions officer might look at the actual score and ask "So what?"

TABLE 2.4. Ordinal (Rank) Data for SAT Scores

Student	SAT	Rank
A	800	1
B	799	2
C	798	3
D	300	4

Interval Data

Interval data are the first of two types of data that are called *quantitative* or *continuous*. By this, we mean that a data value can hypothetically fall anywhere on a number line within the range of a given dataset. Test scores are a perfect example of this type of data since we know, from experience, that test scores generally range from 0 to 100 and that a student taking a test could score anywhere in that range.

As suggested by its name, the number line below represents the possible range of test scores and is divided into equal increments of 10. Given that, the 10-point difference between a grade of 70 and a grade of 80 is the same as the difference between a grade of 10 and a grade of 20. The differences they represent however, shown in Table 2.5, are not relative to one another. For example, if a person scores 80 on the exam, his grade is four times larger than that of a person who scores 20 on the exam. We cannot say, however, that the person who scores 80 knows four times more than the person who scores 20; we can only say that the person with the 80 answered 80% of the questions correctly and that the person scoring 20 answered 20% of the questions correctly. We also cannot say that a score of zero means they know nothing about the subject matter, nor does a score of 100 indicate absolute mastery. These scores only indicate a complete mastery or ignorance of the questions on the examination.

TABLE 2.5. Test Scores

0	10	20	30	40	50	60	70	80	90	100

The point is that even though the score data are measured in equal intervals on this type of scale, the scale itself is only a handy way of presenting the scores without indicating the scores are in any way related to one another. Other examples of interval-level data include temperature, aptitude scores, and intelligence quotients. These ideas are shown in Table 2.6.

TABLE 2.6. Interval Data for Temperatures

School	Fahrenheit temperature	Comment
University of Buffalo	Zero	This doesn't mean there is no temperature in Buffalo; this is just an arbitrary value where the temperature in Buffalo falls on the scale. There could be a value below zero.
University of Virginia	25	This doesn't mean that Virginia has a temperature and Buffalo doesn't. The two temperatures are relative to the scale and nothing else.
University of Alabama at Birmingham	50	Birmingham is not twice as warm as Virginia. Just like above, this is simply where the temperature for Birmingham falls on the scale.
University of Texas at El Paso	100	El Paso is not four times as hot as Virginia, nor is it twice as hot as Birmingham. Having lived there, I know this is actually not that hot!

Ratio Data

Ratio data are also classified as quantitative or continuous data. Ratio data differ from interval data because they do have an absolute zero point and the various points on the scale can be used to make comparisons between one another. For example, weight could be measured using an interval scale because we know that the relative difference between 20 pounds and 40 pounds is the same as the relative difference between 150 pounds and 300 pounds; in both cases the second value is twice as large as the first. There are, however, three important distinctions between interval and ratio data.

First, a value of zero on a ratio scale means that whatever you are trying to measure doesn't exist; for example, a swimming pool that is zero feet deep is empty. Second, ratio data allow us to establish a true ratio between the different points on a scale. For example, a person who owns 600 shares of a company has twice as many shares as a person owning 300 shares; the same person would own six times as many shares as a person owning 100 shares. Third, this added degree of precision allows us to use ratio scales to measure data more accurately than any of the other previously mentioned scales. Other examples of ratio-level data include distance and elapsed time; one of my favorite examples, annual income, is shown in Table 2.7.

TABLE 2.7. Ratio Data for Income

Job	Annual income	Comment
Student	$0.00	This student is making absolutely nothing—he has a lack of income. That's OK; he can study now and earn money later.
Textbook author	$25,000	Believe me; we are only in this for the love of writing. This is a very optimistic estimate!
College professor	$50,000	This is another profession where you better not be in it for the money. At the same time, look at the bright side; you're making twice as much as a textbook author.
Medical doctor	$250,000	This is better. Ten times as much as the author and five times as much as the college professor. Did I mention it takes longer to get a PhD than it does to get an MD?
Chairman of Microsoft	$1,000,000,000	This is where we want to be, a billion dollars. The college professor's salary times 20,000! It is even 4,000 times what the medical doctor makes! For those of you majoring in education and psychology, forget it. It is probably too much trouble keeping up with that much money anyway.

Data Types—What's the Good News?

The first bit of good news is that we're making a lot of progress in our six-step model. We can identify an independent variable and its levels, as well as identify a dependent variable and the type of data we are collecting. This is shown in Figure 2.4.

There is one more bit of good news that will help us in our journey toward becoming a proficient consumer of statistics. Once we are able to identify the different types of data, there is one general rule that you need to keep in mind that will really help narrow down the type of statistical tool you're going to use. Generally speaking, if the data you collect are quantitative, we will use *parametric statistics* when we begin our inferential decision making. These include many of the names you might be familiar with—for example, the *t*-test and the analysis of variance. If the data are nominal or ordinal, we will use *nonparametric statistics* to help us make decisions. With the exception of the chi-square test, you may have never heard of most of them since we tend to

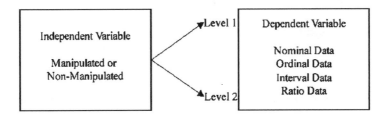

FIGURE 2.4. The dependent variable and types of data that may be collected.

use quantitative data more frequently. Again, these are general rules, and we will see minor exceptions to them later. For now, we can feel comfortable with the summarization provided in Table 2.8.

TABLE 2.8. Relationship between Data Type and Appropriate Statistical Tests

Type of data	Quantitative?	Parametric or Nonparametric statistics?
Nominal (categorical)	No	Nonparametric
Ordinal (rank)	No	Nonparametric
Interval	Yes	Parametric
Ratio	Yes	Parametric

Summary of the Dependent Variable and Data Types

The dependent variable in a hypothesis represents the "effect" we're trying to measure. In order to do so, we have four types of data—nominal, ordinal, interval, and ratio; the last two of these can be thought of together as quantitative data. Each of these data types has unique characteristics, and understanding these characteristics will ultimately help us identify the appropriate statistical tool we need to test our hypotheses. For now, however, we'll learn how to "describe" our data numerically and graphically. This will include discussing common measures such as the average of a dataset but will also include things—like the normal distribution—that some of us may have never heard of. Like everything we have talked about so far, though, these concepts are easy to understand and are integral in our quest to be good consumers of statistics.

Measures of Central Tendency

Measures of central tendency are just that—tools that help us determine the "center" or midpoint of data we have collected. This is essential because it not only allows us to better understand one set of data, but it also lets us begin comparing two or more sets of data to each other. As we go along, I will show you the basic formulas that underlie numeric descriptive statistics; learning to compute these "by hand" will help you better understand and apply the results. Later in the chapter we'll see how to do the same calculations the easy way—by using a computer! For now, let's create a dataset to use an example.

Table 2.9 shows data from a sample of students in a typical classroom (remember, we call all possible participants in a group the *population* and a subset of the population is a *sample*). Our sample represents things a teacher might be interested in knowing.

As you can see by looking at the headers, the contents of the columns are self-explanatory, but let's take a closer look at Table 2.9 and decide what type of data they represent. First, look at the column labeled Gender. Since the valid values for this field are M for male students and F for female students, what type of data are we collecting? That's easy; since the value of the field represents a category a given student falls into, we know it is nominal-level data. What about the field called Ethnicity? It is also an example of nominal data since we generally say a student falls into only one ethnic category. In this case, the categories are AA (African American), NHW (non-Hispanic White), and H (Hispanic). The next column, Class Rank, is obvious. In this case, we are ranking our students based on their math exam grade. We can see that the highest grade has a rank of one, the next highest rank a two, and so on. Notice that in this case there were no ties in the rankings—every student has a unique score.

TABLE 2.9. Dataset for Nominal, Ordinal, Interval, and Ratio Data

Student number	Gender	Ethnicity	Class rank	Math exam	Absences
1	F	H	1	92	7
2	F	NHW	2	91	4
3	M	AA	3	90	0
4	M	NHW	4	89	8
5	M	AA	5	88	3
6	F	NHW	6	87	3
7	F	H	7	86	3

We could also rank Absences by adding a column to Table 2.10 showing how the students rank, from highest to lowest, in terms of their number of absences. In this case, we can see that the highest ranking, representing eight absences, goes to student 4. After that, you'll see that student 1 has seven absences and is ranked second. This goes along well until we get to students 5, 6, and 7, each of whom has three absences. Since they all have an equal number of absences, they are all tied for a ranking of fourth. The lowest number of absences is reported for student 3; therefore that student is ranked seventh.

Now, what type of data does "math exam" represent? In order to answer that question, let's first assume that achievement scores range from the traditional 0 to 100. Given that, we can narrow down our choices to interval- or ratio-level data. The next question is, "Does a math exam have an absolute zero point?" From our previous discussion, we know the answer is "no." A student may make a zero on the exam, but that does not mean she knows absolutely nothing about math; nor does 100 indicate absolute mastery of the topic. We also cannot make comparisons or ratios using the data; a person scoring 80 on a test does not necessarily know twice as much about math as a person who scores 40; they simply scored twice as much on that given exam. Knowing that, the data are interval in nature.

TABLE 2.10. Dataset for Nominal, Ordinal, Interval, and Ratio Data Ranked by Absences

Student number	Gender	Ethnicity	Class rank	Math exam	Absences	Absence rank
1	F	H	1	92	7	2
2	F	NHW	2	91	4	3
3	M	AA	3	90	0	7
4	M	NHW	4	89	8	1
5	M	AA	5	88	3	4
6	F	NHW	6	87	3	4
7	F	H	7	86	3	4

Finally, let's look at the field titled Absences. We know these data are quantitative in nature, but this time an absolute value of zero is possible; a student may have never missed class. We can also make meaningful comparisons between the values; a student who has been absent four times has been absent four times as many as a student who has been absent once; this means the data are measured on a ratio scale. Remember, however, it is generally not important to know if the data are either interval or ratio level. Just knowing whether the data we are collecting are quantitative is enough.

The Mean, Median, and Mode—Measures of Central Tendency

A measure of central tendency helps determine the center of a set of dataset. We have already mentioned one of the most common measurements, the *mean* or average, but it is not the only measure of this type. Two other closely related measures of central tendency are the *median* and the *mode*. Before talking about them, let's take a closer look at how we compute the mean.

The Mean or Average

To begin computing a mean score, let's start by totaling the Math exam data from Table 2.10; this leaves us with 623. To determine the mean score, we simply divide this value by the total number of scores, 7, giving us an average of 89. If you ever want to impress your friends, however, there is a technical formula for calculating the mean. Before we get into it, however, there might be several characters in the formula that you might not recognize; let's go through them first.

In the formula below, the symbol $\bar{x}$ (referred to as "x-bar") stands for the actual mean value that will be calculated. The upper-case Greek letter sigma (i.e., Σ) means to sum everything following it. In this case, the x represents the individual values in our sample so the top line of the formula says, "add up all of the values in the sample." The lower-case n on the bottom stands for the number of items in the sample; here we have collected data for seven students. Putting this all together, the formula for the mean of a sample is

$$\bar{x} = \frac{\Sigma x}{n}$$

Now, we can use the values in our table above to fill in the formula:

$$\bar{x} = \frac{92 + 91 + 90 + 89 + 88 + 87 + 86}{7}$$

This gives us:

$$\bar{x} = \frac{623}{7}$$

By dividing 623 by 7, we can see that the average math exam score is 89, just as we computed above:

$$\bar{x} = 89$$

Remember, in reality, we very rarely know the population mean, but, on any of the rare occasions where you might know all of the values in a population, we can compute it for a population using the same formula. The only difference is that we would substitute the lower-case Greek letter mu (i.e., μ) for $\bar{x}$ and an upper-case N as the number of members in the population. The equation would then look like this:

$$\mu = \frac{\Sigma x}{N}$$

Before we move forward, let me point out something here that might easily be overlooked. Did you notice that we used a lower-case x to refer to the value we computed for a sample and then we used a Greek letter to refer to that same value for a population? That idea holds true throughout. If we're calculating a value for a population, we'll use upper-case Greek letters; for samples, we'll use lower-case letters. The ability to recognize and understand these seemingly insignificant things is a big part of becoming a good consumer of statistics!

We Do Not Calculate the Mean for Nominal or Ordinal Data

Calculating the average for quantitative data is commonplace, but it does not make sense to calculate it for nominal or ordinal data. For example, can we calculate the average gender? The real question is, "Why would we want to calculate an average gender? A person is either a male or a female!" Despite already knowing that, let's try calculating the mean for gender and see where it takes us.

First, we already know that we must have numeric values to compute the mean of data we have collected. In this case, since gender is not represented by numeric values, let's go so far as to replace all the values of M and F with numeric values of 1 and 2, respectively; we can see this in Table 2.11.

TABLE 2.11. Nominal Data Transformed to Numeric Values

Gender	Numeric gender
F	2
F	2
M	1
M	1
M	1
F	2
F	2

We could then use these values to enter into our equation for a sample mean:

$$\bar{x} = \frac{2+2+1+1+1+2+2}{7}$$

Using these data, we would compute an average of 1.57. Is this meaningful to us? No, because it doesn't tell us anything meaningful about our data.

The same holds true if we try to calculate the mean for ordinal data. For example, let's use the data for class rank from Table 2.10 and calculate the mean. Just as before, we would add all the values of class rank leaving us with 28. When we divide that by the 7, the number of values in the dataset, it leaves us with an average of 4. This is simply the ranking that falls in the center of the dataset; it does not tell you anything about the other rankings. Trying to calculate the mean of ordinal data would be akin to reporting the results of a 11-car race by saying the average finish was sixth—it just doesn't make sense! The point is, while the mean is a good way to determine the central point of a dataset containing interval- or ratio-level data, it tells us nothing at all about nominal- or ordinal-level data.

The Median

The median is the score that falls in the exact center of a sorted dataset and can be used with ordinal, interval, or ratio data. Because there is no commonly agreed upon symbol for either the population or sample median, we will use "Mdn" in this book. To see an example of a median, let's use the absences from our original table shown again in Table 2.12; if we sorted them, lowest to highest, we can see that the 3 is right in the middle.

We don't use the median too often for quantitative data, but it really helps when we are trying to find the center of a dataset that is ordinal in nature. We have already alluded to what the median represents when we were attempting to find the average value of the field called class rank. As we saw, when we arranged the data values into ascending or descending order, spotting the center point was easy. For class rank, shown in Table 2.13, it is easy to see that the median score belongs to a non-Hispanic White male with a math exam grade of 89.

TABLE 2.12. Median Absence

Absences	Rank
0	1
3	2
3	3
3	4
4	5
7	6
8	7

TABLE 2.13. Median Ethnicity Based on Class Rank

Ethnicity	Gender	Class rank	Math exam
H	F	1	92
NHW	F	2	91
AA	M	3	90
NHW	M	4	89
AA	M	5	88
NHW	F	6	87
H	F	7	86

In cases where you have a larger dataset and you cannot readily see the median, there are formulas to help us locate the median of a dataset. There are two ways to do this; the formula you use depends on whether you have an even or odd number of values in your set of data.

LOCATING THE MEDIAN OF A DATASET WITH AN ODD NUMBER OF VALUES

In order to find the midpoint of a dataset with an odd number of values, add 1 to the total number of items in the dataset and then divide the total by 2. Using the data above showing ethnicity and math exam information, we can find the midpoint of the data using the following steps:

1. $\text{Midpoint} = \dfrac{\text{Total number in dataset} + 1}{2}$

2. $\text{Midpoint} = \dfrac{7+1}{2}$

3. $\text{Midpoint} = \dfrac{8}{2}$

4. Midpoint = 4

Remember, this simply means that 4 is the midpoint of the dataset; we have to go to that point in the table to determine the actual median value. For example, in this case, if you look at the Math exam column in the table and count 4 up from the bottom or 4 down from the top, you'll see, based on the math scores, that the actual median value is 89.

LOCATING THE MEDIAN OF A DATASET WITH AN EVEN NUMBER OF VALUES

When we had an odd number of values in our dataset, it was easy to find the middle; it is the value where there are just as many data values above it as there are below it. In cases where there is an even number of values, our logic has to change just a bit. For example, let's use the data from Table 2.10 and add information to the bottom, shown in Table 2.14, about a student who scored 50 on the math exam.

TABLE 2.14. Ranked Math Exam Data

Student number	Math exam
1	92
2	91
3	90
4	89
5	88
6	87
7	86
8	50

Looking at this table, we can see that there is no middle record, so we cannot readily point out the median of the dataset. In order to get around that, we have to use the two records that fall in the middle of the dataset—in this case the two records with the scores of 89 and 88. In order to find the median we would add these values together and take the average; doing so would give us a median of 88.5.

We can use the same formula as before to find our midpoint but will wind up with a fractional number:

1. $\text{Midpoint} = \dfrac{\text{Total number in dataset} + 1}{2}$

2. $\text{Midpoint} = \dfrac{8 + 1}{2}$

3. $\text{Midpoint} = \dfrac{9}{2}$

4. $\text{Midpoint} = 4.5$

The resulting answer, 4.5, shows the position in the dataset of the median. Since there are no fractional values in our table, it's necessary to average the two numbers in the fourth and fifth position. That again means you use 88 and 89. The resultant average, just like before, is 88.5; this is our median. Another thing we need to notice is that, while the test scores are ranked highest to lowest, the actual scores are not relevant to one another. For example, look at the difference between the students ranked first and second; we can see a difference of one point; the difference between the seventh and eighth is 36 points. This means that, although the data are ranked, the differences are not relative and computing a mean score tells us nothing about rank.

The median is also a good tool to use if our data are quantitative but are badly *skewed*; this simply means that the values on one side of the mean are more spread out than the values on the other side of the mean. We will talk about skewness in detail later, but here's an example to help us get a basic understanding. Suppose we are interested in moving to a neighborhood where a realtor tells us there are five homes and the average home price is $100,000. Upon hearing that, we decide to visit the neighborhood where we find there are, indeed, five houses. Four of the houses are worth $10,000 each and one is worth $460,000. This is shown in Table 2.15.

TABLE 2.15. Home Values in Order of Cost

House 1	House 2	House 3	House 4	House 5	Average
$10,000	$10,000	$10,000	$10,000	$460,000	$100,000

While we might be disappointed with what the realtor told us, it was true; the average home in the neighborhood is worth $100,000. However, since there are many more values below the mean than there are above the mean, it would have been better if the realtor had told us that the median value was $10,000. That would have been a more meaningful indicator of what the average home in the neighborhood is worth.

WE DO NOT CALCULATE THE MEDIAN FOR NOMINAL DATA

The median is the best measure of central tendency for ordinal-level data, as well as badly skewed quantitative data, but we cannot use it with nominal data. For example, Table 2.16 presents a list showing the highest college degree earned by employees of a marketing company. In this case, we can see that the midpoint in the list represents a person with an MBA, but what does that tell us? Absolutely nothing; it tells us the center point of an ordered list and that's it. As far as we know, everyone on the list has an MBA.

TABLE 2.16. Position of Median Degree Earned

Degree earned
BS
BS
BS
BS
MBA
MBA
MBA
MS
PhD

The Mode

Up to this point we have talked about the mean and the median; both can be used to help us examine the central tendency of a dataset. As we saw, the mean cannot be used with nominal and ordinal data, and the median cannot be used with nominal data. Fortunately we have another tool, the mode, which can be used with all types of data. The mode is nothing more than the value that occurs most often in a dataset. Like the median, there is no one standard symbol, so we will use *Mo*.

If we look at the Absences field from our original dataset (shown as Table 2.17), we can see that our mode is 3 since it occurs most often; we can further see that F is our modal value for gender since we have four females and three males.

TABLE 2.17. Determination of Modal Values for Absences and Gender

Student number	Gender	Ethnicity	Math exam	Class rank	Absences
1	F	H	92	1	7
2	F	NHW	91	2	4
3	M	AA	90	3	0
4	M	NHW	89	4	8
5	M	AA	88	5	3
6	F	NHW	87	6	3
7	F	H	86	7	3

The idea of the mode is straightforward, but we have to be aware of a small issue that arises when we have an equal number of any given value. For example, suppose we added another student, whose ethnicity is AA, to the data; what then would our mode be for Ethnicity? Since we would then have three NHWs, three AAs, and two Hs, it appears we do not have a mode since no one value occurs more often than the others. Actually, since NHW and AA occur an equal number of times, we have two modes; we call this a *bimodal* dataset. In other instances, you might find three, four, five, or even more modes in a dataset. Instead of getting complicated and trying to use a different name for each distribution, we just refer to any dataset having more than one mode as *multimodal*.

So, after all of that, we have three measures of central tendency: the mean, the median, and the mode. Table 2.18 helps clarify exactly when to use each.

TABLE 2.18. Relationship between Data Type and Measure of Central Tendency

	Nominal	Ordinal	Interval	Ratio
Mean	No	No	Yes	Yes
Median	No	Yes	Yes	Yes
Mode	Yes	Yes	Yes	Yes

Using SPSS to Analyze Our Data

First, before we go any further, let me remind you that, in the Introduction, I told you this book is designed to help you become a good consumer of statistics. I also said we would be using examples from statistical software to make our lives easier. Knowing that, there are many good software packages we could use and, if we used the same data, we would get the same answers. You'll find that some software is easier to use than others, there are many different abbreviations that mean the same thing from package to package and some are more sophisticated than others. I've found that IBM® SPSS® (i.e., Statistical Package for the Social Sciences) is the easiest to use and comprehend. Because of that, we'll look at how to actually use the software as well as examples of output from it to explain the concepts we're covering. Let's start with Table 2.19 showing data collected from 10 patients.

TABLE 2.19. Patient Data

Gender	Systolic blood pressure	Anxiety	Exercise (minutes)
F	180	100	0
M	175	55	60
F	160	80	10
F	158	25	45
F	135	65	30
M	135	90	15
M	132	100	45
F	120	40	50
F	105	10	40
F	90	0	60

Within the table you see four columns; the first column shows valid values for Gender—F and M. Knowing that this is nominal data, we could easily compute a mode of 7 since there are seven females represented in the table. When looking at the remaining variables, the second column represents systolic blood pressure. We have used blood pressure to rank the patients from highest to lowest; this allows us to see that the person with the highest blood pressure is female, so is the person with the lowest blood pressure. From our earlier discussion, we know it is appropriate to compute both the mode and median for ordinal data; here the mode is 135 since it occurs twice. Since our dataset contains an even number of cases, we would average the fifth and sixth values of blood pressure to determine the median; in this case, it is also 135. Although the data are ordinal, we could also compute a mean simply because the data being used for the ranking are numeric.

The third column, labeled Anxiety, represents a person's self-perceived anxiety on a scale from 0 to 100. Knowing that, is this an interval- or ratio-scale item? We can determine that by answering two questions. First, does anxiety have an absolute zero point? In this case, it does not. Although some of the patients have a score of zero, it doesn't mean they do not experience anxiety in their lives; it only means they have no anxiety as measured by the survey or scale used as a measurement instrument. Second, is a person with a score of 80 twice as anxious as a person with a score of 40? They are not; the scores only represent each person's anxiety level relative to the scale. Now, knowing the answers to both of those questions leads us to determine that the data are on an interval scale. Because of that, we can compute a mean of 56.5, a median of 60, and a mode of 100.

Next we have the number of minutes that each person exercised daily. Using this data, can ratios be made? Yes, a person who exercised for 60 minutes exercised twice as long as a person who exercised for 30 minutes. There is also an absolute zero—if the person did no exercise at all, their score would be zero. Knowing both of these things, we can feel safe in saying that the data are ratio level. As we did with interval data, we can compute a mean, in this case, 35.5 and a median of 42.5. We can also see that the data are bimodal since we have two occurrences each of 45 and 60.

Everything we've computed so far is shown in Table 2.20.

TABLE 2.20. Central Tendency Based on Data Type

	Gender	Systolic	Anxiety	Exercise
Mean			56.5	35.5
Median		135	60	42.5
Mode	F	135	100	45 and 60

Entering the Data into SPSS

Entering our data into SPSS is a fairly straightforward process. It's based on a spreadsheet format, much like EXCEL or any of the other spreadsheets we've all used at one time or another. When we start SPSS, we will see the screen shown in Figure 2.5. There are a number of commands listed at the top of the spreadsheet, and we'll get to them shortly. For now, however, let's click on the Variable View button at the bottom left of the screen. This will take us to the next page, shown in Figure 2.6, where we can define the variables we will use for each of the values in our dataset.

Each variable in our dataset will be defined within a row of the table. We will use the Name column to identify each variable and then use the remainder of the columns in that row to define the characteristics of that variable. Figure 2.7 shows the spreadsheet after it has been set up to work with our data. The first row represents the Gender. It is a string variable Type with a maximum Width of one character; obviously there are no decimal points. The label "Gender" will be used when referring to this variable on output produced by SPSS. In this case, by clicking on the actual variable name, using the Variable Labels dialog box, we have created two Values. If the actual data is F, the value will show as "Female" on any reports we create; a value of M will show as "Male." We have no entry in the Missing column, which means the system will not automatically enter values if a data value is missing; Column and Align indicate that there are eight spaces, with the value left justified, on any output we create. The measure, of course, is nominal, and the Role label indicates this field will be used as input.

The next variable, Systolic, represents systolic blood pressure and is numeric with a width of 3. No decimal points are allowed; the Label is "Systolic Pressure" on any

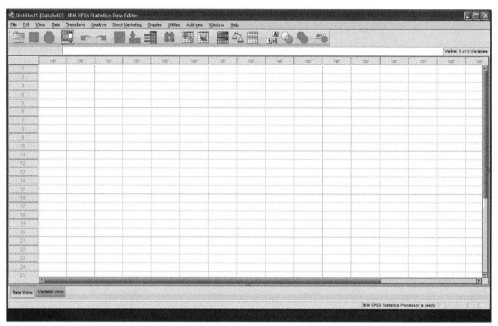

FIGURE 2.5. SPSS Data View spreadsheet.

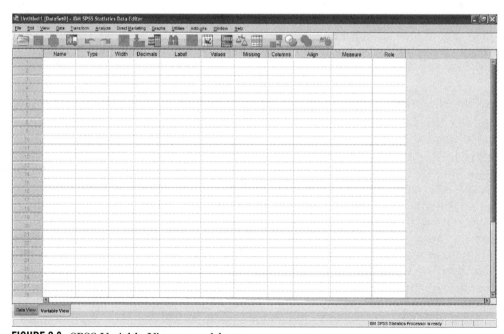

FIGURE 2.6. SPSS Variable View spreadsheet.

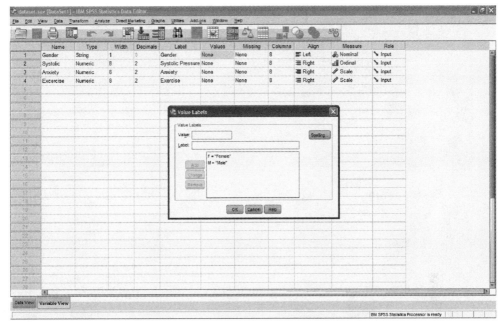

FIGURE 2.7. SPSS Variable View spreadsheet with variables entered.

output and, because the range of blood pressure is so large, there are no Values for the individual variables. The Missing, Column, and Align categories work the same as for the Gender value, but, in this case since we're ranking the cases by their systolic blood pressure, the Measure is ordinal.

The remaining two columns show levels of Anxiety and the amount of Exercise. In this case, the data are quantitative, either interval or ratio; SPSS simply labels them both as "scale" data. The remaining columns are self-explanatory. From here, we need to analyze our data, so let's go back to the Data View screen and enter the data from above.

We've entered the data exactly as in the earlier table. Now, it's time to actually analyze the data. First, as seen in Figure 2.8, we'll select the Analyze command, Descriptive Statistics and Frequencies. This will be followed by the screen shown in Figure 2.9. In this case, in the Frequencies: Statistics box we will select Mean, Median, and Mode, followed by Continue; we will then select OK in the left box. SPSS will then produce the output shown in Figure 2.10.

It's readily apparent that the values shown in this box are exactly equal to those we computed earlier. Three things bear mentioning however. First, as we just said, systolic blood pressure is ordinal; we've used it to rank the cases from a high of 180 to a low of 90. Despite that, since the values for blood pressure are numeric, SPSS computes the mean, median, and mode. Second, we can see there is a lower-case "a" next to the Mode for Exercise. This only occurs when the data are multimodal and the software displays the lowest value; this is explained in the note below the table. Finally, since Gender is nominal, none of the descriptive statistics other than N (i.e., the number of values) are produced.

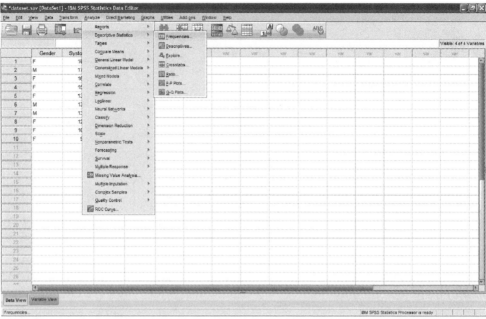

FIGURE 2.8. Selecting the Descriptives and Frequencies commands on the SPSS Data View spreadsheet.

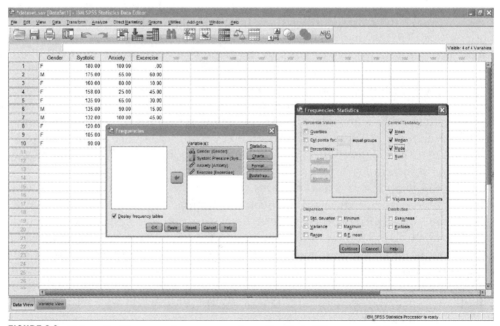

FIGURE 2.9. Using the Frequencies command on the Data View spreadsheet.

Statistics

		Gender	Systolic Pressure	Anxiety	Exercise
N	Valid	10	10	10	10
	Missing	0	0	0	0
Mean			139.00	56.5000	35.5000
Median			135.00	60.0000	42.5000
Mode			135	100.00	45.00[a]

a. Multiple modes exist. The smallest value is shown

FIGURE 2.10. Descriptive statistics generated by the Frequencies command.

Summary of the First Part of Step 4: Identify and Describe the Dependent Variable

In this section we've covered the first part of Step 4; we learned to identify independent variables and their levels, dependent variables, and the different types of data and, finally, we learned how to compute measures of central tendency both manually and using SPSS. In the next chapter, we'll continue with Step 4 of our six-step model by looking at measures of dispersion and relative standing. Since we already know how to locate the center of a dataset, these tools will help us determine how the data are spread out around the central point. For now, let's make sure we completely understand everything we've covered in this chapter.

Do You Understand These Key Words and Phrases?

categorical data	continuous data
dependent variable	independent variable
interval data	levels of the independent variable
manipulated (experimental) independent variable	measures of central tendency
mean	median
mode	nonmanipulated (quasi) independent variable
multivariate	nonparametric statistics
nominal data	ordinal data
numeric descriptive statistics	quantitative data
parametric statistics	rank data
ratio data	univariate

Do You Understand These Formulas?

Mean of a population:

$$\mu = \frac{\Sigma x}{N}$$

Mean of a sample:

$$\overline{x} = \frac{\Sigma x}{n}$$

Midpoint for a sample:

$$\text{Midpoint} = \frac{n+1}{2}$$

Quiz Time!

Let's start by looking at the following hypotheses. Read each one and then identify the independent variable and its levels; then explain why the levels are either manipulated or nonmanipulated. Following that, identify the dependent variable and the type of data it represents.

1. There will be a significant difference in motivation scores between students in online programs and students in traditional programs.

2. There will be a significant difference in the number of males and females working as computer programmers in corporate America.

3. There will be a significant difference in the number of females in computer science classes between students in the United States, France, and Russia.

4. There will be a significant difference in weight gained during their freshman year between students who live at home and students who live away from home.

5. There will be a significant difference in first-year salaries between graduates of Ivy League schools and graduates of state universities.

6. Administrative assistants who work in cubicles are significantly less productive than administrative assistants who work in enclosed offices.

7. Primary-care patients who are treated by an osteopathic physician will have significantly fewer health problems than primary-care patients who are treated by an allopathic physician.

8. Truck drivers who check their tire pressure frequently will have significantly higher miles-per-gallon than truck drivers who do not check their tire pressure frequently.

9. Insurance companies that use computer-dialing services to call prospective clients will have a significantly lower number of sales than insurance companies that use live agents to call prospective clients.

10. The rankings of favorite sporting activities will be significantly different between Mexico, the United States, and Canada.

Let's check to make sure you understand how to compute the basic measures of central tendency using the data in Table 2.21. Let me warn you, be careful! Some of these answers are not as obvious as they might appear! You'll be able to check your answers at the end of the book.

1. What is the average height?

2. What is the median class rank?

3. What is the mode of shoe size?

4. What is the average class?

5. What is the median height?

6. What is the mode of height?

7. What is the median shoe size?

8. What is the mode class rank?

9. What is the mode of class?

10. What is the median class?

11. What is the average class rank?

12. What is the average shoe size?

TABLE 2.21. Data for Quiz Time!

Height	Class	Class rank	Shoe size
60	FR	1	8
66	FR	10	7
69	SO	5	8
66	SR	13	11
69	JR	6	8
68	JR	12	12
64	FR	2	7
68	SO	14	9
70	SR	8	12
69	JR	3	11
70	FR	15	9
73	SR	6	13
72	SR	9	8
65	SO	11	8
72	SO	4	13

CHAPTER 3

Measures of Dispersion
and Measures of Relative Standing

Introduction

In this chapter we'll continue with Step 4 of our six-step model by learning how to compute and understand *measures of dispersion* and *measures of relative standing*. When we measure dispersion, we're simply looking at how spread out our data-set is. Measures of relative standing allow us to determine how far given values are from each other as well as from the center of the dataset. As usual, we'll learn how to compute everything manually, but then we'll use SPSS to show us how easy it really is.

■ Measures of Dispersion

When we're talking about the dispersion of a dataset, we're asking a simple question, "How spread out is the dataset we plan to use?" For example, I might tell you I own paintings that average about $300 each; I might also tell you that I paid as little as $25 for some and as much as $1,000 for others. Based on that, you could say that the amount I've paid for paintings is widely dispersed around the mean. In fact, if we subtract $25 from $1,000, we can see that the amount I've paid for pictures could fall anywhere between these two values; in this case a range of $975. Computing the *range* is the most common of the measures of dispersion, but we'll also look at the *standard deviation* and *variance*. All three measures are critical for the accurate computation of inferential statistics.

The Range

As we just saw, the range tells us how far apart the largest and smallest values in a dataset are from each other; it is determined by subtracting the smallest value from the largest value. For example, in Table 3.1, when we subtract 25 from 75, we have a range of 50; we call this our *computed range*. In this case, we also have a mean score of 50.

subtract start from finish

TABLE 3.1. Range of Data Values from 25 to 75

25	30	35	40	45	50	55	60	65	70	75

While this is very easy to compute, we have to be careful in its interpretation when we are interested in the dispersion of the data. Remember, the purpose of the measures of dispersion is to understand how the data are spread out around the center of the dataset. Using the range to help us understand this is very imprecise and can be very misleading. For example, what does a range of 50 tell us? What can we learn about how the data are dispersed around the center of the data by computing it? Again, let's use the data in Table 3.1 and look at some of the issues we might encounter.

We already know that both our mean and range are 50. In this case, all we can tell by using the range as a measure of dispersion is that the data are evenly spread out around the average. This can be seen very clearly; the smallest value is 25 points below the mean, the largest value is 25 points above the mean, and the data values are spaced at equal increments of 5 points each.

We have to be careful, though, because we can often compute the same range for two datasets that have the same mean score and range but their dispersion will be completely different. For example, as in the previous example, the lowest value in Table 3.2 is 25 and the largest is 75; this leaves us with a range of 50.

TABLE 3.2. Range of Data Values from 25 to 75 Clustered around the Mean

25	42	43	44	45	50	51	52	53	54	75

In this case, although we can see that the range is fairly large; most values are clustered around the mean. This isn't to say that the range cannot be used effectively; it is perfectly acceptable for a given task while remembering its limitations. For example, investors might want to look at the range of values that a particular stock sold for on a given day, or a college professor might be interested in knowing the overall spread of test scores for an examination. In both of these cases the results are meaningful, but, generally speaking, the practical application of the range is often limited.

The Possible Range of the Data

There is a manner by which the range can be used in a very meaningful fashion. As we noted above, the computed range is the result of subtracting the smallest number in the dataset from the largest number. We can also establish a *possible range* that is nothing more than the entire range of values a dataset might include. For example, the possible range on the Graduate Management Admissions Test (GMAT) runs from 200 to 800 and, since there are 100 senators in the United States Congress, the possible range of the number of "yes" votes on any piece of legislation would be 0 to 100. We can also establish our own possible range by simply deciding the boundaries we are looking for; we'll use shopping mall sales pitches as an example.

We've all experienced those people who approach you in the mall asking for you to try one product or the other; if you're like me, you pretend you don't notice them and get away as quickly as possible. Other people, who I guess are more willing to part with their hard-earned money than I am, stop to talk and try out the new product. In this case, suppose our marketers are interested in knowing the appeal of a new cologne to males in the 20 to 40 age bracket; when we subtract 20 from 40, we wind up with a possible range of 20. While that isn't terribly interesting by itself, we can use the possible range and the computed range together to help in decision making.

For example, if the employee of the marketing company asked each male trying the new product their age, the marketing company could use these data to determine the computed range of mall patrons who actually tried the product. In this case, let's say the youngest person who rated the product was 20 and the oldest was 32; this would result in a computed range of 12. If we were to compare our computed range of 12 to our possible range of 20, we would see that our computed age range is somewhat spread out since it accounts for 60% of the possible range. At the same time, in order to really understand the appeal of the cologne to their targeted age group, they would need to get opinions from the remaining 40%, all of whom would fall in the 33 to 40 age range. We can see this entire idea demonstrated in Table 3.3.

TABLE 3.3. Comparison of 60% Computed Range to the Possible Range

Computed range = 12 (60%)																				
20	21	22	23	24	25	26	27	28	29	30	31	32	33	34	35	36	37	38	39	40
Possible range = 20 (100%)																				

Let's use the same possible range but, this time, let's say the youngest person trying the product was 28 and the oldest was 34; this would give us a computed range of 6. In this case, since our possible range is 20, we've only accounted for 30% of our possible range—not a very wide dispersion. In order to get responses from the entire possible range, the marketers would need to go out and focus their efforts on males in the 20 to 27 age range as well as those in the 35 to 40 age range. Again, this is demonstrated in Table 3.4.

TABLE 3.4. Comparison of 30% Computed Range to the Possible Range

								Computed range = 6 (30%)												
20	21	22	23	24	25	26	27	28	29	30	31	32	33	34	35	36	37	38	39	40
Possible range = 20 (100%)																				

As you can see, the only thing the range tells us is how spread out the data are and what percentage of the possible range can be accounted for by the computed range. In the next section, we'll discuss the standard deviation, a tool that is not only very accurate in measuring dispersion but one that lies at the center of many of the inferential statistics we will be using later in the book.

The Standard Deviation

In many cases we need a tool that will help us determine exactly how far a value is from the center of the dataset. The standard deviation allows us to measure the distance from the mean of a dataset in terms of the original scale of measurement. In other words, if we have a dataset consisting of a range of values, height for example, we can use the standard deviation to tell us how far a given height is, in terms of original scale of measurement (e.g., inches or feet) from the mean of the dataset. We will use the standard deviation as an integral part of several of the inferential statistics we will learn to compute later in the book, but, for now, the concepts underlying the standard deviation are easy to understand.

First, since the standard deviation is computed in terms of the original scale, there is an easy way to think about it. The standard deviation simply approximates the average distance any given value in the dataset is away from the mean of the dataset. Let's use the heights of nine employees in Table 3.5 to demonstrate this idea.

TABLE 3.5. Range of Heights Used to Compute a Standard Deviation

Height in inches								
64	65	66	67	68	69	70	71	72

The first thing you'll notice is that the data are sorted and range from the smallest value on the left (i.e., 64), to the largest value on the right (i.e., 72). The mean, 68, falls right in the center of the dataset. Although we won't go into the calculations right now, the standard deviation for this data is 2.74; again, that is an approximation of the average distance each person's height is from the average height. How do we apply this to the overall idea of dispersion? Simple: the larger the standard deviation, the more spread out the heights are from the mean; the smaller the standard deviation, the closer each average score is to the mean. Let me show you what I mean.

First, in Table 3.6 let's change the data we just used so that the values above and below the mean are more spread out; I have illustrated this with the arrows moving away from the mean in both directions. While the mean remains the same, the standard deviation increases to 5.48. This tells us that the approximate average distance any value in the dataset is from the mean is larger than when the data values were grouped closer to the mean.

TABLE 3.6. Range of Heights to Compute a Larger Standard Deviation

Height in inches								
60	62	64	66	68	70	72	74	76
←				Mean				→

The exact opposite happens if we group the data closer to the mean of the dataset. As can be seen in Table 3.7, the mean is again 68, but the values are much more closely clustered around the mean. This results in a standard deviation of 1.58, much smaller than the original standard deviation of 2.74.

TABLE 3.7. Range of Heights to Compute a Smaller Standard Deviation

Height in inches								
66	66	67	67	68	69	69	70	70
————————————▶				Mean	◀————			

Making Estimations and Comparisons Using the Standard Deviation

Using the standard deviation, unlike the range, we can determine how far each value in a dataset is from the mean; we don't have to simply rely on ideas such as "small" and "large" when talking about dispersion. Later in the book we'll use statistical tools and something called the normal distribution to develop this idea further, but, for now, we can use the standard deviation alone to make comparisons. For example, if we're interested in comparing the number of sick days our employees take on a yearly basis, we might use the data in Table 3.8.

TABLE 3.8. Using the Standard Deviation to Compare Values in One Set of Data

Annual sick days								
2	5	5	6	6	6	7	7	10
				average				

We can see that the average employee calls in sick 6 days each year. If we computed the standard deviation, we would find that the average number of sick days taken is 2.12 days away from the mean. For clarity's sake, we'll round that off to 2.00 and say that the average number of days away from the mean ranges from 4.00 (the mean of 6 minus one standard deviation) to 8.00 (the mean of 6 plus one standard deviation). Anything outside of this is not in the average range. In this case, the person who used only two sick days is below average. Obviously, this is a good thing unless they are coming to work sick when they should be staying at home! On the other end, we might want to talk to the person taking 10 days a year; this might be a bit excessive.

We can also make comparisons of datasets using the standard deviation. For example, suppose we are a new fan of professional football in the United States and want to know as much about it as possible. One thing that interests you is the equality between the two divisions—the National Football Conference (NFC) and the American Football Conference (AFC). Let's use Table 3.9 to represent the total points scored by each team during a season in order to compare the two conferences (yes, I know there are more than nine teams in each conference, but this page is only so wide).

TABLE 3.9. Football Scores Used to Compare Two Sets of Data

NFC	304	311	315	320	325	340	345	350	360
AFC	100	105	298	305	340	410	510	642	710

In Table 3.10, we can see that the NFC teams score, on average, 330 points per season while the AFC teams score 380 points per season; their range is 56 and 610, respectively. At this point, we might decide to head to Vegas; it's apparent that the

AFC teams are much better than the NFC teams—maybe there's some money to be made! Before we bet all of our hard-earned cash, however, let's look a little deeper into how the team scores are dispersed within the two leagues. I've already computed the mean, average range and standard deviation for our data and we'll verify it using SPSS in just a bit. For now, let's look at the standard deviation of the scores for each team to get an idea.

TABLE 3.10. Using the Standard Deviation to Compare Values between Football Conferences

Conference	Average points	Standard deviation	Average score range
NFC	330	19.4	310.6–349.4
AFC	380	213.11	166.9–593.11

First, the standard deviation for the NFC teams is 19.4, while the same value for the AFC teams is 213.11. This means, on average, that the NFC teams score somewhere between 310.6 and 349.4 points per season while the AFC teams score, on average, between 166.9 and 593.11 points per season. What does this tell us? The large standard deviation means the AFC scores are very, very spread out—there are some really good AFC teams and some really poor AFC teams. NFC teams, while they have a lower overall average, are fairly evenly matched. What does this mean? When an AFC team plays an NFC team, do we always bet on the AFC team because of their higher overall average? No, it would probably be better to look at the individual teams when we're placing our wagers!

Computing the Standard Deviation for a Sample

The basic logic of the standard deviation is fairly straightforward. As I said, we'll use it later in the book to make more precise measurements, but, for now, let's move forward and learn how to compute it. Do you remember back in the first of the book where I said all you need is basic math for most statistical calculations? Even though the following formula may look a bit complicated, that's still my story and I'm sticking to it!

$$s = \sqrt{\frac{\Sigma(x - \bar{x})^2}{n - 1}}$$

Although there are quite a few symbols in this equation, it is actually very easy to compute. First, let's make sure we know what we're looking at:

1. s stands for the sample standard deviation.
2. n is the number of values in our sample.
3. x is each individual value in our sample.
4. $\bar{x}$ is the mean of our sample. *average*
5. Σ means to add together everything following it.
6. $\sqrt{}$ means to take the square root of the value under it.

Now that we understand all of the symbols, let's go step by step through the computation; after that, we'll use it with our data to see what the actual standard deviation is:

1. $(x - \overline{x})^2$ means to subtract the average score of the entire dataset from each of the individual scores; this is called the *deviation from the mean*. Each of these values is then squared.

2. Σ is the Greek letter sigma and indicates we should add together all of the squared values from Step 1.

3. $n - 1$ is the number of values in the dataset minus one. Subtracting one from the total number helps adjust for the error caused by using only a sample of data as opposed to the entire population of data.

Let's go back and verify the computations for the NFC scores above by going though this step by step in Table 3.11.

TABLE 3.11. Computing the Standard Deviation for Football Scores

NFC scores	Mean of NFC scores	Deviation from the mean $(x - \overline{x})$	Deviation from the mean squared $(x - \overline{x})^2$
304	330	304 – 330 = –26	–26 * –26 = 676
311	330	311 – 330 = –19	–19 * –19 = 361
315	330	315 – 330 = –15	–15 * –15 = 225
320	330	320 – 330 = –10	–10 * –10 = 100
325	330	325 – 330 = –5	–5 * –5 = 25
340	330	340 – 330 = 10	10 * 10 = 100
345	330	345 – 330 = 15	15 * 15 = 225
350	330	350 – 330 = 20	20 * 20 = 400
360	330	360 – 330 = 30	30 * 30 = 900
$\overline{x}$ = 330		Σ = 0	Σ = 3012

In the first column, we see each of the NFC scores in our sample. When we subtract the sample mean, shown in the second column, from each of those values, we see the difference shown in column three; this is our deviation from the mean. Notice, if the observed value is less than the mean, the difference will be negative. The fourth column shows the result when we square each of the values in the third column (remember Math 101; when we square a negative value, it becomes a positive value).

The sum of all of the squared values, 3012, is shown at the bottom of column four. Remembering that $n - 1$ is the total number of values in our dataset, minus 1, we can now include all of these values in our formula:

$$s = \sqrt{\frac{3012}{8}} \qquad n = 9 - 1 = 8$$

What's left to do? Easy, first we need to divide 3012 by 8:

1. $s = \sqrt{376.5}$
2. $s = 19.4$

Computing the square root of 376.5 leaves us with a standard deviation of 19.4—believe it or not, it's really that easy! We will verify what we've done using SPSS in only a few pages, but first let's discuss a minor inconvenience.

WHY DID WE DO WHAT WE JUST DID?

When we first began our discussion of the standard deviation, we said we could think of it as the average distance that each value is from the mean of the dataset. Given that, when we look back at columns three and four in the table above, the obvious question becomes, "If we're interested in knowing the average distance each value in our dataset is from the mean, why do we even need the fourth column? Why don't we just add the deviations from the mean and divide them by the number of values in the sample?"

Unfortunately, it's not that easy. As you can see in the column labeled "Deviation from the Mean," if we summed all of these values, the answer would be zero. Because the sum of $x - \bar{x}$ is an integral part of our equation, that means we would always wind up with a standard deviation of zero. We avoid that by squaring each of the values; this, of course, means that any deviation from the mean less than zero becomes a positive value. After squaring all of those values, we then divide the sum of those values by $n - 1$, giving us an approximation of the average distance from the mean.

At this point, we run into yet another problem! Since we've squared each of the values, we are no longer measuring the standard deviation in terms of the original scale. In this case, instead of a scale representing the points scored by each division, we wind up with the squared value of each division. In order to transform this value back into the scale we want, all we need to do is take the square root of the entire process; that gets us back to where we want to be.

Computing the Standard Deviation for a Population

As you've probably guessed, if we can compute the standard deviation for a sample, we can do the same for a population. The formula, essentially, remains the same; there are only three small differences:

$$\sigma = \sqrt{\frac{\Sigma(x - \mu)^2}{N}}$$

As you can see, since we are now computing a population parameter, we have changed the symbol for the standard deviation to the lower-case Greek letter sigma (i.e., σ). We have also included the Greek letter mu (i.e., μ), which represents the mean of the population. Since we're no longer worried about the inherent error caused by sampling, we can drop the idea of subtracting 1 from our denominator and replace it with an upper-case N. The overall concept, however, is the same. The population standard deviation simply shows the dispersion of the dataset when we know all of the possible values in the population.

The Variance

The variance is a third tool that is used to show how spread out a dataset is and interpreting it is easy. Generally speaking, if the variance value is large for a given dataset, the data values are more spread out than if the variance value is small. You're probably thinking, "That sounds too simple, what's the catch?" The good news is there is not a catch. Let's look at the formula for calculating the variance of a sample and then put some numbers into it to show you what I mean:

$$s^2 = \frac{\Sigma(x - \bar{x})^2}{n - 1}$$

If it seems that we've seen some of this before, you're right. Other than a couple of minor changes, this is nothing more than the formula for the standard deviation for a sample. We're now using s^2, the symbol for the sample variance, instead of s, which is the symbol for the sample standard deviation. The fact that we're squaring the s value only means that we're no longer taking the square root of our computations, so we have to remove the square-root symbol. In other words, if we know the standard deviation, all we need to do is multiply it by itself to compute the variance. For example, since we know the standard deviation for our NFC football scores is 19.40, all we need to do is multiply that by itself and wind up with a variance of 376.5. That means, of course, that if we know the variance of a set of data, its square root is the standard deviation. The obvious question, then is, "Why do we even need to compute the variance since it's just another measure of dispersion? Can't we just use the standard deviation in its place?"

The answer to those questions is pretty simple; in most cases we can use the standard deviation in our statistical inferential decision making. Certain statistical tests, however, use the variance as the tool for measurement.

Interpreting the Variance

Before we start talking about how to interpret the variance, let's create a dataset to work with. In Table 3.12, the top row of numbers contains values ranging from 3 to 9; the average for this data is 6, the standard deviation is 2.16, and the variance is 4.67. The second row in the table shows each value's deviation from the mean. For example, the deviation of the value 3 below the mean of 6 is –3 while the deviation of 9 above the mean is +3. The rightmost column shows that the sum of all deviations from the mean is zero; that stands to reason simply because of the manner in which the mean is calculated.

The third row shows how far each value is from the mean in terms of the standard deviation. For example, the value –3 is –1.39 standard deviations below the mean. We determined this by dividing the observed value by the standard deviation; for example, we can see that –3/2.16 = –1.39. Again, since the standard deviation is based on the mean, the sum of those values, shown in the rightmost column of Table 3.12, is zero.

TABLE 3.12. Computing Distances from the Mean with a Small Amount of Dispersion

	3	4	5	6	7	8	9	Average = 6 Standard deviation = 2.16 Variance = 4.67
Distance from the mean	−3	−2	−1	0	1	2	3	Sum = 0
Standard deviations from the mean	−1.39	−.926	−.463	0.00	.463	.926	1.39	Sum = 0

In order to interpret these data, you have to look at them from several perspectives.

First, just as is the case with the range and standard deviation, the greater the variance, the more spread out the data are. Unlike the range and standard deviation, however, the variance can be smaller than, equal to, or larger than the actual range of the data. In this case, our range (6) is greater than our variance (4.67); this indicates that the values are not very spread out around the mean, but that's not always the case.

For example, suppose the following data represent the number of times an hour our neighbor's dogs bark; our imaginary neighbor has seven dogs, therefore we have seven values. If we computed the range and variance for the data, we would find that the variance of 1108.33 is much larger than the range of 85. As shown in Table 3.13, this means the number of barks per dog is highly dispersed around the mean of 50; some dogs bark quite a bit, others much less so.

TABLE 3.13. Computing Distances from the Mean with a Large Amount of Dispersion

	5	15	30	60	70	80	90	Average = 50 Standard deviation = 33.29 Variance = 1108.33
Distance from the mean	−45	−35	−20	10	20	30	40	Sum = 0
Standard deviations from the mean	−1.35	−1.05	−.601	.300	.601	.901	1.20	Sum = 0

Second, the variance can never be negative simply because it is the standard deviation squared. Unlike the situation when we use the standard deviation while computing inferential statistics, this means we will not be canceling out like values above and below the mean of a dataset. This will allow us, later in the text, to use specific statistical tests based on the idea of comparing variance between two groups to test a hypothesis. For now, however, just remember that the larger the variance, the more spread out the data values are away from the mean.

Third, we have to remember that both the standard deviation and variance measure dispersion around the center of the dataset, but there is a distinct difference. Since the variance is the squared standard deviation, this means it is no longer representative of the original scale. For example, if we weigh a group of students and wind up with an average of 100 and a standard deviation of 10, it means a person weighing 90 pounds would be one standard deviation, measured in pounds, below the mean. The variance of the dataset, since it's the standard deviation squared, 100 in this case, simply shows relative position from the center of the dataset and it is not in the original scale. Again, all we know is that a larger variance value means that the data are more spread out.

The Variance of a Population

Again, since we can compute the variance for a sample, we can compute the variation for a population if we know all of the population parameters. The only differences between the formula for a population and the formula for a sample are three different symbols:

$$\sigma^2 = \frac{\Sigma(x-\mu)^2}{N}$$

The symbol for a population's variance is σ^2 (i.e., sigma squared). In order to compute it, we subtract the population mean mu (μ) from each of the observed values; we then compute and square the sum of those values. Following that, we divide that sum by the number of values in the population (N) giving us the variance for the population.

Using the SPSS Software to Verify Our Measures of Dispersion

Let's use our SPSS software again to make sure our calculations are correct. Since we already know how to use the Variable View screen, let's go directly to Data View, shown in Figure 3.1; you can see that we've already entered the data under a variable named Score and selected Analyze, Descriptive Statistics, and Frequencies.

Once Frequencies is selected, as you can see in Figure 3.2, you can specify the variables you want to analyze, as well as indicate which values you want SPSS to generate. In the Dispersion box, we're selecting range, standard deviation, and variance.

When we select Continue and OK, we get the output shown in Figure 3.3; as you can see, our standard deviation, variance, and range all match what we computed manually when examining the NFC scores.

■ Measures of Relative Standing

After learning to identify the center of a dataset and determine the dispersion of the values within the dataset, we now need to discuss measures of relative standing. These tools will allow us to look at the relationship between individual values within our data and a given measure of central tendency. The five most common measures of relative standing are *percentiles*, *quartiles*, *z scores*, *T-scores*, and *stanines*.

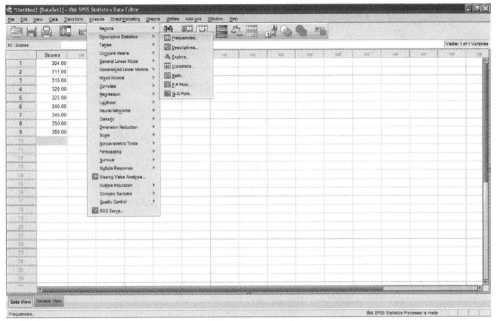

FIGURE 3.1. Selecting the Frequencies command on the Data View spreadsheet.

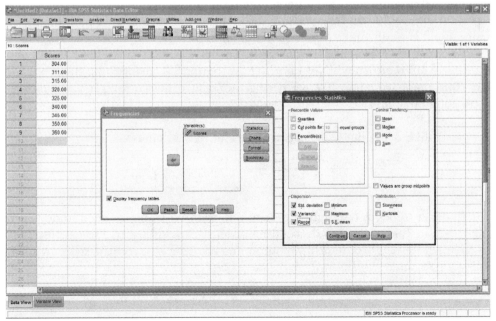

FIGURE 3.2. Using the Frequencies command to generate the standard deviation, variance, and range.

Statistics

Scores

N	Valid	9
	Missing	0
Std. Deviation		19.40361
Variance		376.500
Range		56.00

FIGURE 3.3. Output from the Frequencies command.

Percentiles

Many of us are used to hearing the word percentile, especially when discussing test scores, ability, or aptitude. For example, we might hear that Bob, one of our friends, scored in the 80th percentile on a physical fitness test. What does that mean though? Is this good or bad?

Actually, while this sounds pretty good, we would not really know anything about Bob's physical fitness unless we looked at his actual score on the physical fitness test. His percentile score is simply the percentage of people scoring below that value. With that definition, all we know is that Bob scored higher than 80% of the others taking the same fitness test. Bob might be in horrible shape, but he's better off than 80% of the others! Let's look at a detailed example.

Suppose we have a group of 11 newly hired attorneys in our company and, while they all passed the bar exam, there were some really low scores. One of the young attorneys, a real go-getter, knew his score of 79 was fairly low and was trying to determine a way to "stand out in the crowd." Having heard that statisticians are wizards at making bad things look better, the young attorney approached me and told me his dilemma. While I disagreed with the "wizard" part, I did tell him there is another way he could look at his score. He could compare himself to the rest of the newly hired attorneys and determine, using a percentile score, where he ranked in relation to them. Although he would still have exactly the same score on the bar exam, his percentile score might actually make him look better than he really is.

Knowing that, let's use the following scores to represent the bar scores of all of our newly hired attorneys. Not knowing any better, I'll just assume the scores on the bar exam range from 0 to 100, with our group of attorneys having scores ranging from 70 to 100 (Table 3.14).

TABLE 3.14. Bar Examination Scores for New Attorneys

Bar scores for new attorneys										
100	78	79	76	75	74	75	74	72	70	70

In order to begin computing percentiles, the first thing we need to do is arrange the scores in Table 3.14 into descending order; this is shown in Table 3.15. This gives us the data we need to determine our young attorney's percentile score by using the following formula:

$$\text{Percentile} = [(\text{Below} + 1/2 \text{ Same})/N] * 100$$

TABLE 3.15. Bar Scores Ranked from Highest to Lowest

	100
	79
	78
	76
Bar scores	75
highest	75
to lowest	74
	74
	72
	70
	70

In this equation, "Below" represents the number of scores less than the one we're interested in; "Same" means all values equal to the one we're interested in, and N, as usual, means the total number of values in our dataset. Notice, N is upper-case here; the 11 scores we have refer to the entire population of new hired attorneys.

In this case, our attorney's score is 79; there are nine values less than 79 and only one greater than 79. We can substitute those values, along with N, into our formula, and a few simple calculations will lead us to our percentile score:

1. Percentile = [(9 + 1/2 1)/11] * 100
2. Percentile = [(9 + .5)/11] * 100
3. Percentile = (9.5/11) * 100
4. Percentile = .863 * 100
5. Percentile = 86.3

At this point, it's necessary to drop anything to the right of the decimal point; we don't round it off, but rather just drop it. This means things are looking better for our young friend. Although his actual score was 79, he represents the 86th percentile. This means his score is greater than 86% of the people taking the bar exam. This could possibly make him look better in the eyes of his new employer; that's what he is betting on.

When It Comes to Percentiles, No One Is Perfect

When I tell my students there is no such thing as a 100th percentile, I always get a blank stare from some of them. I know they're thinking, "Wait, if someone scored 100 on a test where that was the highest score possible, shouldn't they be in the 100th percentile?" Let's answer their question by imagining a set of scores ranging from 1 to 100 with only one occurrence of each. If we wanted to compute the percentile for the only person scoring 100, we would have 99 scores below it, 100 would only occur once, and there would be 100 values. Let's substitute those values into our equation.

1. Percentile = [(99 + 1/2 1)/100] * 100
2. Percentile = [(99 + .5)/100] * 100
3. Percentile = (99.5/100) * 100
4. Percentile = .995 * 100
5. Percentile = 99.5
6. Percentile = 99th

Again, some people find this perplexing; the person made a perfect score but was in the 99th percentile; what happened? It's easy when you think about our definition. What is the only score that is greater than 99% of the other scores? Easy, it's 100. There is no way that a person could outscore 100% of people taking the bar exam; that means they would have to outscore themselves. Impossible!

The Median as a Percentile

Although you didn't know it at the time, we have already seen another example of a percentile, but we called it the median. From our earlier definition and what we know now, another name for the median is the 50th percentile. We can further divide our dataset into four *quartiles*, each representing 25% of the dataset. The first quartile, or Q1, is the 25th percentile, Q2 is the 50th percentile (or median), and Q3 is the 75th percentile.

We can make this clearer with an example, but, before we do, let me warn you that there are at least two different ways to compute quartiles. Statisticians tend to argue over which one is most representative (i.e., statisticians have nothing better to do than argue about mundane topics). For our purposes, we'll use these guidelines:

▶ Our definition of the median, represented by Q2, will remain the same; it's the 50th percentile.

▶ Quartile 1, represented by Q1, is the median of the lower half of the data, not including the value Q2.

▶ Quartile 3, or Q3, is the median of the upper half of the data, again not including the value Q2.

Let's demonstrate starting with Table 3.16. Therein we have an odd number of values in our dataset, so our median (Q2) is 5.

TABLE 3.16. Computing Q2 for an Odd Number of Values

1	2	3	4	5	6	7	8	9
				Median Q2				

As we just said, we're going to compute Q1 as the median of all values less than the median; remember, this is the same as the 25th percentile. In this case, the median will be the average of 2 and 3, or 2.5. Q3 will be computed in the same manner; it is the median of all values above the median; in this case the average of 7 and 8, or 7.5. These are marked in Table 3.17.

TABLE 3.17. Computing Q1, Q2, and Q3 for an Odd Number of Values

1	2	3	4	5	6	7	8	9
	2.5 is 25th percentile or Q1			5 is the 50th percentile (i.e., the median or Q2)		7.5 is the 75th percentile or Q3		

In this table, we had an odd number of values; Table 3.18 shows us a dataset with an even number of values:

TABLE 3.18. Data for an Even Number of Values

10	15	20	30	40	46	52	58	60	62

Knowing we have an even number of values, the median, Q2, is the average of the middle two numbers; remember that 10 is the number of values in the dataset:

$$\text{Midpoint} = \frac{10+1}{2}$$

Our results would show the median is in the 5.5th position. In this case, the median, Q2, is 43 [i.e., $(40 + 46)/2 = 43$]. This would also mean that Q1 is the median of all values to the left of 43. Remember, the value 40 is in our original data, and it is less than 43, so we have to include it. Given that, we have an odd number of values; 10, 15, 20, 30, and 40; we simply identify the one in the middle, 20, as our 25th percentile or Q1. In order to compute Q3, the 75th percentile, we would use 46, 52, 58, 60, and 62 and locate the middle value. Here Q3 is equal to 58. All of this is shown in Table 3.19.

TABLE 3.19. Computing Q1, Q2, and Q3 for an Even Number of Values

10	15	20	30	40	46	52	58	60	62
		20 is the 25th percentile or Q1		43 is the 50th percentile (i.e., the median or Q2)			58 is the 75th percentile or Q3		

Computing and Interpreting z Scores

Percentiles help us understand the relative standing of a value when compared to all other values in a dataset. In many instances, however, statisticians want to know exactly how far a given value is away from the mean or from another value in the dataset. We saw earlier that we can determine these differences in terms of the standard deviation, but that was somewhat tedious. Let's both simplify and expand that concept by using the standard deviation to compute a z score:

$$z = \frac{x - \bar{x}}{s}$$

Here we are using the symbols for the sample mean (i.e., $\bar{x}$) and the sample standard deviation (i.e., s) within the formula. Let's use the data in Table 3.20. We have a mean of 10, and to make life simple, let's use a standard deviation of 1 to calculate the z score for an observed value of 12.

TABLE 3.20. Data for Computing z Scores

7	8	9	10	11	12	13

We can compute the z score in three steps:

1. $z = \frac{12 - 10}{1}$

2. $z = \frac{2}{1}$

3. $z = 2$

In this case, our z score is 2; this means our value, 12, is two standard deviations above the mean. We might also find that the z score we compute is negative. For example, using the same standard deviation and mean score, along with an observed score of 7, we would compute a z score of –3. This means our observed score is 3 standard deviations below the mean of the dataset. We can see this as follows:

1. $z = \frac{7 - 10}{1}$

2. $z = \frac{-3}{1}$

3. $z = -3$

Both of these examples are shown in Table 3.21.

TABLE 3.21. z Scores

Value	7	8	9	10	11	12	13
z score	–3	–2	–1	0	+1	+2	+3

In most instances, our z score will not be a whole number; it's quite common to compute a fractional z score. For example, let's use a mean of 15, a standard deviation of 3, and an observed value of 20:

1. $z = \dfrac{20 - 15}{3}$

2. $z = \dfrac{5}{3}$

3. $z = 1.67$

This would indicate that the observed value, 20, is 1.67 standard deviations above the mean.

Using z Scores for Comparison

Besides telling us the relative position of a value within a dataset, the z score can be used to compare values in two different datasets. An example will make this clear.

Suppose a student made 65 on the math section of an achievement test and an 81 on the reading section of the same test. The student's parents, upon seeing these scores, might think their child was doing well in reading but not as well in math. As we learned in our discussion of percentiles, just knowing the raw score on exams of this type is sometimes not enough; we need to know how well the student did in relation to the other students taking the exam. We could do this by breaking the score down into percentiles, but that would involve sorting all of the students' scores and then figuring out all of the different percentile values. Instead of doing this, we can use the z score to report the child's scores in both math and reading.

In order to do that, let's use a sample mean of 50 and a standard deviation of 5 for the math test. Using the child's score of 65, we would compute a math z score of +3; that means the child scored 3 standard deviations above the mean. Using a sample mean score of 90 and a standard deviation of 3, our z score for the reading section would be −3 [i.e., (81−90)/3]; this means the child scored 3 standard deviations below the mean. By looking at this we could tell, contrary to the impression made by the raw scores, our student was well above average in math but well below average in reading. This is shown in Table 3.22.

TABLE 3.22. Comparison of z Scores and Raw Scores

	z = −3	z = −2	z = −1	z = 0	z = +1	z = +2	z = +3
Math	35	40	45	50	55	60	65
Reading	81	84	87	90	93	96	99

When we get into the section of graphical descriptive statistics, we will revisit the subject of z scores. We will find that we can use a predefined table to assign actual percentile values to our z scores, thereby eliminating the problem of sorting all of the test scores and then determining overall percentile values.

Computing and Interpreting T-Scores

Sometimes trying to explain z scores relative to the average of a group is rather difficult. For example, some people just can't comprehend what you are trying to say when you tell them a student has a score of +2 or a –3.07. To address this problem, we can use the T-score, which is nothing more than the z score presented in another form.

Before we learn how to compute it, let me point out something that confuses some people. Later in the book we are going to learn to use a t-test, and it will involve computing a t value; they are not related to what we are computing here. In this case, we have an upper-case T; the t-test will use a lower-case, italicized t. In order to compute the T-score, use the following formula:

$$T = (z * 10) + 50$$

You can see that all we need to do to compute T is to multiply our z value by 10 and then add 50 to it. If we used the z score of +1 we just computed using the math score, the T-score for that same value would be 60; when our z score was –3, our T-score would be 20 (i.e., –30 + 50).

When you compute the T-score, the average will always be 50 with a standard deviation of 10. For example, in the prior example we had an average math score of 50. This means that if we had an observed value of 50, the z score would be zero. It also means that when we are computing the T-score using a z score of zero, the answer is still zero. If we add zero to 50, the mean of our T-score distribution is 50. You can see that in Table 3.23.

TABLE 3.23. Comparison of the Mean, z Score, and T-Score

Observed value	Mean	z score	T-score
50	50	0	50

Now, let's use our observed score of 65. Since it is 3 standard deviations above the mean, our z score is +3; this means our T-score is 80. When we have an observed value of 35, our z score is –3; this translates into a T-score of 20. Both of these examples are shown in Table 3.24.

TABLE 3.24. Using the Mean, z Score, and T-Score to Compare Values in Two Datasets

Mean	Observed value	Standard deviation	z score	T-score
50	65	5	+3	80
50	35	5	–3	20

At this point, if we wanted to show these test scores to a parent, we could explain that the T-scores range from 20 to 80, with an average of 50. This would provide them with information in a format they are more used to seeing.

Stanines

Stanine (short for "standard nine") scores divide the set of values we are looking at into nine groups. Stanine scores are frequently used in education, especially to report standardized test scores and to compare groups of students. Computing a stanine score is very easy; simply multiply the z score by 2 and add 5:

$$\text{Stanine} = (z * 2) + 5$$

For example, let's use a dataset with a mean score of 40 and a standard deviation of 8. If we have an observed value of 56, our z score is 2.

1. $z = \dfrac{56 - 40}{8}$

2. $z = \dfrac{16}{8}$

3. $z = 2$

When we insert our z score of 2 into the stanine formula above, we wind up with a stanine score of 9:

$$\text{Stanine} = (2 * 2) + 5$$

Here we need to check to determine if the result is fractional; if so, round it up to the next number. If the value is greater than 9, you just use 9; this only occurs in a very few cases.

In other cases, you might be working with a z score that is not even. For example, we'll leave our observed value as 56 and our mean as 50, but let's change our standard deviation to 7; this would leave us with a z score of .857.

1. $z = \dfrac{56 - 50}{7}$

2. $z = \dfrac{6}{7}$

3. $z = .857$

We could then substitute our z score into the stanine formula:

$$\text{Stanine} = (.857 * 2) + 5$$

This would result in a stanine score of 6.714 which, when we removed everything to the right of the decimal point, would result in an actual score of 6. By looking at this example, educators can see that a student with a high stanine number is doing better than a student with a lower stanine number and, in many instances, will use these scores for ability grouping. For example, it is common to see children with stanine scores of 1 through 3 included in a low-ability group, students with scores of 4 through 6 in an intermediate group, and students with the highest scores placed in an advanced group.

Putting It All Together

In order to get a better feel for the relationship between each of the measures of relative standing, look at Table 3.25. You can see that we have taken the mean and standard deviation of a dataset and, using an observed value, computed the z score, the T-score, and the stanine score for that value. As you examine this table, three things will become obvious:

1. When an observed value is greater than the mean, the z score will be greater than zero. That will cause the T-score to be greater than 50 and the stanine to be greater than 5.

2. When an observed value is equal to the mean, the z score will equal zero. This will cause the T-score to equal 50 and the stanine to equal 5.

3. When an observed value is less than the mean, the z score will be less than zero. This will cause the T-score to be less than 50 and the stanine to be less than 5.

This is all seen in Table 3.25.

TABLE 3.25. Comparing Standard Deviations, z Scores, T-Scores, and Stanines

Mean	Observed value	Standard deviation	z score	T-score	Stanine score
20	15	5	−1	40	3
80	80	5	0	50	5
55	60	5	1	60	7

Using SPSS for T-Scores and Stanines—Not So Fast!

Up to this point, we have been able to verify our results using the SPSS software. Unfortunately, at this point, we run into somewhat of a snag. First, SPSS will not easily compute T-scores or stanines, but we're OK with that. A good consumer of statistics needs only to be able to recognize and interpret those values. The second part of our dilemma is that, while SPSS will compute a z score, it is presented in spreadsheet format; it is easy to do but presents the results in a manner we're not expecting. Knowing that, the following SPSS spreadsheet contains the same values we used earlier to compute percentiles and quartiles; you can see I have selected Analyze, Descriptive Statistics, and, because it gives us more detailed output, I have chosen the Explore option, shown in Figure 3.4.

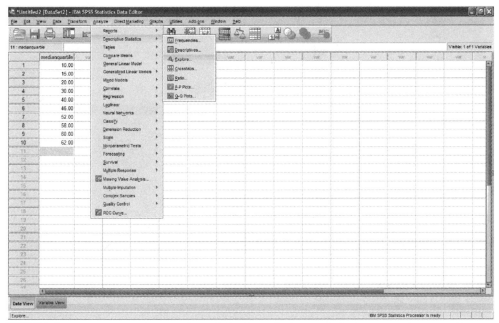

FIGURE 3.4. Using the Explore option of the Descriptive statistics command.

Once Explore is selected, we can choose the variables we want to analyze and the statistics we want to compute; in this case, in the Display field at the bottom of the Explore box I've chosen to show Statistics only; in the Explore Statistics box, I've selected Descriptives and Percentiles (Figure 3.5).

Pressing Continue and OK will show us the descriptive statistics, but let's focus on the percentiles in Figure 3.6. As I said earlier, statisticians love to argue about mundane topics and there is no one agreed-upon way to compute quartiles. Because of that, SPSS uses two different computations: weighted average and Tukey's hinges. Although I didn't label it as such, in our manual calculations we used Tukey's formula. Because of that, you can see that the values we computed for the median, as well as the first and second quartiles (i.e., the 25th and 75th percentiles), match exactly. Besides this formula, what else was Tukey famous for? If you said that he coined the word "software," you're right!

Finally, let's see how SPSS computes z scores for us. Let's start with the same dataset but change the variable name to zscore to avoid confusion; we'll then select Analyze and Descriptive Statistics (Figure 3.7). Again we'll get the Descriptives screen, but this time we're going to do something a bit differently. First, within the Descriptives box, we'll check the box in the lower left that says "Save standardized values as variables"; then we'll select any of the statistics we want in the Options box. In this case, we've asked for the mean score and the standard deviation (Figure 3.8).

Once we select Continue and OK, SPSS will do two things. First, it will compute and display a mean score of 39.3 and a standard deviation of 19.45964 for the variable we called z score; this is shown in Figure 3.9.

More importantly, when we return to the Data View screen, shown in Figure 3.10,

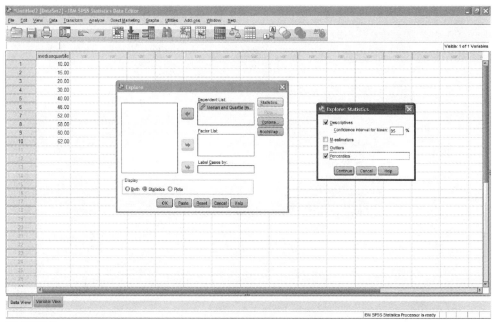

FIGURE 3.5. Using the Explore option to calculate descriptive statistics and percentiles.

we'll see that SPSS has saved the actual computed z score for each value in the column labeled Zzscore. In short, it has done nothing more than put a "z" in front of whatever we named the original variable. For example, we can see that the z score for 62 is 1.16652. This means that 62 is 1.16652 standard deviations above the mean of 39.3. We can check that with a bit of basic math:

Percentiles

Percentiles	Weighted Average(Definition 1) Median and Quartile	Tukey's Hinges Median and Quartile
5	10.0000	
10	10.5000	
25	18.7500	20.0000
50	43.0000	43.0000
75	58.5000	58.0000
90	61.8000	
95	.	

FIGURE 3.6. Output from the Frequencies command showing percentiles.

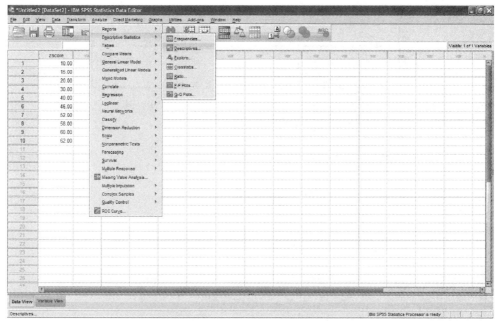

FIGURE 3.7. Using the Descriptive statistics option.

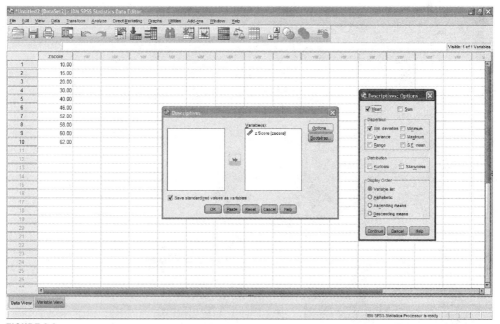

FIGURE 3.8. Using the Descriptive statistics option to compute the mean and standard deviation.

Descriptive Statistics

	N	Mean	Std. Deviation
z Score	10	39.3000	19.45964
Valid N (listwise)	10		

FIGURE 3.9. Output from the Descriptive statistics option showing the mean and standard deviation.

1. $z = \dfrac{62 - 39.3}{19.45964}$

2. $z = \dfrac{22.7}{19.45964}$

3. $z = 1.17$

Again, we've shown that we're as good as a computer; let's wrap this chapter up.

Summary

In this chapter, we continued talking about Step 4, learning to understand how spread out our dataset is, using measures of dispersion; we also learned how to determine the relative position of a value within a dataset when compared to a given measure of central tendency. We'll continue with Step 4 of our six-step model

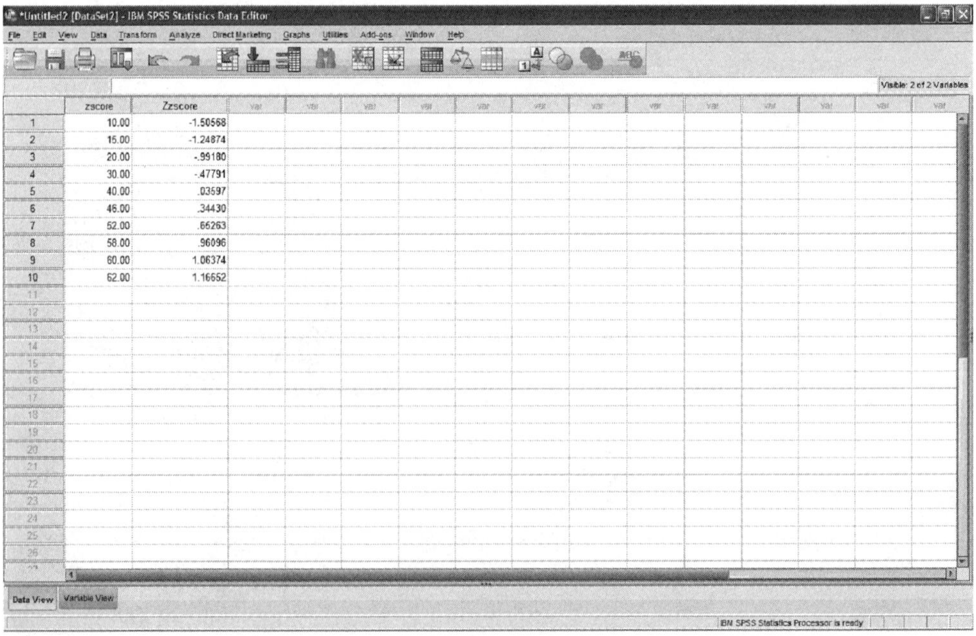

FIGURE 3.10. Computed z scores created as a new column in the Data View spreadsheet.

by using graphical tools to examine our data. Just like everything else we've talked about up to this point, these tools are essential in deciding which inferential statistical tool we will use to analyze our data.

Do You Understand These Key Words and Phrases?

computed range	deviation from the mean
measures of dispersion	measures of relative standing
percentile	possible range
quartile	range
standard deviation	stanine
T-score	variance
z score	

Do You Understand These Formulas?

Computed range:
Largest *observed* value minus the smallest *observed* value

Percentile:
Percentile = [(Below + 1/2 Same)/N] * 100

Population standard deviation:
$$\sigma = \sqrt{\frac{\Sigma(x - \mu)^2}{N}}$$

Population variance:
$$\sigma^2 = \frac{\Sigma(x - \mu)^2}{N}$$

Possible range:
Largest *possible* value minus the smallest *possible* value

Sample standard deviation:
$$s = \sqrt{\frac{\Sigma(x - \bar{x})^2}{n - 1}}$$

Sample variance:
$$s^2 = \frac{\Sigma(x - \bar{x})^2}{n - 1}$$

Stanine:

$$\text{Stanine} = (z * 2) + 5$$

T-score:

$$T = (z * 10) + 50$$

z score for a sample:

$$z = \frac{x - \bar{x}}{s}$$

Quiz Time!

Table 3.26 shows a dataset representing the ages of employees working for three different companies. Assuming a minimum working age of 18 and mandatory retirement at 65:

TABLE 3.26. Ages for Employees Working for Three Companies

Company A	35	25	44	62	55	22	31	41	36	65
Company B	47	37	25	24	22	31	33	35	27	36
Company C	20	40	40	35	42	47	50	52	54	60

1. What is the possible age range of each company?

2. What is the possible age range of all companies combined?

3. What is the computed age range of each company?

4. What is the computed age range of all companies combined?

5. What does the computed versus actual age range tell you about the dispersion computed in Questions 1–4?

6. What is the standard deviation of age in each company?

7. What does the standard deviation tell you about the dispersion of age in each company?

8. What is the standard deviation of age in all companies combined?

9. What is the variance of age in each company?

10. What is the variance of age when all companies are combined?

11. What does the variance of age when all companies are combined tell you about their dispersion?

Let's check to make sure we completely understand these important topics.

1. You can see in Table 3.27 that there are 10 sets of mean scores, observed scores, and standard deviations. For each of these, compute the z score and T-score. Following that, you should rank the mean values and compute the stanines.

TABLE 3.27. Data for Computation of z Scores, T-Scores, Stanines, and Ranked Mean Scores

Mean	Observed value	Standard deviation	z score	T-score	Stanine	Ranked mean score
30	33	2.00				
48	52	5.00				
55	54	3.00				
71	77	7.00				
14	8	2.70				
23	35	5.00				
61	48.6	2.90				
100	114	6.33				
81	78.5	1.55				
47	60.00	12.0				

2. Intelligence tests have shown that the average IQ in the United States is 100 with a standard deviation of 15. What z score would enable students to qualify for a program where the minimum IQ was 130?

3. Let's assume that average income of families in the United States is $22,000; we also know that families whose annual income is less than 2 standard deviations below the mean qualify for governmental assistance. If the variance of the incomes in America is $9,000, would a family making $19,500 a year qualify?

CHAPTER 4

Graphically Describing the Dependent Variable

Introduction

As we have already said, there are two ways of describing data—numerically and graphically. We spent the last two chapters talking about the numeric methods: measures of central tendency, measures of dispersion, and measures of relative standing. In this chapter we will focus on basic graphical tools, many of them already familiar to you. Let me say, right off the bat, that we are only going to look at examples of graphical statistics that were generated using SPSS. While many may disagree, I think creating the graphs "by hand" is a waste of time. I learned to do that in a basic statistics class years and years ago, and I have not had to do it since. We have the software to create the graphs for us; why not use it?

■ Graphical Descriptive Statistics

What is the old saying about one picture being worth a thousand words? You'll find how true this really is once you learn how to represent datasets graphically. Like many other topics in statistics, you'll find there are many different ways we could graphically describe our data; many of them are dependent on the type of data we have collected. In this chapter, we are going to focus on a few commonly used techniques. As mentioned earlier, we will use SPSS to create our graphs for us.

Graphically Describing Nominal Data

Before we get started, let's create an example of a nominal dataset. In Table 4.1 we can see the classification for 36 college students. In the table, FR stands for freshman, SP for sophomore, JR for junior, and SR for senior. By counting each occurrence, we can see the table has entries for 6 freshmen, 18 sophomores, 6 juniors, and 6 seniors. If we counted the values, we would see that the modal value is SP, as it occurs in the data more often than the other values. Since the data are nominal, it wouldn't make sense to try to compute the median or mean.

TABLE 4.1. Class Data

FR	FR	FR	FR	FR	FR	SP	SP	SP	SP	SP	SP
SP	SP	SP	SP	SP	SP	SP	SP	SP	SP	SP	SP
JR	JR	JR	JR	JR	JR	SR	SR	SR	SR	SR	SR

Although a table like this one is a perfectly acceptable way to describe our data graphically, it doesn't concisely show what we want to see. In order to try to get a better understanding, we could break it down further by creating a frequency table (Table 4.2) showing the count of each value as well as the percentage of the total that each frequency represents.

TABLE 4.2. Frequency Distribution of Class Data

Class	Frequency	Percentage	Cumulative percentage
FR	6	16.7	16.7
SP	18	50.0	66.7
JR	6	16.7	83.3
SR	6	16.7	100.00
Total	36	100.00	

Figure 4.1 shows how we could use SPSS to verify this by entering the data into SPSS, then select Analyze, Descriptive Statistics, and Frequencies. Selecting Frequencies brings us to the next step (Figure 4.2). By selecting Class as our Variable, checking the box titled "Display Frequency Tables" in the lower left, and clicking on OK, we get the output shown in Figure 4.3; as expected, it matches what we computed manually in Table 4.2.

Pie Charts

Another graphical tool that we've all seen used quite often is a *pie chart*. It is nothing more than a picture that looks like a pie, with each slice representing how often a given value occurs in the dataset. Using the same data from the frequency table, if you look closely at Figure 4.4, you will see options for many types of graphs. We will look at many of these graphs in the following pages but, for now, select Graph, Legacy Dialogs, and Pie. After selecting Pie, the dialog box shown in Figure 4.5 will appear. We want our Slices to represent the number of cases and we've defined our cases as Class. Entering OK results in the pie chart shown in Figure 4.6.

Doesn't this really make a difference? It becomes immediately clear, with just a glance, that there are as many sophomores as there are all the other classes put to-

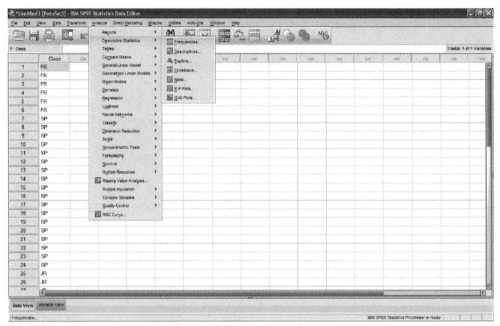

FIGURE 4.1. Using the Frequencies command.

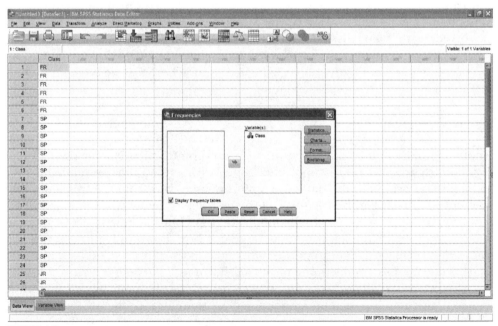

FIGURE 4.2. Using the Frequencies command for class (nominal) data.

Class

		Frequency	Percent	Valid Percent	Cumulative Percent
Valid	FR	6	16.7	16.7	16.7
	JR	6	16.7	16.7	33.3
	SP	18	50.0	50.0	83.3
	SR	6	16.7	16.7	100.0
	Total	36	100.0	100.0	

FIGURE 4.3. Output from Frequencies command for class data.

gether. It is also very obvious that there are an equal number of students in each of the other three classes.

This verifies what we just saw in our frequency table. The percentage for freshmen, juniors, and seniors is the same (i.e., 16.7% each), and the number of sophomores is the same as the other three groups combined (i.e., 50%). Obviously, the table showing the frequency count is easy to read, but if you want to get your point across, the picture is far more dynamic.

Bar Charts

Another easy way to present data is using a *bar chart*. As I said earlier, when we looked at the box where we indicated we wanted to create a pie chart, there were options for

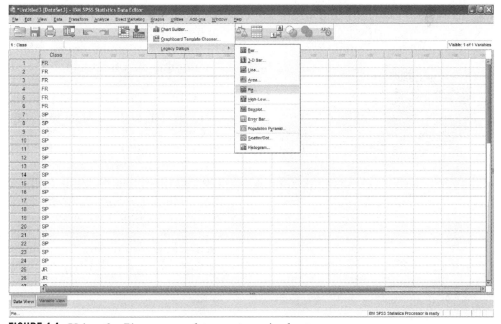

FIGURE 4.4. Using the Pie command to create a pie chart.

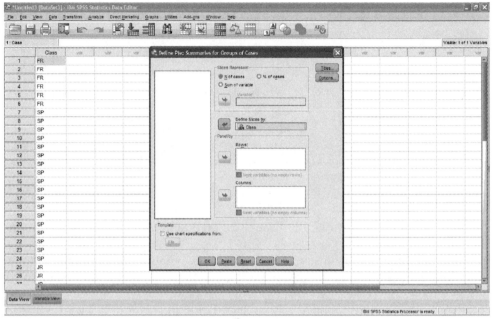

FIGURE 4.5. Setting up the Pie command to use class (nominal) data.

creating several other types of graphs. Rather than present all of those input screens again, Figure 4.7 shows us what SPSS would generate if we had selected Bar instead of Pie.

On the bar chart, the actual count for each of the variables is displayed; you can tell that by looking at the values on the left side of the graph. We can verify what we saw on the pie chart and in the descriptive statistics; the number of freshmen, juniors, and seniors is the same, while the number of sophomores is equal to the number in each of the other three groups combined.

Had we wanted to, we could have created a bar chart that showed the percentages each of these values represented. If we had, the values on the left side of the chart would have ranged from 0 to 100 and the values for each group would have represented their percentage out of 100 rather than their actual count. Of course, the relative sizes of the bars for each group would have been the same. As was the case with the pie chart, this bar chart really makes the data "come to life."

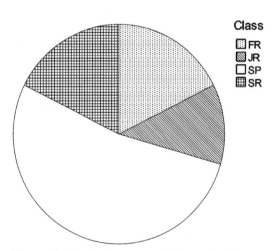

Class
- FR
- JR
- SP
- SR

FIGURE 4.6. A pie chart created from the class data.

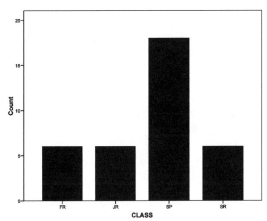

There is one additional thing to note here. Later in this chapter we'll talk about a *histogram*, and you'll see that it's very similar to the bar chart. The difference between the two is one of scale; the bar chart is used to present nominal data, while the histogram will be used when we're working with quantitative data.

FIGURE 4.7. A bar chart created from the class data.

Graphically Describing Quantitative Data

Although it is possible to use pie charts, frequency tables, and bar charts to display quantitative data, it is far more appropriate to use other tools we are about to discuss. Before we do, however, let's create a dataset we can use as we are creating the new graphs. In Table 4.3, we're showing data for a company that instituted a new tuition policy 10 years ago. The company is now noticing that, as the amount of money spent each year goes up, the number of resignations goes up as well.

TABLE 4.3. Data Showing a Positive Relationship between Tuition and Resignations

Year	Tuition spent	Resignations
1	$32,000	20
2	$38,000	25
3	$41,000	28
4	$42,000	26
5	$50,000	25
6	$60,000	30
7	$57,000	40
8	$70,000	45
9	$80,000	45
10	$90,000	90

The company is interested in seeing if a relationship exists between these two variables. It's afraid that some employees might be using the tuition for a better education only to prepare themselves for a move to another company where they will make more money! Later in the book, we will use one of two numeric tools, Pearson's or Spearman's correlation coefficient, to understand this better, but, for now, a scatterplot will give us a good feel for what's happening.

Scatterplots

A scatterplot is used to show the relationship between paired values from two datasets. In order to create the scatterplot, we will prepare the SPSS spreadsheet shown in Figure 4.8 to include the three variables.

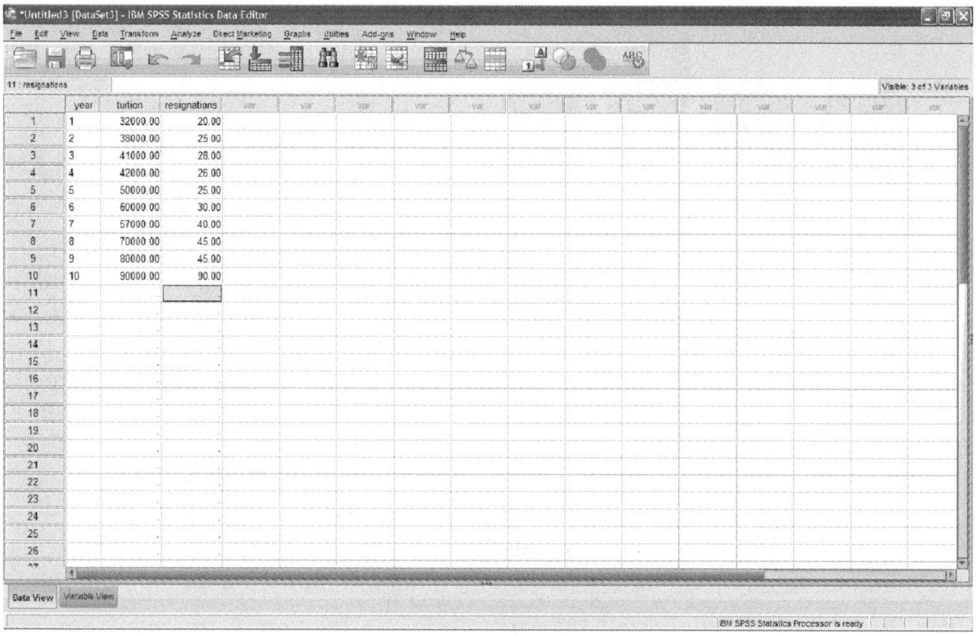

	year	tuition	resignations
1	1	32000.00	20.00
2	2	38000.00	25.00
3	3	41000.00	28.00
4	4	42000.00	26.00
5	5	50000.00	25.00
6	6	60000.00	30.00
7	7	57000.00	40.00
8	8	70000.00	45.00
9	9	80000.00	45.00
10	10	90000.00	90.00

FIGURE 4.8. Data for year, tuition spent, and resignations in the Data View spreadsheet.

Creating a scatterplot takes several steps, so let's go through it carefully to ensure we understand each one. First, although not shown, we would select Graphs, Legacy Dialogs, and Scatter/Dot. Then, as shown in Figure 4.9, we'll ask for a Simple Scatterplot. We then can select Tuition to be on the *x*-axis (i.e., the horizontal axis) and Resignations to be on the *y*-axis (i.e., the vertical axis). Notice, in Figure 4.10, that we're not including Year because it is not pertinent to the scatterplot itself. After clicking on OK, SPSS creates the scatterplot shown in Figure 4.11.

The left and bottom sides of the box are labeled with the possible range of Resignations and Tuition. Since our data didn't indicate to the software what the possible range actually is, SPSS created it by including a small amount above and below the actual values. The possible range of values for Resignations goes from about 15 to 100, while the possible range of Tuition goes from about $32,000 to $90,000.

SPSS then paired the variables, by year, and plotted them on the chart. For example, in the first year of the program, there were 20 resignations and the company spent $32,000 on tuition reimbursement. To plot that point on the chart, SPSS went out to the right on the *x*-axis to a point equivalent to $32,000 and then up the *y*-axis to a point equivalent to 20. The intersection of these two values is marked with a small circle. This process is then repeated for each of the sets of values.

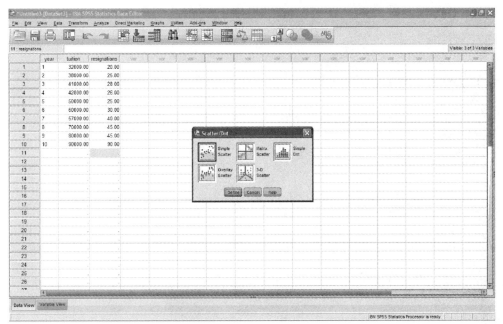

FIGURE 4.9. Selecting the Scatterplot command.

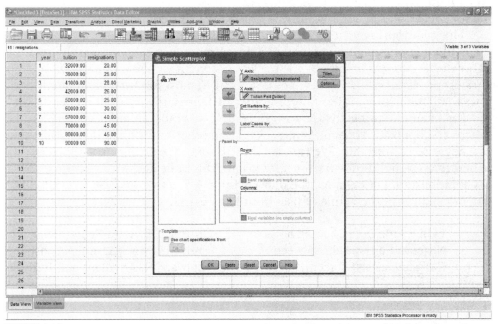

FIGURE 4.10. Setting up the Scatterplot command to use the tuition and resignations data.

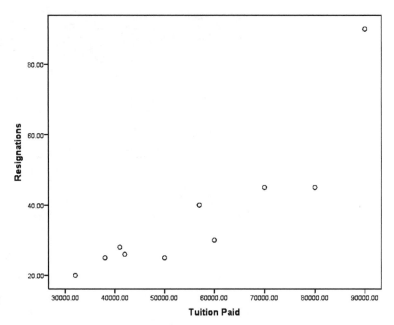

FIGURE 4.11. Output from Scatterplot command.

Although it is not necessary, SPSS can plot a *line of best fit* through the data points. This line, shown in the graph below, shows the trend of the relationship between the plotted variables. Later in the book we will compute the actual *correlation coefficient* between the two variables, but just by looking at Figure 4.12, we can see that the company might have a reason to be concerned. It does seem that the amount of tuition money spent and the number of resignations are related.

This plot represents a positive relationship; as one value goes up, the other goes up. If the line of best fit was exactly 45 degrees, we would know that the relationship between the two sets of variables is perfect in that, if we knew one value, we could accurately predict the other. This does not occur in most instances, however; we will see lines that look nearly perfect, and, in other instances, the line will be nearly flat.

For example, using the data in Table 4.4, we could create a scatterplot and see exactly the opposite. As one value goes up, the other tends to go down; this is known as a negative relationship. We can use these data to create Figure 4.13. Here, the line of best fit goes slightly down from left to right; this indicates we have a somewhat negative relationship. Let's change the data in Table 4.4 to that in Table 4.5. SPSS shows a flat, but slightly positive, line of best fit in Figure 4.14.

TABLE 4.4. Data Showing a Negative Relationship between Resignations and Tuition

Tuition	Resignations
$32,000	40
$38,000	35
$41,000	28

TABLE 4.5. Data Representing a Moderately Negative Relationship between Resignations and Tuition

Tuition	Resignations
$32,000	20
$38,000	35
$41,000	58

$42,000	25	$42,000	25
$50,000	20	$50,000	22
$60,000	22	$60,000	72
$57,000	15	$57,000	4
$70,000	25	$70,000	15
$80,000	10	$80,000	90
$90,000	30	$90,000	5

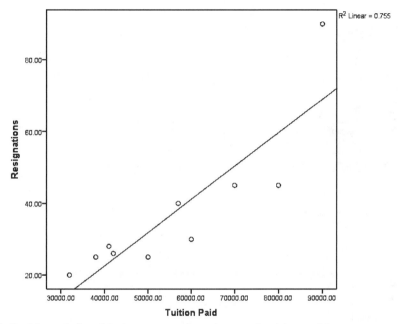

FIGURE 4.12. Positive relationship between resignations and tuition paid.

Histograms

A histogram is really nothing more than a bar chart used to chart quantitative data. We'll use the data in Table 4.6, collected from our coworkers, when we asked them how many minutes they usually take for lunch. Using these data, SPSS would produce the histogram shown in Figure 4.15.

TABLE 4.6. Minutes Taken for Lunch

30	35	35	40	40	40	45	45	45	45	50	50
50	50	50	55	55	55	55	55	55	60	60	60
60	60	65	65	65	65	70	70	70	75	75	80

In a histogram, the numbers across the bottom of the chart represent the actual values in our dataset; it begins with 30 on the left and goes to 80 on the right. The

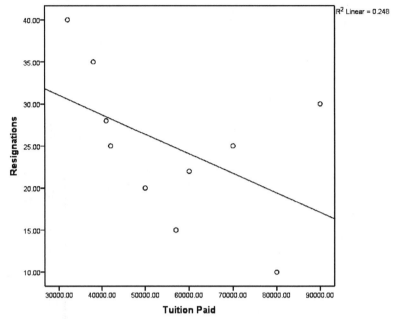

FIGURE 4.13. Negative relationship between resignations and tuition paid.

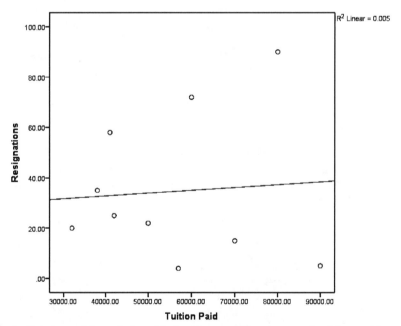

FIGURE 4.14. A slightly positive relationship between resignations and tuition paid.

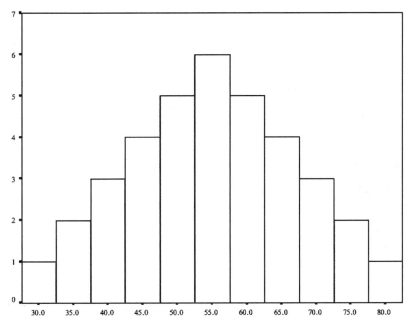

FIGURE 4.15. Output from the Histogram command.

values on the left side of the chart are the number of times a given value appears in the dataset. As you can see, the value 30 appears once in the dataset, and the bar representing it stops at the line showing the value of 1. We could double-check ourselves by seeing that the value 50 occurs 5 times in both the table and the bar chart. We'll use the histogram extensively in the next chapter to discuss the *normal distribution*, but for now, let me give you a word of warning about graphical statistics.

Don't Let a Picture Tell You the Wrong Story!

Like we said, a picture tells a thousand words; unfortunately, it's also been said that there are three kinds of lies—lies, damn lies, and statistics. Given that, it probably won't surprise you that statistics have been used, in many cases, to try to mislead, misinform, or manipulate an unsuspecting victim. This seems to be especially true when graphical statistics are used.

For example, let's imagine a local environmental group denouncing the effects of humans on the natural habitat of brown bears in the Rocky Mountains. As an example, they compare the life expectancy of the brown bears to their cousins above the Arctic Circle. They first explain that the average life expectancy of all bears ranges from 0 to 30 but adamantly believe that the following graph clearly shows that brown bears have a far shorter life expectancy than polar bears. Based on this, they demand the government spend more money on environmental conservation; their results are shown in Figure 4.16.

Although the environmentalists certainly seem to have a case, do you notice any-

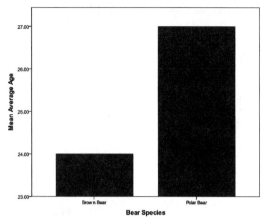

FIGURE 4.16. Histogram showing the relationship between ages of bears.

thing suspicious about this graph? If not, look on the left side of the chart. As you can see, the possible values range from 23 to 27. This means that the bars showing average life expectancy are not actually proportionate to the possible range of 0 to 30; instead they are proportionate to a much shorter possible range. This makes the difference between the average life expectancy of brown bears and the average life expectancy of polar bears look much more dramatic than it really is. If we were to plot the average life expectancy using the true possible range, the chart would look like that shown in Figure 4.17.

That makes quite a bit of difference, doesn't it? While the first graph might support the environmentalists' concerns, the second graph shows there is really very little difference in the average life span of brown and polar bears. This isn't to say that we shouldn't worry about a shorter life expectancy of brown bears, but we should worry about graphs like this where the scale has been changed; it happens more often than you would believe!

Let's look at another example. In this case, United States Olympics officials are desperately begging for more money. After all, they claim, their athletes won the fewest medals in the recent summer games; this is shown in Figure 4.18. In this case, it appears that the Olympic Committee may have a valid concern; the United States does have the smallest "piece of the pie." Unfortunately, however, the committee seems to be trying to pull the wool over our eyes. By looking at the list of "countries" shown in the upper-right corner we can see that, instead of being compared to other individual countries, the United States is being compared to each continent and the multiple countries they represent; it's no wonder it looks like the United States is lagging behind! Again, while this might sound like an outlandish example (i.e., our Olympic Committee would never do this), I've seen graphs far more misleading than this being sold as the truth.

That being said, always keep a careful eye out for misleading statistics; they're everywhere. Being able to identify them and understand how they're being misused will not only make you a hero among your friends, but it will also help you become a better consumer of statistics.

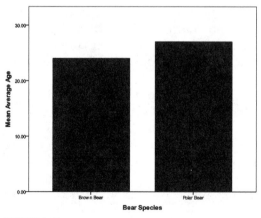

FIGURE 4.17. Adjusted histogram showing the relationship between ages of bears.

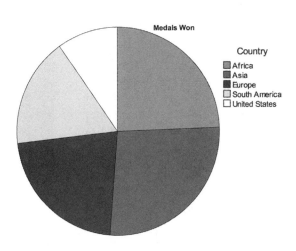

Medals Won

Country
■ Africa
■ Asia
■ Europe
☐ South America
☐ United States

FIGURE 4.18. Olympic medals won.

Summary of Graphical Descriptive Statistics

To this point, we've covered the very basic graphical tools for describing numeric data; we are used to seeing pie charts, bar charts, histograms, and scatterplots. News reports, financial documents, and advertisements, for example, are quick to use these tools to make sure their audience "gets the point." We're going to move forward now and use a histogram to demonstrate the idea of the normal distribution. As I've said about so many things, this will really help us as we move forward with the idea of inferential decision making.

The Normal Distribution

As you've seen, there are quite a few tools for graphically describing numeric data. While each of the topics we've covered is important in its own right, at this point, we need to start tying everything together. Let's start by discussing the relationship between a histogram and a *normal distribution*.

The Histogram and the Normal Distribution

In many instances, if we plot data values from a relatively large sample or population, the data form a "mound" or "bell"-shaped distribution clustered about the center of the dataset. For example, we saw that earlier with the data describing the number of minutes employees take for lunch. While all data distributions aren't shaped accordingly, when they are, we call it a *normal distribution*. By definition, in datasets that are normally distributed, the mean, median, and mode are equal, and the remaining values are symmetrically distributed around the mean. When that happens, we wind up with the distribution we saw earlier; it's shown here again in Figure 4.19.

We can demonstrate the idea of a normal distribution in Figure 4.20 by tracing a line over the tops of each of the bars. When we do, we can see that our data distribution is, in fact, bell-shaped. In this case, we have a normal distribution. The mean, median, and mode are all equal, and the other values are distributed symmetrically around the center. It would be nice to have a normal distribution every time we collect data; that would allow us to use a discrete set of statistical procedures, called parametric statistics, to analyze our data. You may already be familiar with the names of many of these statistical tests: the *t-test* and the *analysis of variance (ANOVA)*, for example. Although using these parametric tools will be the primary focus of the latter part of this book, we have to keep two things in mind.

First, remember that histograms are used to plot quantitative data. As we talked

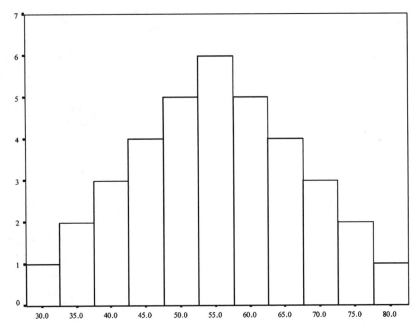

FIGURE 4.19. Histogram of a normal distribution.

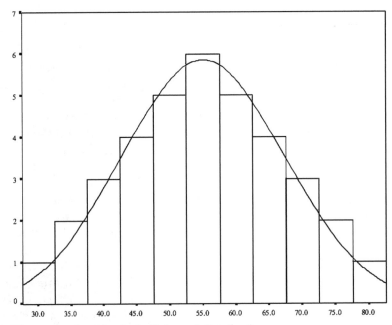

FIGURE 4.20. Histogram showing the bell-shaped distribution.

about earlier, when we are working with nominal- or ordinal-level data, we are forced to use a less powerful set of tools called nonparametric statistics. Like the parametric statistics, you may already be familiar with some of them; the *chi-square test*, for example, is the most widely recognized.

Second, the use of parametric statistics is based on quantitative data that are normally distributed. In most cases the distribution doesn't have to be perfectly bell-shaped since the inferential statistics are powerful enough to allow for some latitude within the data. In certain instances, however, a researcher might find a set of quantitative data where the distribution is so distorted that parametric statistics will not work. Right now we will not get into the underlying reasons, so let's just leave it at this. In this book, when we have nominal or ordinal data, we will use nonparametric statistics; if we have quantitative data, we will use parametric statistics.

Things That Can Affect the Shape of a Distribution of Quantitative Data

Although it would be nice, a perfect normal distribution rarely happens. Think about it—if we are looking at the heights of students in an elementary class, we could compute the average. How often, however, would the heights less than average and the heights greater than average be exactly equally distributed? You might have a few more shorter students than taller students or vice versa; either way it would affect how the distribution would look. This type of distortion is very common and very rarely leads to having a data distribution where both sides of the curve are perfectly symmetrical. This isn't a big concern in most instances because, like I said, the inferential tests we will use can compensate for problems with the shape of a distribution up to a certain degree. Knowing that, let's look at a few things that can affect the way a distribution looks.

Skewness

Skewness means that our data distribution is "stretched out" to one side or the other more than we would expect if we had a normal distribution. When we have more values greater than the mean, we say the distribution is *positively skewed* or *skewed to the right*. When we have more values than expected less than the mean, we say that it is *negatively skewed* or *skewed to the left*. Let's take a look at each.

POSITIVE SKEWNESS

Let's add these values—90, 90, 90, 100, 100, and 105—to the lunch-time data we have been using. This means we have more data points on the high end of the scale than we do on the lower end. In Figure 4.21, you can see that this causes the normal curve to be far more stretched out on the right side than it is on the left side. When this occurs, we say the dataset is *positively skewed* or skewed to the right.

In a dataset that is positively skewed, the mean score will be larger than the median score. Since we already know how to use SPSS to create graphs, let's just look at the SPSS printout shown in Figure 4.22; in it you can see that the mean is 60.83 while

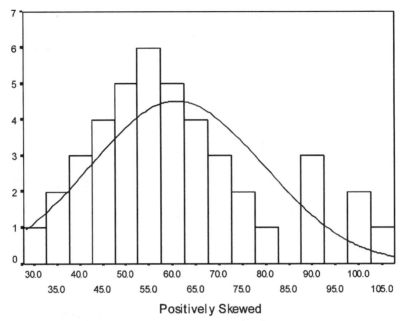

FIGURE 4.21. Positive skewness.

the median is 57.5. Larger degrees of skewness would cause this difference to become even greater. Besides just looking at the data distribution and the difference between the mean and median, we also have statistical formulas to help us to determine the degree of skewness. For example, the output below includes a field labeled "skewness." Its value, either positive or negative, indicates how skewed the data distribution is; the further away from zero it is, the greater the skewness is in one direction or the other.

In this case, we have a skewness statistic of .734. Since this value is greater than zero, it verifies that the dataset is positively skewed. This only becomes problematic, however, if the value is greater than +2. Anything larger would tell us that the distribution is so skewed that it will interfere with our ability to use parametric inferential statistics. In those cases we would have to use a nonparametric alternative.

Before we move on, let me point out a couple of other things in Figure 4.22 you might have already noticed about the SPSS output. First, for some reason, it doesn't include the modal value. This isn't usually an issue as we can easily determine it by looking at the graphical data distribution, and there are also other commands in SPSS that will compute the mode for you. Second, the printout includes many descriptive statistics that we haven't discussed up to this point. A lot of these statistics, such as the Minimum and Maximum, are common sense; others will be covered later in the text.

NEGATIVE SKEWNESS

To demonstrate negative skewness, let's add these values to the original lunch-time dataset: 5, 10, 10, 20, 20, and 20. By doing that, we can see in Figure 4.23 that the distribution is more spread out on the left side than it is on the right.

			Statistic	Std. Error
Scores	Mean		60.8333	2.85599
	95% Confidence Interval for Mean	Lower Bound	55.0655	
		Upper Bound	66.6011	
	5% Trimmed Mean		60.0926	
	Median		57.5000	
	Variance		342.581	
	Std. Deviation		18.50895	
	Minimum		30.00	
	Maximum		105.00	
	Range		75.00	
	Interquartile Range		21.25	
	Skewness		.734	.365
	Kurtosis		.041	.717

FIGURE 4.22. Descriptive statistics showing positive skewness.

In Figure 4.24, you can see the mean, 49.17, is less than the median, 52.50. As you might have expected, the relationship between the mean and the median is opposite that of the positively skewed distribution. The more negatively skewed the distribution is, the smaller the mean value will be when compared to the median value.

The negative skewness in the diagram is supported by a skewness index of –.734

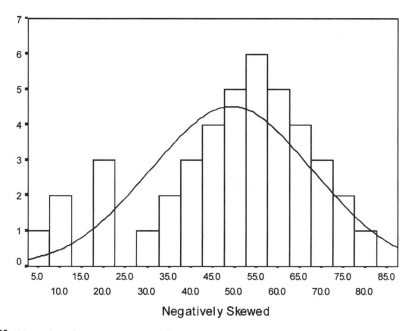

Negatively Skewed

FIGURE 4.23. Negative skewness.

			Statistic	Std. Error
Scores	Mean		49.1667	2.85599
	95% Confidence Interval for Mean	Lower Bound	43.3989	
		Upper Bound	54.9345	
	5% Trimmed Mean		49.9074	
	Median		52.5000	
	Variance		342.581	
	Std. Deviation		18.50895	
	Minimum		5.00	
	Maximum		80.00	
	Range		75.00	
	Interquartile Range		21.25	
	Skewness		-.734	.365
	Kurtosis		.041	.717

FIGURE 4.24. Descriptive statistics showing negative skewness.

which, although it is negative, is not less than –2 (i.e., it is between –2 and 0). As was the case with a positive skewness value less than +2, this means the normality of the data distribution is not significantly affected; while this distribution is not perfectly normal, it is still close enough to use parametric statistics.

Kurtosis

Kurtosis refers to problems with the data distribution that cause it to look either more peaked or more spread out than we would expect with a normal distribution. *Platykur-tosis* occurs when we have more values than expected in both ends of the distribution, causing the frequency diagram to "flatten" out. *Leptokurtosis* happens when we have more values in the center of the dataset than we might expect, causing the distribution to appear more peaked than it would in a normal distribution.

PLATYKURTOSIS

Once again, let's modify the original dataset we used. This time, however, instead of adding values to either end of it, let's remove one each of the values 50, 55, and 60 in order to help us understand platykurtosis. If we look at our descriptive statistics in Figure 4.25, we can see that the mean and median both equal 55, just as we saw in the normal distribution in Figure 4.19. In this case, however, the data values are not symmetrical around the mean. Instead we can see in Figure 4.26 that the normal curve is plotted slightly above the higher value. That means our dataset is flatter than would be expected if the distribution was perfectly bell-shaped.

As was the case with skewness, SPSS will calculate an index of kurtosis which, as seen in Figure 4.25, is –.719. This value, since it is negative, verifies what we saw in the graph. Much like the skewness numbers, however, the kurtosis number would have to be less than –2.00 to significantly affect the normality of the dataset.

Descriptives

			Statistic	Std. Error
Scores	Mean		55.0000	2.21906
	95% Confidence Interval for Mean	Lower Bound	50.4799	
		Upper Bound	59.5201	
	5% Trimmed Mean		55.0000	
	Median		55.0000	
	Variance		162.500	
	Std. Deviation		12.74755	
	Minimum		30.00	
	Maximum		80.00	
	Range		50.00	
	Interquartile Range		20.00	
	Skewness		.000	.409
	Kurtosis		-.719	.798

FIGURE 4.25. Platykurtosis descriptive statistics.

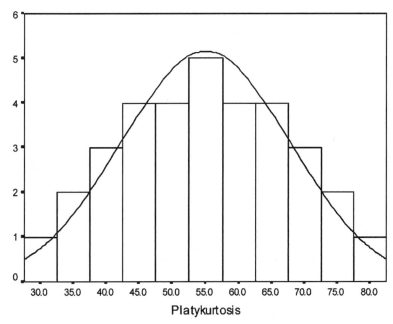

FIGURE 4.26. Platykurtosis.

LEPTOKURTOSIS

In other instances, there may be more values around the center of the distribution than would be expected in a perfectly normal distribution. This causes the distribution to be taller in the center than we would expect; this condition is called leptokurtosis. In Figure 4.27, we again have the mean, median, and mode equal to 55 (remember, we have to look at the actual diagram in Figure 4.28 to determine the mode),

			Statistic	Std. Error
Scores	Mean		55.0000	1.78174
	95% Confidence Interval for Mean	Lower Bound	51.3829	
		Upper Bound	58.6171	
	5% Trimmed Mean		55.0000	
	Median		55.0000	
	Variance		114.286	
	Std. Deviation		10.69045	
	Minimum		30.00	
	Maximum		80.00	
	Range		50.00	
	Interquartile Range		10.00	
	Skewness		.000	.393
	Kurtosis		.334	.768

FIGURE 4.27. Leptokurtosis descriptive statistics.

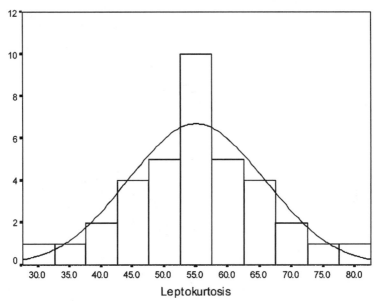

FIGURE 4.28. Leptokurtosis.

but the distribution isn't normal; you can see that we've added several occurrences of the value 55 as well as taken away some of the lower and higher values. As odd as it looks, however, the index of kurtosis for this distribution is only .334, which is still well within the range allowed by parametric statistics; these descriptive statistics are shown in Figure 4.27.

The Empirical Rule

Does the histogram in Figure 4.29 look familiar? It should; we have already used it to demonstrate a normal curve. This time, however, I have added some of the computed inferential statistics to the side of the chart. We can see that the mean is 55 and we've computed a standard deviation of 12.25. We can use these two values, along with the *empirical rule*, to better understand the dispersion of our data.

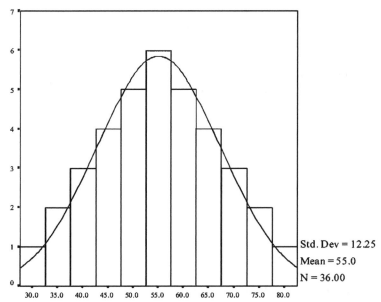

FIGURE 4.29. Normal distribution.

The first thing the empirical rule tells us is, if the distribution is approximately normally distributed, at least 68% of the values in the dataset fall within ±1 standard deviation of the mean. Using the descriptive statistics on the right side of Figure 4.29 means we know that just over two-thirds of the scores fall somewhere between 42.75 and 67.25. The rule goes on to say that about 95% of the scores will fall within ±2 standard deviations of the mean, 30.5 to 79.5, and nearly all of the scores (99.7%) will fall between ±3 standard deviations of the mean, 18.25 to 91.75. This also tells us that anything falling outside of these boundaries is an anomaly; values less than 18.25 or greater than 91.75 occur less than .30% of the time. We will go into more detail about this later, but for now let's look at an example of what we have just described.

You may not know it, but the average height for a male in the United States is 5 feet 9 inches (69 inches); the standard deviation for this distribution is 3 inches. This means that about 68% of men are somewhere between 66 and 72 inches tall (i.e., 69 ±1 standard deviation), about 95% of men are somewhere between 63 and 75 inches tall (i.e., 69 ±2 standard deviations), and nearly all men are somewhere between 60 and 78 inches tall (i.e., 69 ±3 standard deviations). The remaining .3% of men are either less than 60 inches or taller than 78 inches. Knowing that explains why we tend to really notice men that are very, very short or very, very tall; there are not too many of them.

Let's Tighten This Up a Little

You probably noticed that in the last section, I kept saying "about" while referring to a certain percentage from the mean. I did that to get you used to the concept of the empirical rule; now let's look at the exact percentages. What the empirical rule actually says is that, in a mound-shaped distribution, at least 68.26% of all values fall within ±1 standard deviation from the mean, at least 95.44% of all values fall between the mean ±2 standard deviations and nearly all (99.74%) values fall between the mean and ±3 standard deviations. Using our example of height, this now means only .26% of all men fall outside the mean ±3 standard deviations; they are even rarer than we thought!

Let's look at Figure 4.30. There you can see that I used *SD* for standard deviation instead of a symbol. I did this because the empirical rule applies to both mound-shaped samples and populations, so it wouldn't be appropriate to use either of the symbols discussed earlier.

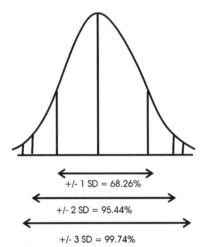

+/- 1 SD = 68.26%

+/- 2 SD = 95.44%

+/- 3 SD = 99.74%

FIGURE 4.30. The empirical rule.

It makes sense, by looking at this figure, that we can see percentage differences between the mean and any value above or below the mean. Although it isn't as clear conceptually, we can also look at percentages between any two values in the distribution, be they both above the mean, both below the mean, or one on either side of the mean. For example, we know that 68.26% of all values fall within ±1 standard deviation of the mean. It naturally follows that half of these values, 34.13%, would fall between the mean and +1 standard deviation, and half of the values, 34.13%, would fall between the mean and –1 standard deviation. You can see that in Figure 4.31.

If we wanted to determine the difference between the mean and –3 standard deviations, we would go through the same process. As shown in Figure 4.32, we know that the region between ±3 standard deviations is 99.74%; if we divide that in half, we know that 49.87% of the values lie in that range.

We can also measure areas under the curve that include values above and below the mean. For example, what percentage of values lies between 1 standard deviation

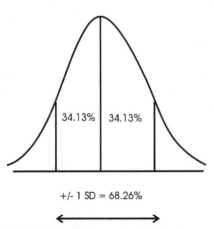

34.13%   34.13%

+/- 1 SD = 68.26%

FIGURE 4.31. Mean ±1 standard deviation.

below the mean and 3 standard deviations above the mean? All we would have to do is add the percentage of values between the mean and –1 standard deviation, 34.13%, to the percentage of all values that lie between the mean and +3 standard deviations, 49.87%. We would wind up with 84% of the values; this can be seen in Figure 4.33.

Understanding the relationship between standard deviations and the percentage of values they represent allows us to tie together something we learned earlier. For example, if we have a z score of +1, we know this represents one stan-

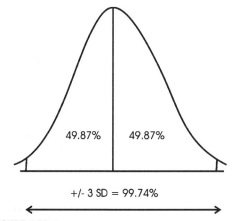

FIGURE 4.32. Mean ±3 standard deviations.

dard deviation above the mean. Given that we also know the percentage of values between the mean and a z score of +1 is 34.13%, the same would hold true if we had a z score of –1. If we had a z score of +2, it would represent 47.72% of the values between it, and the mean and a z score of –3 would include 49.87% of the values between it and the mean.

This works well if you have a whole number of standard deviations, but what happens if you have a fractional value? We can deal with that situation using Table 4.7, the *area under the normal curve table* or, as it is sometimes called, a *critical values of z table*; this table is also shown in Appendix A.

Although the table may look confusing, do not be deceived; it is actually very easy to use. Look at the leftmost column. These are the first two digits of any z score (i.e., the whole number and the first decimal position). In order to find the percent-age of values that lie between the mean and a given z score, go down this col-umn until you find the row where the value corresponds to your z score. For example, if we wanted to verify that the percentage for a z score of 1 is 34.13%, go down that column until you get to the row containing 1.0. Immediately to the right of that, you'll see the value .3413; this is the decimal notation for 34.13%. All you have to do is move the decimal point in the table two places to the right to get the percentage. In this case, you can see we are in the column where the heading is 0.00; this means we have a

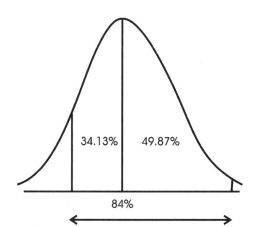

FIGURE 4.33. The area between –1 standard deviation and +3 standard deviations.

z score of exactly 1.00. If we had a z score of 1.04, we would have to go to the right, in the same row, until we get into the column under the header .04. In that case, the percentage would be .3508 or 35.08%.

Decimal Values versus Percentages

Although most z tables are shown in decimal format, many people find it easier to work with percentages instead of decimals, so that's how we will do it throughout the book. Again, all you have to do to change the decimal values to percentages is move the decimal point two spaces to the right. You can also multiply the decimal value by 100 and get the same result.

TABLE 4.7. Area under the Normal Curve (Critical Value of z)

z score	0.00	0.01	0.02	0.03	0.04	0.05	0.06	0.07	0.08	0.09
0.0	.0000	.0040	.0080	.0120	.0160	.0199	.0239	.0279	.0319	.0359
0.1	.0398	.0438	.0478	.0517	.0557	.0596	.0636	.0675	.0714	.0753
0.2	.0793	.0832	.0871	.0910	.0948	.0987	.1026	.1075	.1103	.1141
0.3	.1179	.1217	.1255	.1293	.1331	.1368	.1406	.1443	.1480	.1517
0.4	.1554	.1591	.1628	.1664	.1700	.1736	.1772	.1808	.1844	.1879
0.5	.1915	.1950	.1985	.2019	.2054	.2088	.2123	.2157	.2190	.2224
0.6	.2257	.2291	.2324	.2357	.2389	.2422	.2454	.2486	.2517	.2549
0.7	.2580	.2611	.2642	.2673	.2704	.2734	.2764	.2794	.2823	.2852
0.8	.2881	.2910	.2939	.2967	.2995	.3023	.3051	.3078	.3106	.3133
0.9	.3159	.3186	.3212	.3238	.3264	.3289	.3315	.3340	.3365	.3389
1.0	.3413	.3438	.3461	.3485	.3508	.3531	.3554	.3577	.3599	.3621
1.1	.3643	.3665	.3686	.3708	.3729	.3745	.3770	.3790	.3810	.3830
1.2	.3849	.3869	.3888	.3907	.3925	.3944	.3962	.3980	.3997	.4015
1.3	.4032	.4049	.4066	.4082	.4099	.4115	.4131	.4147	.4162	.4177
1.4	.4192	.4207	.4222	.4236	.4261	.4265	.4279	.4291	.4306	.4319
1.5	.4332	.4345	.4357	.4370	.4382	.4394	.4406	.4418	.4429	.4441
1.6	.4452	.4463	.4474	.4484	.4495	.4505	.4515	.4525	.4535	.4545
1.7	.4554	.4564	.4573	.4582	.4591	.4599	.4608	.4616	.4625	.4633
1.8	.4641	.4649	.4645	.4664	.4671	.4678	.4686	.4693	.4699	.4706
1.9	.4713	.4719	.4726	.4732	.4738	.4744	.4750	.4756	.4761	.4767
2.0	.4772	.4778	.4783	.4788	.4793	.4798	.4803	.4808	.4812	.4817
2.1	.4821	.4826	.4830	.4834	.4838	.4842	.4846	.4850	.4854	.4857
2.2	.4861	.4864	.4868	.4871	.4875	.4887	.4881	.4884	.4887	.4890
2.3	.4893	.4896	.4998	.4901	.4904	.4906	.4909	.4911	.4913	.4916
2.4	.4918	.4920	.4922	.4925	.4927	.4929	.4931	.4932	.4934	.4936
2.5	.4938	.4940	.4941	.4943	.4945	.4946	.4948	.4949	.4951	.4952
2.6	.4953	.4955	.4956	.4957	.4959	.4960	.4961	.4962	.4963	.4964
2.7	.4965	.4966	.4967	.4968	.4969	.4970	.4971	.4972	.4973	.4974
2.8	.4974	.4975	.4976	.4977	.4977	.4978	.4979	.4979	.4980	.4981
2.9	.4981	.4982	.4982	.4983	.4985	.4984	.4985	.4985	.4986	.4986
3.0	.4987	.4987	.4987	.4988	.4988	.4989	.4989	.4989	.4990	.4990

Determining the Area under a Curve between Two Values

Now, let's work more with the table to make sure we have the idea. What's the area under the curve between the mean and a z score of 2.44? Go down the leftmost column until you get to the row labeled 2.4 and then to the right until you're in the column under ".04." We can see the value is .4927 or 49.27%. What about the area between the mean and a z score of .03? Go over the row labeled 0.00 until you get to the column labeled .03. The value for the area under the curve is .0120 or 1.2%.

Finally, how about a z score of –1.79? Although the z score is negative, the table works the same. In this case, we would go down the leftmost column until we get to 1.7 and then across that row until we come to the column under .09; we can see the value there is .4633 or 46.33%.

In certain cases, we find it necessary to determine the percentage of data values that fall between two given z scores; we call this "determining the area under a curve." The general procedure is to use our table to determine the percentage for two z scores you're interested in and then, by addition or subtraction, determine the percentage of values between them. A couple of examples make this clear.

Let's use z scores of 1.10 and 2.21. We can see the percentage value for a z score of 1.1 is 36.43% and the z score of 2.21 is 48.64%. Since both values are positive, we subtract the smaller value from the larger value and find a difference of 12.21%. In other words, 12.21% of the values under the curve lie between a z score of 1.1 and a z score of 2.21. This is shown in Figure 4.34.

Now, let's assume we have a z score of –2 and a z score of +2.2, what's the percentage between them? Using the table, we see the value for 2.2 is 48.61% and the value for 2 is 47.72%. Unlike the prior example, here we have one positive value and one negative value. In instances such as this you have to add the two percentages together. In this case, when we add the two values together, we find that 96.33% of the values under the curve are contained in the range from a z score of –2 and a z score of +2.2. This is shown in Figure 4.35.

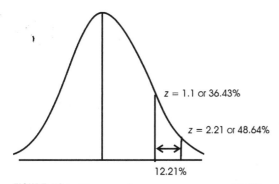

FIGURE 4.34. The area between z scores of 1.10 and 2.21.

Finally, let's look at a case where both z values are negative; let's use –1 (34.13%) and –3 (49.87%). In this case, since both values are on the same side of the mean, we have to again subtract the smaller number from the larger. We can plot them on our graph, shown in Figure 4.36, and we find that 15.74% of values fall between our two z scores.

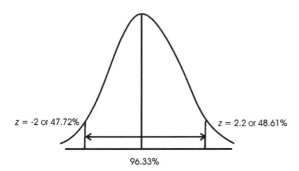

FIGURE 4.35. The area between z scores of –2 and +2.2.

Summary of the Normal Distribution

Understanding the normal distribution is the key to inferential testing since many of the classic statistical tests are based on the assumption that the data are normally distributed. As we saw, quantitative data that result in a perfect normal distribution have the same mean, median, and mode and the data values are symmetrically dis-

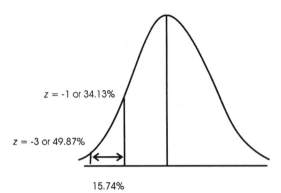

$z = -1$ or 34.13%

$z = -3$ or 49.87%

15.74%

FIGURE 4.36. The area between z scores of -1 and -3.

tributed on both sides of these measures of central tendency.

In instances where we have more values greater than the mean than expected, we see that the distribution is positively skewed, whereas datasets with more scores below the mean than expected are negatively skewed. In other instances, our data may be skewed on both ends, resulting in a normal distribution that is flatter than expected; we say this distribution is platykurtotic. Distributions that have more values in the center of the dataset than expected are called leptokurtotic. All of these aberrations can be controlled for by the robust nature of many of the parametric statistical tests. Sometimes, however, the distribution is so skewed or kurtotic that nonparametric tests must be used when evaluating our hypotheses. Keep in mind, however, that this is the exception and not the rule.

Finally, we used z scores to help determine the distance between two data points in the normal distribution and tied everything together by studying the empirical rule. All of this together, especially the understanding of distances under the curve, will be critical when we begin testing hypotheses based on data we have collected.

Do You Understand These Key Words and Phrases?

area under the normal curve table

empirical rule

graphical descriptive statistics

kurtosis

line of best fit

normal distribution

platykurtosis

scatterplot

skewed right

bar chart

frequency distribution table

histogram

leptokurtosis

negatively skewed

pie chart

positively skewed

skewed left

skewness

Quiz Time!

Since we said that we always use a computer to help us when we're graphically describing our dependent variable, we are in somewhat of a jam. Knowing we can't do that, let's use the following questions to understand conceptually where we are. When you are finished, check your answers at the end of the book.

1. If I wanted a quick and easy way to present the frequency of values in a dataset containing nominal data, which graphical tool(s) would I use? Why?

2. Using math data on the y-axis and the reading data on the x-axis, as shown in Table 4.8, what trend would the line of best fit show us if we plotted the data on a scatterplot?

TABLE 4.8. Math and Reading Scores

Math score	80	60	75	50	77	45	100
Reading score	20	100	75	90	59	65	95

3. If we plotted the salary data in Table 4.9 on the y-axis and years of education on the x-axis, what would our plot look like?

TABLE 4.9. Years of Education and Salary

Education	10	12	14	14	16	16	18
Salary in thousands	20	30	40	37	59	55	70

4. Explain the difference between histogram and measures of tendency? Can this be used with nominal data? Why? Quantitative data? Why?

5. What are the requirements for a dataset to be called a perfect normal distribution?

6. If a dataset had a mean of 45 and a median of 40, would it be skewed? Why?

7. If we plot a dataset with a mean of 60 and a median of 55, would it be either platykurtotic or leptokurtotic? Why?

8. Does a dataset with an equal mean and median always mean the data are normally distributed? Why?

9. Do we always have to use nonparametric statistics when we're working with datasets that are skewed or kurtotic?

10. Discuss the relationship between the mean, median, and skewness. How would we know if it is problematic for inferential decision making?

11. On each row in Table 4.10 there are two z values. Using these, determine the percentage of the curve for each and then calculate the difference between the two.

TABLE 4.10. Values for Quiz Time Question 11

z value 1	Area under the curve for z 1	z value 2	Area under the curve for z 2	Difference
3.01		2.00		
−2.50		3.00		
−1.99		−0.09		
1.50		−1.50		

CHAPTER 5

Choosing the Right Statistical Test

Introduction

Finally, we're to the point where we will choose the statistical test we need to test our hypothesis. Before we can, though, we've got to pass through a roadblock. In order to move forward, we need to go into a lot more detail about hypothesis testing. First, we will discuss the overall concept of inferential decision making and learn to manually compute a few of the basic statistical tests necessary for hypothesis testing; we'll also use SPSS to verify our results. Once we've done that, we'll be ready to move forward, select the test we need, and go on to Step 6.

The Very Basics

As we know, inferential statistics are used to help us make decisions based on data we have collected. Before we move forward, however, let's think about the word "inferential" for a minute. What does it really mean? First, it's easy to see that it's based on the word "infer"; that means we're going to make a decision based on evidence we collect. Since, in statistical decision making, we are generally using a sample of data from a population, this means we are drawing a conclusion based on limited evidence. As we saw earlier, that also means there is always a risk of making an incorrect decision based on chance or random error. What does that mean to us? Simple: inferring something is a long way from proving something! As we said way back in our words of wisdom, we never try to prove anything with inferential decision making.

Therefore, when we use sample data to make inferences about a population, we want to answer one or both of the following questions:

1. Can we estimate population parameters based on sample data? For example, if we know a sample statistic such as the mean, can we estimate the mean of the population?

2. Can we use sample data to test a belief about the population the sample was drawn from? For example, if we hypothesize that a population value, such as the mean, is equal to a given value; can we use the sample data to support that assertion?

In order to start answering these questions, let's start by talking about the *central limit theorem* and the *standard error of the mean*.

The Central Limit Theorem

A theorem is nothing more than a statement or an idea that has been demonstrated or proven to be true. For example, we use the Pythagorean theorem in geometry all the time; we use it to compute the length of an unknown side of a right triangle if we know the length of the other two sides. The theorem has been computed and verified for hundreds of years, so it's apparent that Pythagoras knew what he was talking about!

In developing the central limit theorem, French mathematician Pierre-Simon Laplace proved that if you take samples from a population, compute the mean of each sample, and then plot those means on a histogram, the distribution will be approximately normal. In fact, if you plot enough samples, the distribution will be normally distributed, and the mean of the distribution will be equal to the mean of the population from which the samples were drawn. I'm sure we can all appreciate the work Pierre put into this, but what does it mean? Let's use an example to help understand it.

Suppose you are teaching a section of Psychology 101 and give a final exam to all 50 of your students. Based on the results of the exam, you are interested in estimating the mean score of the population of all 1,000 students at your university who are taking the same course. While you are thinking about how to do it, you suddenly realize four things:

1. If you knew the final exam scores for the population of all students in your university you wouldn't have a problem. You could easily add up the scores for all 1,000 students in the population, divide that by 1,000, and wind up with the average score for the population.

2. Your class of 50 students is only a sample of the population of 1,000 students, and your class's average score may or may not be exactly equal to the mean score of the population.

3. Based on what we already know, the mean scores from the other samples (i.e., classes of 50 students) are going to be exactly equal to the population mean or, much more likely, they are going to fluctuate around the mean.

4. The more samples you select and plot, the closer the overall mean of the samples will be to the mean of the population. If you take an infinite number of samples, the overall mean of the samples will be exactly equal to the population mean.

In other words, if we are going to attempt to make inferences about a population parameter based on a sample statistic, it becomes readily apparent that understanding the fluctuation of the values is the key to making such decisions. The central limit theorem helps us to do exactly that.

The Sampling Distribution of the Means

In order to demonstrate this idea, imagine that we could take an infinite number of random samples from the population of Psychology 101 students. Each of these samples would include 50 students, the same as the number in your class. If we calculated the mean of each of these samples and plotted them on a histogram, something interesting would occur; the more mean scores we plotted, the more the plot would start looking like a normal distribution.

For example, suppose you use the values in Table 5.1 to represent the mean scores from your first 10 samples of 50 students. Since we've already seen how to compute graphical descriptive statistics, I've gone ahead and plotted the following data; when you look at the histogram, you can see that the distribution looks very spread out. First, the curve that is superimposed on it is very flat, and, second, although it is not shown, the standard deviation of 4.49 is fairly large given the narrow range of 14. As we saw earlier, this also means the data are skewed on both ends of the distribution.

TABLE 5.1. Data for 10 Samples

85	87	87	92	92
92	92	95	97	99

As we can see in Figure 5.1, because the distribution is flatter than a normal distribution, it is also platykurtotic. Let's continue by using the data in Table 5.2 to represent the mean scores for 24 samples.

We can see in Figure 5.2, by using a larger number of mean scores, the distribu-

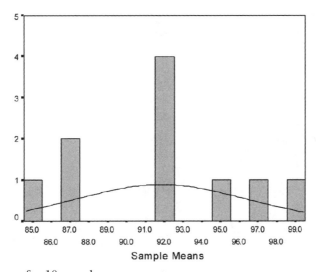

FIGURE 5.1. Histogram for 10 samples.

tion is starting to look different. The curve is more bell-shaped, and the standard deviation (3.46) is getting smaller. Since the standard deviation is a measure of dispersion, the data distribution is less spread out, less skewed, and less platykurtotic.

TABLE 5.2. Data for 24 Samples

86	86	87	87	89	90
91	91	91	92	92	92
92	92	92	93	93	93
94	95	97	97	98	98

TABLE 5.3. Data for 50 Samples

86	87	87	88	88	88	89
89	89	89	90	90	90	90
90	91	91	91	91	91	91
92	92	92	92	92	92	92
93	93	93	93	93	93	94
94	94	94	94	95	95	95
95	96	96	96	97	97	98

Now, let's use the set of 49 sample means shown in Table 5.3, plot them, and see what happens. This is really starting to make a difference isn't it? We can see that the curve is becoming very bell-shaped and the standard deviation, in this case, 2.86, is getting smaller (Figure 5.3).

We could keep adding data values and plotting them on our distribution but, suffice it to say, the process we're following is shown in Figure 5.4.

By looking at the distribution of sample means in Figure 5.4, you can see it looks like a normal distribution. It is, but unlike the normal distributions we have seen so far, this one has three special qualities.

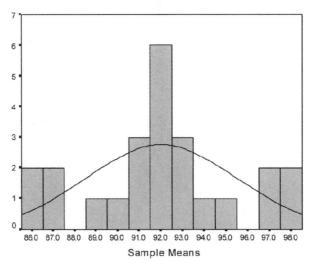

FIGURE 5.2. Histogram for 24 samples.

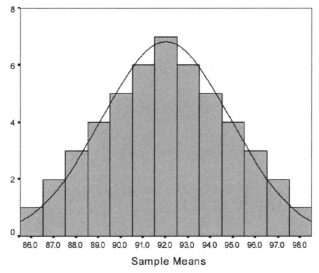

FIGURE 5.3. Histogram for 50 samples.

1. The distribution has a special name. Since we are plotting the mean scores from repeated samples, it is called the *sampling distribution of the means*.

2. The mean of the *sampling distribution of the means* is called the *mean of means* and is shown as $\mu_{\bar{x}}$. The more samples you plot on a histogram, the closer the mean of means gets to the mean of the population from which we take the samples.

3. When we compute a standard deviation for our sampling distribution of the means, we have to keep in mind the error inherent in the sampling process. Because of that, instead of calling it the standard deviation, we now call it the *standard error of the mean* or SEM or $\sigma_{\bar{x}}$.

Figure 5.5 gives us the overall feel for the central limit theorem.

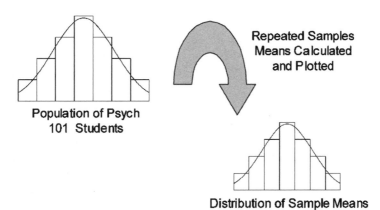

FIGURE 5.4. Creating a distribution of sample means.

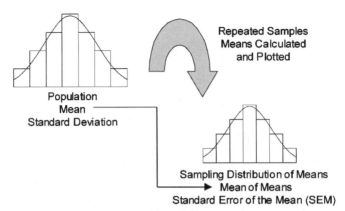

Population
Mean
Standard Deviation

Repeated Samples
Means Calculated
and Plotted

Sampling Distribution of Means
Mean of Means
Standard Error of the Mean (SEM)

FIGURE 5.5. Comparison of the mean between a population and a sampling distribution of means.

Three Important Notes

At this point, three things have to be stressed: sample size, the number of samples, and the shape of the population distribution.

First, in the definition we said that a sample size needs to be sufficiently large. While there is some disagreement as to what "sufficiently large" means, it is generally agreed that a sample size of 30 is large enough.

Second, I used the word "repeated" to describe the number of samples taken from the population. As we said earlier, measures from samples, such as the mean, cluster around the same measure of the population. They create a bell-shaped distribution with a mean equal to that of the population, only if we have an adequate number of samples. We demonstrated that with our three examples:

1. When we plotted only 10 sample means, we saw a flat, spread out distribution.

2. When we plotted 24 sample means, we saw a distribution that looked more like a normal distribution.

3. When we plotted 49 sample means, we had a nice, bell-shaped distribution with the majority of scores clustering around the mean of the population. Many of the scores were exactly equal to the mean, but most were a little higher or a little lower; this we know is due to sampling error. Had we continued plotting mean scores, eventually we would have a perfectly normal distribution with a mean of means exactly equal to the population mean.

Third, although it was implied that the distribution of scores for the entire population of PSY101 students was normally distributed, it doesn't have to be a normal distribution for the central limit theorem to apply. As long as a sufficient number of samples are taken and the samples are large enough, when the means of the samples are plotted, a normal curve will emerge.

The Empirical Rule Applies to the Sampling Distribution of the Means

Having a normal distribution emerge as a result of the central limit theorem brings with it some good news; the empirical rule applies. However, since we're plotting the means from multiple samples, instead of measuring distances from the mean using the standard deviation, we'll use the standard error of the mean (SEM). As a result, we can safely say that at least 68.26% of all sample means fall within plus or minus (±) one SEM of the mean, 95.45% will fall between the mean (±) two SEMs, and almost every sample mean (99.73%) will fall between the mean (±) three SEMs. At this point you might be thinking, "This sounds good, but how can we compute the standard error of the mean? I do not have access to the entire population and cannot take an infinite number of samples and plot their means." Luckily we have a very easy formula that allows us to do that:

$$\sigma_{\bar{x}} = \frac{\sigma}{\sqrt{n}}$$

We have seen most of this before. The symbol on top of the equation is sigma (i.e., σ), the symbol for the population standard deviation; below that is the square root of our sample size. The standard error of the mean, the value we are computing, is represented by a lower-case sigma followed by the symbol for a sample mean (i.e., $\sigma_{\bar{x}}$).

Instead of using the actual population standard deviation, just to make it easier to understand, let's use a nice whole number, in this case 14. Along with our sample size of 49, we can easily compute the SEM:

1. $\sigma_{\bar{x}} = \dfrac{14}{\sqrt{49}}$

2. $\sigma_{\bar{x}} = \dfrac{14}{7}$

3. $\sigma_{\bar{x}} = 2$

Keep in mind that we probably won't know the population standard deviation; we simply used 14 in this case to illustrate what we are doing. If the population standard deviation is not known, we can use the sample standard deviation in its place. When we do this, it is no longer called the standard error of the mean; instead we call it the *sample standard error of the mean*. This means our formula changes slightly because we have to replace the symbol for the population standard deviation (i.e., σ) with the symbol for a sample standard deviation, that is, s, as well as replace the symbol for the population SEM (i.e., $\sigma_{\bar{x}}$) with that of the sample standard error of the mean (i.e., $s_{\bar{x}}$):

$$s_{\bar{x}} = \frac{s}{\sqrt{n}}$$

You'll find that the sample standard error of the mean is a very good estimate of what we would have computed had we known the population value.

Before We Go Any Further

Since we will use these formulas and terms extensively in the next section, let's put together a reference list (Table 5.4) we can easily refer to.

TABLE 5.4. Symbols Related to the Sampling Distribution of the Means

Sampling distribution of the means	The resulting distribution when repeated samples are drawn from a population and their mean is calculated and plotted. Given enough samples, this will result in a perfect normal distribution.
$\mu_{\bar{x}}$	Mean of the sampling distribution of the means; also known as the "mean of means."
$s_{\bar{x}} = \dfrac{s}{\sqrt{n}}$	$s_{\bar{x}}$ represents the sample standard error of the mean; s represents the sample standard deviation, and n is the number of data values. This is the same as the standard deviation in a normal distribution of values except in this case we are using a sampling distribution of means.
$\sigma_{\bar{x}} = \dfrac{\sigma}{\sqrt{n}}$	$\sigma_{\bar{x}}$ represents the population standard error of the mean, sigma (i.e., σ) represents the population standard deviation, and n is the number of data values. This is the same as the standard deviation in a normal distribution if you know all of the values in a population.

Putting It All Together

Using the information from our example, we now have enough information to start putting things together. If we computed the mean score of our 49 samples, we would find that it is 92. From our discussion, we know that the mean of the sampling distribution of the means (i.e., the mean of means) will be very close to the mean of the population (i.e., 92); in this case I've made them exactly equal for the sake of simplicity. In the preceding equation, when we used 14 as the population standard deviation and 49 as our sample size, we calculated a SEM of 2. We know that the empirical rule applies, and we can see everything shown together in Figure 5.6.

If the empirical rule still applies, this means our idea of z scores still applies. We can compute the z score using the same basic formula, but we have to change the values to reflect those of the sampling distribution of the means, not those of a sample. We can see this in the following formula:

$$z = \frac{\left(\bar{x} - \mu_{\bar{x}}\right)}{\sigma_{\bar{x}}}$$

This is just a variation of the z score formula we used earlier. On the top of the equation, we're subtracting the mean of means (i.e., $\mu_{\bar{x}}$) from the mean of an individual plotted sample value (i.e., $\bar{x}$). After that, we divide by the standard error of the mean for the population (i.e., $\sigma_{\bar{x}}$). Remember, if we do not know this value, we can substitute the sample standard error of the mean (i.e., $s_{\bar{x}}$).

We already know that the mean of means (i.e., $\mu_{\bar{x}}$) is 92. We also know, from our example, that the population standard error of the mean (i.e., $\sigma_{\bar{x}}$) is 2. If we have an observed sample mean that is one sample SEM above the mean (i.e., 94), then our z score is 1:

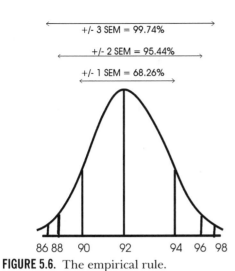

$$\text{+/- 3 SEM} = 99.74\%$$
$$\text{+/- 2 SEM} = 95.44\%$$
$$\text{+/- 1 SEM} = 68.26\%$$

FIGURE 5.6. The empirical rule.

1. $z = \dfrac{(\bar{x} - \mu_{\bar{x}})}{\sigma_{\bar{x}}}$

2. $z = \dfrac{(94 - 92)}{2}$

3. $z = \dfrac{(2)}{2}$

4. $z = 1$

Let's try that again and see if we've got the hang of this. Suppose the mean of our sampling distribution of means is 50 and the mean of the sample we're interested in is 42. In this case, we don't know the population SEM, but the sample SEM is 3; what is our z score? Let's go through the steps:

1. $z = \dfrac{(42 - 50)}{3}$

2. $z = \dfrac{(-8)}{3}$

3. $z = -2.67$

In this case, what does a z score of –2.67 tell us? Simple: the sample mean we're interested in is 2.67 SEMs below the mean of the sampling distribution of the means (i.e., the mean of means).

Now, let's really start tying this together. Based on what we learned earlier, what is the area under the curve between the mean and our z value? If we go to the back of the book and look at the area under the normal curve table, we can see that our z value is equivalent to .4962. This tells us that 49.62% of the values under the curve lie between 42 and 50. Remember, even though our observed value is negative, that doesn't affect the percentage we're interested in; a z score of 2.67 above the mean would result in the same percentage under the curve. All we're interested in is measuring the distance between two points.

Summary of the Central Limit Theorem and the Sampling Distribution of the Means

The concepts in this section are very straightforward. Creating a sampling distribution of the means involves nothing more than selecting a sufficiently large number of samples from a population, calculating their means and then plotting those means on a histogram. Given enough samples, this results in a normal distribution with a mean equal to that of the population (i.e., the mean of means). Since the sampling distribution is normally distributed, the empirical rule still applies, and you can use z scores to determine any given area under the curve. You'll find these tools will be very useful in the next sections when we begin talking about actually *testing* a hypothesis. Table 5.5 recaps everything we have talked about in this chapter and will help you as we move forward.

TABLE 5.5. The Process for Creating a Sampling Distribution of the Means

Original data distribution	Population	An infinite number of random samples with means computed and plotted	Sampling distribution of the means
Type of data	Quantitative		Quantitative
Shape of distribution	Any shape		Bell-shaped
Measure of central tendency	Mean		Mean equal to the population mean (i.e., the mean of means)
Measure of dispersion	Standard deviation		Standard error of the mean
Measure of relative standing	Empirical rule applies		Empirical rule applies

How Are We Doing So Far?

This is a rather long chapter filled with a lot of useful information. Given that, let's not wait until the end of the chapter to check our progress. Work through these exercises and, as usual, check your work in the back of the book.

1. For each of the following sample means, compute the z score using the descriptive statistics given in Figure 5.7. Note that the column labeled "Mean Statistic" is the mean of the sampling distribution of the means and the column labeled "Std. Deviation Statistic" is the population standard error of the mean.

   a. 109

   b. 77

   c. 113

   d. 101

   e. 95

   f. 88

   g. 96

   h. 90

2. In Table 5.6, using a SEM of 2 and a population mean of 92, complete Table 5.6.

Descriptive Statistics

	N	Mean		Std. Deviation	Variance
	Statistic	Statistic	Std. Error	Statistic	Statistic
Score	100	98.0000	.81650	8.16497	66.667
Valid N	100				

FIGURE 5.7. Descriptive statistics for computing *z* scores.

TABLE 5.6. Computing the Distance between *z* Scores

Value 1	*z* 1	Area under the curve for *z* 1	Value 2	*z* 2	Area under the curve for *z* 2	Area between *z* 1 and *z* 2
90.0			95.0			
89.0			90.0			
91.5			93.5			
92.0			96.0			

■ Estimating Population Parameters Using Confidence Intervals

Now that we understand the central limit theorem and the sampling distribution of the means, it's time to put them to work for us. The first thing we'll do is learn to estimate a value in a population (i.e., a population parameter) based on data from a sample (i.e., a sample statistic). Before we do that, however, we need to discuss a very important technical term.

The Alpha Value

As we saw earlier, any time we are dealing with a sample of data we must realize that our sample represents only one of the many samples that could be drawn from a given population. Because of that, we are never exactly sure if statistics calculated from sample data are equal to their counterpart parameter in the population data. For example, if we are looking at a distribution of sample means, 50% of the sample means are going to be at or below the population mean and 50% of the sample means are going to be at or above the population mean. Because of sampling error, our sample mean could be anywhere in that range (Figure 5.8).

Knowing this, what we, as consumers of statistics, have to determine is "what is our cutoff point?" If we are trying to test a hypothesis, how do we decide which sample statistics are different owing to sampling error and which are significantly different from the population parameter?

First, we have to accept the fact that any decision we make may be wrong; after all, we are dealing with probability. Luckily, we have some leeway at this point. We are able, by stating an alpha value, to state a priori (i.e., beforehand) the degree of risk we

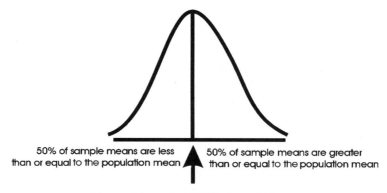

50% of sample means are less than or equal to the population mean

50% of sample means are greater than or equal to the population mean

Mean of Sampling Distribution of Means
Is Equal to the Population Mean

FIGURE 5.8. The population mean and the mean of the sampling distribution of means.

are willing to take when making a decision. The alpha value, often abbreviated using the Greek letter α, is sometimes called the *level of significance*. Its use is probably best explained using an example.

Suppose we are interested in determining if the average final exam score for the students in our Management class (i.e., our *sample*) is significantly different from the average score of other Management students throughout the university (i.e., the *population* of Management students). In order to determine this, we would have to test the following null hypothesis:

■ *There is no significant difference in exam scores between our Management students and other Management students throughout the university.*

Now, let's imagine we have computed an average score of 92 for our students and found the population average to be 80. Because the difference between these mean scores is fairly large, it appears we should reject our null hypothesis (i.e., there appears to be a significant difference between the two groups). Before we make that assumption, though, let's think it through.

We have already said that statistics from a sample, the mean for example, cluster around the same measure in the population. Knowing that, the mean score of our class (i.e., our sample) is nothing more than one of the great number of mean scores fluctuating around the population mean. In our case, all we know about our sample mean is that it is different from the population mean. This difference is either due to sampling error, or it is so large it cannot be attributed to chance; it represents a significant difference between the sample mean and the population mean.

If the first is true, we will fail to reject our null hypothesis—although there is a difference, it is due to chance and is not significant. If the second is true, then we will reject our null hypothesis—it is unlikely we can attribute our results to chance; they represent a significant difference between our sample mean and the population mean.

The question that remains, then, is, "This sounds good on the surface but how do we know when the sample mean and the population mean are different due to chance and when they are significantly different from one another?" In answering that question, I have some bad news and some good news. The bad news is, since we are dealing with a sample of data, we are never 100% sure if the differences are due to chance or represent a significant difference. The good news is, by using our alpha value we can control the risk of making the wrong decision.

Our alpha value allows us to do four things:

1. We can use it to indicate the percentage of time we are willing to make an incorrect decision based on the sample data we have collected. We have to admit that sampling error is going to occur, but we are going to set limits within which we can say that a sample statistic is significantly different from the population parameter due to chance (i.e., sampling error) and times when it is truly significantly different due to reasons other than sampling error (i.e., a "real" difference).

2. We can control the level of risk we are willing to take by using different alpha values (e.g., .10 and .01), although it is generally set at .05 (or 5%) for the types of studies we are most likely to undertake. By doing this we are saying, prior to making even the first statistical calculation, we are willing to take a 5% chance of making an error by rejecting a true null hypothesis.

3. The alpha value can help us create a range called a *confidence interval*. We can use a confidence interval to predict a population parameter using a sample statistic.

4. The alpha value can help us test a hypothesis about the population using data drawn from a sample.

In short, because statistical decision making is based on sampling and probability, we can never be 100% sure that any decision we make is going to be correct. We can, however, use our alpha value to help us feel more comfortable in the decisions we do make.

Type I and Type II Errors

Often alpha is referred to as the *Type I error rate* or the probability of making a *Type I error*; they both mean the same thing. Because we are dealing with samples, we are never 100% sure if the decisions we make are correct. If we reject a null hypothesis when we should not have (i.e., due to chance), then we have made a Type I error. Unfortunately, these things happen from time to time; along with that comes even more good news and bad news. The good news is that we can control for Type I errors using our alpha value, but changing the alpha value, as we will see shortly, may also lead to other problems.

The counterpart of a Type I error is called a *Type II error* and occurs when, because of sampling error, we fail to reject a null hypothesis when we actually should reject it (i.e., we say the values are not significantly different when they really are). In some texts, you will see the Type II error rate referred to as beta; the two are synonymous. A mistake that many people make is to assume that, if our alpha value is .05, then our beta value must be .95–that is not true! Calculating beta is a very complicated process, but, luckily for us, it is often computed as part of the output when using a statistical software package.

Although beta is not often a topic for beginning statisticians, it does play an integral part in computing the *statistical power* of an inferential test. Statistical power, defined as the probability of not making a Type II error (i.e., we will reject the null hypothesis when it is false), is computed by subtracting beta from 1. This value, of course, is affected by the alpha value we use, the actual difference between values we are comparing and the sample size. Generally speaking, any power value over .80 is considered acceptable. Again, this is something that consumers of statistics aren't frequently faced with, but we must be able to recognize and understand terms like this when we see them!

Criticality of Type I and Type II Errors

Understanding the criticality and relationship of Type I and Type II errors is imperative, so here's an analogy that is guaranteed to stick with you.

Suppose you have a null hypothesis that reads:

■ *The defendant is not guilty.*

The research hypothesis would read:

■ *The defendant is guilty.*

In this case, a Type I error (i.e., rejecting the null hypothesis when you should not) would mean that an innocent person would be sentenced to jail because you would be supporting the research hypothesis. Failing to reject the null and not supporting the research hypothesis due to sampling error (i.e., a Type II error) would mean a criminal would be found not guilty and be allowed to go free. Which is worse, an innocent man in jail or a criminal on the street? Of course, we could argue circumstances here, but suffice it to say the first scenario is worse!

Having said all of that, Table 5.7 shows the relationship between null hypotheses and the two types of errors.

TABLE 5.7. Type I and Type II Errors

	If we do not reject the null hypothesis	If we reject the null hypothesis
If null hypothesis is true	Correct decision	Type I error
If null hypothesis is false	Type II error	Correct decision

■ Predicting a Population Parameter Based on a Sample Statistic Using Confidence Intervals

In this section, we are only going to look at making estimations where one sample is taken from a population. In doing so, we will learn to estimate a population parameter (e.g., a population mean) based on a sample statistic (e.g., a sample mean). We can use the same tools to estimate other values (e.g., differences between two independent population means, population proportions, and differences between two dependent populations), but we are going to stick with the basics. The reasoning is simple; we'll focus on the most basic form of estimation because it is easy to understand and the underlying principles are the same for other types of estimations. That being said, let's move forward.

Let's go back to our example using the students in the Management class. This time, however, instead of hypothesizing that our group will be different from the population, we want to use a sample statistic to predict a population parameter. In this case, we are going to use our class average to estimate the average for all of the other Management students.

It would be impossible to ask every student in the university what they scored and then compute the population average, so that option is out of the question. Instead, we can use a *confidence interval* to predict a population parameter, in this case the mean, based on a mean of the sample. A confidence interval is defined as

A range of numbers around a sample statistic within which the true value of the population is likely to fall.

As we have said, any time we have a sample from a population, the sample statistic may or may not be equal to the corresponding population parameter. Because of that, we can never be 100% sure of any estimates made based on the sample data. That being said, we can build a range around the sample statistic, called the confidence interval, within which we predict the population parameter will fall.

To compute the confidence interval you need two things: a sample from the population and an alpha value. We will stick with the standard alpha value of .05 (i.e., 5%). This means that you want to create a range that, based on the data from the sample, will have a 95% (i.e., 100% minus your alpha value of 5%) probability of including the population average. Obviously, it also means there is a 5% probability that the population mean will not fall into the confidence interval. Let's use the following data to put this all together:

1. The sample size is 50 (the number of students in your class).
2. Your average test score for your sample is 92.
3. You are setting alpha to .05—this means we are going to have a 95% confidence interval.

Using this information you can use the following formula to compute the confidence interval you need:

$$\text{Confidence interval} = \overline{x} \pm (z_{a/2})\left(\frac{\sigma}{\sqrt{n}}\right)$$

The individual parts are easy to understand:

▶ $\overline{x}$ stands for the mean of the sample we are dealing with (i.e., 92).

▶ z is the z value for the alpha level we chose (i.e., 1.96). As you can see, we will be dividing our z value by 2; we will talk more about that in a minute.

▶ a is our alpha value (i.e., .05).

▶ $\sqrt{n}$ is the square root of the sample size (i.e., 50) we are dealing with (i.e., 7.07).

▶ σ is our sigma value and represents the population standard deviation.

Here it seems we have a potential problem. This last step in the equation, shown below, is the formula for the standard error of the mean, but we still do not know the population standard deviation (i.e., σ) in order to compute it:

$$\frac{\sigma}{\sqrt{n}}$$

As we have said, if we do not know the population standard deviation we can replace it with the sample standard deviation and get a very good estimate. In this case, however, we'll make it easy and use a population standard deviation of 5 so we can move forward. That will give us everything we need for our formula:

$$\text{Confidence interval} = 92 \pm (1.96)\left(\frac{5}{7.07}\right)$$

At first glance, everything here is pretty straightforward except for the whole idea of the z score for alpha divided by 2. We do that because, if you are trying to create a range that contains a population parameter, obviously you will have fluctuation around that parameter due to sampling error. Since you know this is going to occur, you have to be willing to make a mistake in either direction—you may estimate too high, you may estimate too low, or you may estimate just right.

If we set alpha to .05 and then divide it by 2, we get a value of .025 or 2.5%. That means you are willing to make a 2.5% mistake on either side of the mean. When you subtract this value from the total percentage (50%) on either side, you are now saying your range will include from 47.5% below the mean to 47.5% above the mean (remember 47.5% + 47.5% = 95%, the size of our confidence interval). When you find 47.5 (or .4750) in our area under the normal curve table, you find the corresponding z value of 1.96. To verify this, here is the particular row in the table showing these values. As you can see in Table 5.8, .4750 lies at the intersection of the 1.9 row and the .06 column; obviously, 1.9 and .06 equals 1.96. All of this is shown in Figure 5.9.

TABLE 5.8. Percentage Value for z Score of 1.96

z score	0.00	0.01	0.02	0.03	0.04	0.05	0.06	0.07	0.08	0.09
1.9	.4713	.4719	.4726	.4732	.4738	.4744	.4750	.4756	.4761	.4767

We can now compute the actual confidence interval using what we know up to this point:

1. When alpha = .05, we are looking for a confidence interval that contains 95% of the values (i.e., 1.00 − .05 or 100% − 5%).

2. Since we must establish a range on both sides of the value we are predicting, it's necessary to divide the size of our confidence interval by 2. This leaves us with 95%/2 or 47.5%.

3. The area under the normal curve table shows a z value of 1.96 for 47.5%.

4. The population standard deviation (5) divided by the square root of the sample size ($n = 50$) is .707.

5. Confidence interval = $92 \pm (1.96)(.707)$

6. Confidence interval = 92 ± 1.39

7. Confidence interval = 90.61 to 93.39

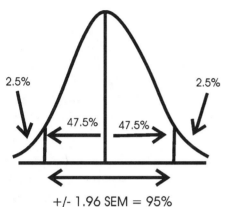

+/- 1.96 SEM = 95%

FIGURE 5.9. Relationship between the SEM and percentages of values in a distribution.

Pay Close Attention Here

Right about here, many beginning statisticians (and if the truth were told, many statisticians who should know better) think a 95% confidence interval means we are 95% sure the confidence interval contains the population parameter of interest. This isn't true. The confidence interval you compute around the sample statistic either contains the population parameter or it doesn't.

Instead, what it does say is, if we were to repeatedly take samples from the same population and calculate the confidence interval, the intervals computed would contain the population parameter 95% of the time. In this case, it means that 95% of the time, the true population parameter would be somewhere between 90.61 and 93.39. While this may sound like semantics, it is a very important point.

In short, there is a 95% probability that the population mean is somewhere between these two values. Again, we have a 5% chance of making an error, but a 95% chance of being right is a pretty darn good bet! This is shown in Figure 5.10.

Let's look at one more case before we move on. Here let's imagine our class average is 70, our population standard deviation is 10, and our sample size is 100:

1. When alpha = .05, we are looking for a confidence interval that contains 95% of the values (i.e., 1.00 – .05 or 100% – 5%).

2. Since we must establish a range on both sides of the value we are predicting, it's necessary to divide the size of our confidence interval by 2. This leaves us with 95%/2 or 47.5%.

3. The area under the normal curve table shows a z value of 1.96 or 47.5%.

4. The population standard deviation (10) divided by the square root of the sample size ($n = 100$) gives us 10/10 or 1.

5. Confidence interval = $70 \pm (1.96)(1.00)$

6. Confidence interval = 70 ± 1.96

7. Confidence interval = 68.04 to 71.96

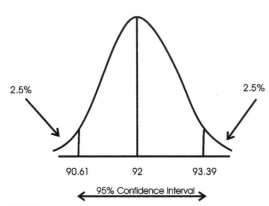

FIGURE 5.10. 95% Confidence interval around a mean of 92.

Here, we're 95% confident that our population average is somewhere between 68.04 and 71.96. Again, we're not saying that 95 out of 100 values fall in this range; this is just the probability that any given value does.

Confidence Intervals for Alpha = .01 and Alpha = .10

At this point, we run into a sticky situation. We just saw that we were able to find the correct percentage for the 95% confidence interval because that meant we were looking for a range of numbers that represented 47.5% above the mean and 47.5% below the mean. We simply went into the body of the area under the normal curve table and found that 47.5% represents a z score of 1.96.

Unfortunately, it is not quite so easy to determine the z scores for the other common alpha values (i.e., .01 and .10). For example, if our alpha value is .10, that means we want to compute a 90% confidence interval; this means 45% on either side of the mean. If we look at another section of the area under the normal curve table, shown in Table 5.9, we can see this creates somewhat of a problem.

TABLE 5.9. Percentage Value for z Score of 1.645

z score	0.00	0.01	0.02	0.03	0.04	0.05	0.06	0.07	0.08	0.09
1.6	.4452	.4463	.4474	.4484	.4495	.4505	.4515	.4525	.4535	.4545

In looking at the table, there isn't an entry for 45% (i.e., .4500) but we do have entries for 44.95% (i.e., .4495 in the table and a z score of 1.64) and 45.05% (i.e., .4505 in the table and a z score of 1.65). Since the value we are looking for, 45.00%, is in the exact middle between 44.95% and 45.05%, we must compute a z score that represents the average between the two. This process is as follows:

1. When alpha = .10, we are looking for a z score for 45% (i.e., 90%/2).

2. The table shows scores of 44.95% (z = 1.64) and 45.05% (z = 1.65).

3. 1.64 + 1.65 = 3.29

4. z = 3.29/2

5. z = 1.645

Let's use the same data as from the first example above; our mean is 92, our population standard deviation is 5, and we have a sample size of 50. This leads us to a confidence interval ranging from

1. Confidence interval = 92 ± (1.645)(.707)
2. Confidence interval = 92 ± 1.16
3. Confidence interval = 90.84 to 93.16

The same holds true when we have an alpha value of .01. In this case, we want to compute a 99% confidence interval that represents 49.5% of the values above the mean and 49.5% of the values below the mean. That means we're looking for a value of .4950 in the area under the normal curve table; let's use Table 5.10 to help us.

TABLE 5.10. Percentage Value for _z_ Score of 2.575

z score	0.00	0.01	0.02	0.03	0.04	0.05	0.06	0.07	0.08	0.09
2.5	.4938	.4940	.4941	.4943	.4945	.4946	.4948	.4949	.4951	.4952

Here we can see there is no entry for 49.5% (.4950), but there are entries for the two values that surround it, .4949 and .4951. Because 49.50% is exactly in the middle between 49.49% and 49.51%, we have to again use the z score that represents the average of the two (i.e., $(2.57 + 2.58)/2 = 2.575$). Again, this process is

1. When alpha = .01, we are looking for a z score for 49.5% (i.e., 99%/2).
2. The table shows scores of 49.49% ($z = 2.57$) and 49.51% ($z = 2.58$).
3. $2.57 + 2.58 = 5.15$
4. $z = 5.15/2$
5. $z = 2.575$

Using this value, we go through the same process to compute our confidence interval:

1. Confidence interval = 92 ± (2.575)(.707)
2. Confidence interval = 92 ± 1.82
3. Confidence interval = 90.18 to 93.82

Let me reiterate just one more time. This only means that we're 99% confident that our population average lies somewhere between 90.18 and 93.82. Remember, although it's close, 99% isn't 100%; because of random error, we can never be absolutely certain!

Another Way to Think about z Scores in Confidence Intervals

It's seems like we've gone to a lot of trouble to find the necessary z scores for these three common confidence intervals, doesn't it? If so, here's the good news. The z

scores we have just computed are obviously going to be constant every time we create a 90%, a 95%, or a 99% confidence interval. Knowing that, I have created a reference chart that's a lot easier to deal with. As you can see, the common confidence interval sizes, along with their corresponding information, are shown in Table 5.11.

TABLE 5.11. z Scores for Common Confidence Intervals

Confidence interval	a	$a/2$	Area between mean and tails	$z\,a/2$
.90	.10	.05	45%	1.645
.95	.05	.025	47.5%	1.96
.99	.01	.005	49.5%	2.575

Tying This All Together

Now that we've seen how to compute confidence intervals for different alpha values, let's use one case to better understand how changing our alpha value can affect our overall decision making. Suppose you are hired as a school psychologist and are interested in finding the average IQ in your school. You do not have time to test all 500 students in your school and you want to work with a representative sample of 30. How do you proceed?

First, let's assume you have a sample size of 30 with a mean IQ of 110; assume further you have a population standard deviation of 15 and an alpha value of .05. Using these values you are able to compute the confidence interval just like we did in the preceding example:

$$\text{Confidence interval} = 110 \pm (1.96)\left(\frac{15}{5.48}\right)$$

1. First, 15 (i.e., the population standard deviation) divided by 5.48 (i.e., the square root of the sample size) gives us 2.74.

2. When we multiply 2.74 by 1.96 (the z score of $\alpha/2$), we get 5.37.

3. When we subtract 5.37 from our mean of 110, we get 104.63 (the lower limit of our confidence interval).

4. When we add 5.37 to 110, we get 115.37 (the upper limit of our confidence interval).

The results, shown in Figure 5.11, indicate there is a 95% probability that the average IQ in the school is between 104.63 and 115.37.

Now, let's lower our alpha level to .01; this means we are going to create a 99% confidence interval (i.e., 100% – 1% = 99%). The calculation is the same except we need to substitute 2.575 for 1.96 (this is the z score for $\alpha/2$ or .005% [i.e., .5%]). Using our information, we can compute our confidence interval using the following formula:

$$\text{Confidence interval} = 110 \pm (2.575)\left(\frac{15}{5.48}\right)$$

In this case, there is a 99% probability that the average IQ in the school is be-

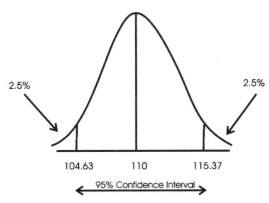

2.5%   2.5%

104.63   110   115.37

← 95% Confidence Interval →

FIGURE 5.11. 95% Confidence interval around a mean of 110.

tween 102.94 and 117.06. By using the same data and decreasing the size of our alpha value, we can see that we have widened the confidence interval (Figure 5.12). In order to give ourselves the greatest opportunity to create a range that includes the population mean, this means we have to use a smaller value for alpha.

Let's do the same thing but increase our alpha value to .10. Again, our formula would look the same except our z value now is 1.645.

$$\text{Confidence interval} = 110 \pm (1.645)\left(\frac{15}{5.48}\right)$$

In this case, we have narrowed our confidence interval down to 105.50–114.50. Here in Figure 5.13, we can see, by using a larger alpha value, we have decreased the range of the confidence interval.

Be Careful When Changing Your Alpha Values

Although you've seen it is easy enough to change your alpha value and compute the appropriate statistics, be careful when you do—you might be creating a problem when you analyze your data. For example, in the prior three examples, we originally set our alpha level at .05 and then saw what the difference in the confidence interval would be if we set alpha to .01 and then to .10. The type of decision and the needed precision of your results are going to dictate the alpha level you choose. This idea is shown in Table 5.12.

If you look closely at this table, you can see that it conceptually makes sense. The size of your alpha value is inversely related to the width of your confidence interval; larger alpha values lead to smaller confidence intervals and vice versa. Of course, this creates havoc when you're trying to estimate a population parameter. Do you try to "cast as large a net" as possible with a small alpha value, or do you try to be as accurate as possible with a large alpha value? Unfortunately, this is the dilemma that statisticians face when dealing with uncertainty. How do we strike a happy medium? Easy—set alpha equal to .05. That's a standard that researchers in the social sciences are

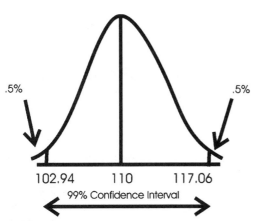

.5%   .5%

102.94   110   117.06

99% Confidence Interval

FIGURE 5.12. 99% Confidence interval around a mean of 110.

familiar and comfortable with. There's still some degree of error, but it's small enough to live with.

TABLE 5.12. The Effect of Alpha on the Size of a Confidence Interval

Confidence interval (CI)	Alpha	z score	Lower end of CI	Mean of CI	Upper end of CI	Width of CI
90%	.10	1.645	105.50	110	114.50	9.0
95%	.05	1.96	104.63	110	115.37	10.74
99%	.01	2.575	102.94	110	117.06	14.12

Do We Understand Everything We Need to Know about Confidence Intervals?

As we did with the section on the central limit theorem, let's go ahead and test our comprehension of the chapter up to this point. In Table 5.13, you are supplied with the sample mean, the alpha value, the number of values in the sample, and the population standard deviations. Use these values to compute the lower and upper limits of the confidence interval.

TABLE 5.13. Learning to Compute the Width and Limits of a Confidence Interval

$\bar{x}$	Alpha	n	σ	Lower limit of CI	Upper limit of CI	Width of CI
100	.10	25	5			
500	.05	50	25			
20	.01	15	3			
55	.05	20	7			
70	.01	22	6			
220	.10	40	10			

■ Testing Hypotheses about a Population Parameter Based on a Sample Statistic

In the last section, we learned to predict a population parameter based on data from a sample. We saw there are many different types of predictions we could make, but we used only one example, the mean, as an example. At this point, we are again going to use one example that will lay the groundwork for the various types of hypothesis testing we'll do throughout the remainder of the book. In this case, we are going to learn to make decisions about a population parameter based on sample information. In order to do that, let's use the following example.

Let's imagine you are the dean of the graduate school of psychology at ABC University. Unlike many other graduate schools in the United States, your students are not required to take the *Graduate Record Examination* (GRE) as part of the admissions process. Critics have suggested that this attracts weaker students and, because of that, they will not score as high on the national board certifications as their peers in schools where the GRE is required.

You admit to yourself that you've never really given it too much thought and, to

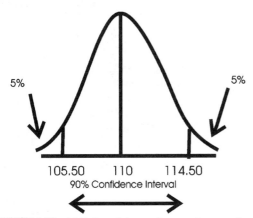

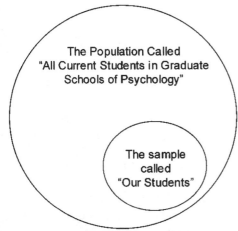

FIGURE 5.13. 90% Confidence interval around a mean of 110.

FIGURE 5.14. The relationship between a population and a sample from the population.

tell the truth, you're really not sure how your students do in comparison to students in other graduate schools. Interested in finding out, you visit one of the statistics professors in your department and tell her what's going on.

After telling her your problem, your professor immediately tells you "In order to conduct your study, you have to remember two things. "First," she says, "our student body is assumed to be nothing more than a sample of the population of all graduate psychology students throughout the United States. There are an infinite number of other samples you could select from that same population." She illustrates what she was saying by drawing Figure 5.14 on the board.

She continues by saying, "What you want to know is if our students are really members of that population or, because of our admissions process, they are significantly different from other samples from that population. In other words, you want to know if the average score our students make on the national certification examination is significantly different from the average score of the other samples." She also draws Figure 5.15 on the board.

"As you can see," she says, "our critics are claiming that our students will not do as well on the certification exams as the other schools, and we do not know if we are doing significantly better or worse than they are. Given that, you need to develop the hypothesis you want to test and then collect the data you need." After thinking about what she said, you decide on the following two-tailed research hypothesis:

■ *There will be a significant difference in scores on the national certification exam between graduate psychology students at ABC University and graduate psychology students at other universities.*

This means you will also have a corresponding null hypothesis:

■ *There will be no significant difference in scores on the national certification exam between graduate psychology students at ABC University and graduate psychology students at other universities.*

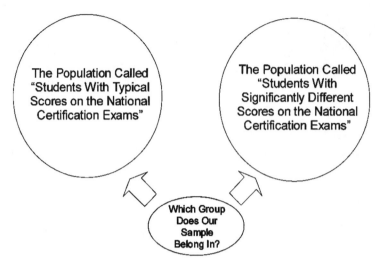

FIGURE 5.15. Different versus significantly different.

Can you tell me why we made the hypothesis two-tailed (i.e., nondirectional)? That's easy—we do not know if our students are doing better or worse than other students in the country. Because our scores could be greater than or less than the population's scores, we have to make the hypothesis nondirectional.

Making a Decision about the Certification Examination Scores

Finally, after all of this time, we're going to actually test our first hypothesis. Before we do, however, let's summarize what we know about populations and samples up to this point. First, a population represents all possible subjects about which we might want to collect data—voters in the United States, every student in a given university, all managers within a corporation who have a graduate degree, and so on. We have also said that, when we take a sample, we are hoping it is representative of the population from which it was taken—smaller samples tend to be less representative, larger samples more so. Finally, we've shown that, when we plot values, the mean for example, from a sample of 30 or more, the distribution is generally bell-shaped or, as we call it, a normal distribution. Having a sample that is normally distributed brings with it some good news when it comes to hypothesis testing:

1. We can test hypotheses to determine if one of the sample means is significantly different from the population mean by using the z score we have already learned to compute. We call this a *one-sample z-test*.

2. We saw earlier that the z score allowed us to report distances from the mean in terms of the standard deviation. The one-sample z-test allows us to do the same but, instead of using the standard deviation, we will use the standard error of the mean.

3. The empirical rule still applies.

4. The principles underlying the z-test are fairly straightforward and are basically the same for most of the other statistical tests we will use.

In our case, let's suppose we have 100 students in our graduate program. Knowing that, as we just said, we can use the z statistic to test our hypothesis. This means using the following values to compute our z value.

1. The mean score ($\bar{x}$) we want to compute a z score for.
2. Either the population mean (μ) or the mean of means ($\mu_{\bar{x}}$).
3. The population standard error of the mean ($\sigma_{\bar{x}}$).

Let's start with the actual equation:

$$z = \frac{(\bar{x} - \mu)}{\sigma_{\bar{x}}}$$

For the sake of our example, let's assume the population mean is 800 (i.e., the average score for graduate students at all universities), the SEM is 10, and the mean value for our students is 815. We can substitute those into our formula:

$$z = \frac{(815 - 800)}{10}$$

This would result in a z score of 1.5. From this point forward, this will be called our *computed value of z*. In order to test our hypothesis, we have to compare this value to a *critical value of z*. This is nothing more than a value from the area under the normal curve table we used earlier. The given value we choose will be based on the alpha value we're using and the type of hypothesis we've stated (i.e., directional or nondirectional). Let's use an example to better understand this.

Determining the Critical Value of z

We saw, earlier in the book, that a normal curve has a distribution of z scores with a mean of zero. Fifty percent of all possible z scores fall at or below the mean of zero, and 50% fall at or above the mean of zero. These z values represent all of the possible critical values of z. Our problem is to figure out which one to use.

In order to do this, we have to decide on our alpha value, so let's use the traditional .05 (i.e., 5%). We then have to subtract that from the range of all possible z values (i.e., 100%) to give us the range of possible z values we can use to test our hypothesis. This means our possible range is 95% of all possible z values (i.e., 100% – 5% = 95%).

Next we have to use the type of hypothesis we have stated to determine the distribution of all of the possible critical values of z. If we have a two-tailed hypothesis, we have to distribute the z scores equally under the curve; if we have a one-tailed hypothesis we have to distribute them according to the direction of the hypothesis. For now, let's look at the nondirectional (i.e., two-tailed) hypothesis; we will do the same for a one-tailed hypothesis shortly.

If we already know that we want to build a range that includes 95% of all possible z values and that range has to be equally distributed on either side of the mean, it is easy to figure out what the range would be. We would equally divide the 95%, meaning 47.5% of the values would fall on one side of the mean and 47.5% of all values would fall on the other. We would then have 2.5% of all possible values left over on either side. We can see that in Figure 5.16.

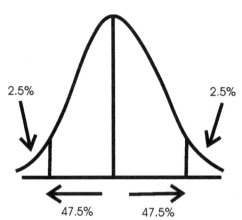

FIGURE 5.16. Equally dividing 95% of z values under the curve.

At this point, things start coming together. If we have 2.5% of all possible values left over on either side of the mean, that means we have a total of 5% left over. Now, where have we heard that value before? Isn't the 5% that is left over the same as our alpha value (i,e., .05 or 5%)? Sure it is! That means, instead of going through all of this trouble, it's much easier to remember that, when we have a two-tailed hypothesis, all we have to do is divide alpha by 2 to determine the percentage of all possible values on either side of the mean.

Now that we have identified an area that contains 95% of all values of z, our job is to determine exactly which value we want to use as our critical value. In order to do that, we need to use the area under the normal curve table.

As you'll remember, this table only shows values for one side of the curve. That's fine. If we know values for one side of the curve, it is easy to figure out what the corresponding value would be on the other side; it is just the numeric opposite. Although we've gone through all of this before back when we talked about confidence intervals, let's look at Table 5.14; this is a small part of the critical value of the z table and will help us determine the critical z value we need.

TABLE 5.14. Percentage Value for a z Score of 1.96

z score	0.00	0.01	0.02	0.03	0.04	0.05	0.06	0.07	0.08	0.09
1.9	.4713	.4719	.4726	.4732	.4738	.4744	.4750	.4756	.4761	.4767

We can see that the value .475 (i.e., 47.5%) lies at the intersection of the row labeled 1.9 and the column labeled .06. If you add these two values together, it gives us a z value of 1.96. This means, for 47.5%, our critical value of z on the right side of the mean is 1.96 and the critical z value on the left side of the mean is –1.96. Putting all of this together, we can see that a range of 95% (i.e., 47.5% on either side of the mean) encompasses a range of critical z scores from –1.96 to +1.96. This is shown in Figure 5.17.

We're Finally Going to Test Our Hypothesis!

We are now ready to test our hypothesis. Let's use the same values we did earlier: the population mean is 800, the SEM is 10, and our university's mean score is 815. That

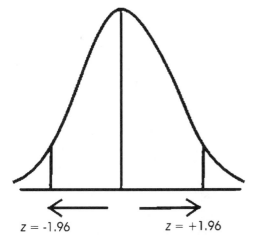

gave us a computed z score of 1.5. All we have to do to determine if our group is significantly different is to see if our computed z score falls within the 95% range of all possible critical z scores. In this case, it is easy to see that it does; 1.5 is between −1.96 and +1.96. This means we fail to reject our null hypothesis; there is no significant difference between our scores and the scores of the other universities throughout the nation.

FIGURE 5.17. The relationship of z scores and percentages.

Believe it or not, it is as easy as that! We will plot what we have done in just a minute, but for now let's look at another example.

Let's use the national mean score of 800 again, but this time let's use a mean score of 780 and a SEM of 10. What would be our computed value of z? Putting those values into our formula we get:

$$z = \frac{(780 - 800)}{10}$$

This gives us a z score of −2. Since this doesn't fall between −1.96 and +1.96, we will reject the null hypothesis. In this case, our score of 780 is significantly less than the national average of 800. Given this, perhaps your critics are right; maybe it is time to implement the GRE.

In Figure 5.18, I've used a population mean of 800 and SEM of 10 and plotted a few more mean values for our students. This, along with the corresponding z scores, will help solidify this idea.

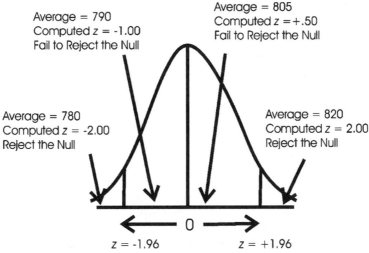

FIGURE 5.18. Using the z score to reject or fail to reject a null hypothesis.

▦ Testing a One-Tailed Hypothesis

Much like the two-tailed hypothesis, the logic behind testing a one-tailed hypothesis is quite straightforward. As its name implies, we will be using our alpha value on only one side of the distribution, but that is the only difference. Let's use the following case to explain what I mean.

Imagine you're the dean of a school of education where the average professor's annual salary is $73,000. One day, one of your more outspoken faculty members comes into your office and asks, "How do you ever expect us to stay here? We work harder, teach more classes, and have a larger dissertation load than faculty at other universities. Unless you can show us we are making significantly more than the average faculty salary at those other places, we are going on strike!" After saying this, the professor leaves your office in a huff, slamming the door behind him. After careful consideration, you decide the professor might have a point. He is pretty good and he does a lot of fine work; perhaps an investigation is called for.

What's the first thing you do? It is easy. Since he wants you to show that your faculty is earning more than the national average, you state the following research hypothesis:

▦ *Your faculty's average annual salary will be significantly higher than the national average salary of faculty members.*

The corresponding null hypothesis would read:

▦ *There will be no significant difference between your faculty's average annual salary and the national average salary of faculty members.*

In order to begin investigating this, you call the human resources department and find that the average salary in your department is $74,000. After a little investigation, they tell you the national average is $70,000 with a SEM of $2,000. This means your faculty's salary, $74,000, has a z score of +2:

1. $z = \dfrac{(74000 - 70000)}{2000}$

2. $z = \dfrac{(4000)}{2000}$

3. $z = +2$

Is that large enough to consider the difference significant? Let's compare the computed z score to the critical z score and find out. Before we do, however, let's take a minute to consider where we are going conceptually.

In the prior example, when we were using a two-tailed hypothesis, we had to divide our alpha value by 2 and mark off 2.5% on each end of the distribution. In order to determine the critical z score, we had to take the remaining 47.5% (remember, 50%

on either side of the distribution, minus 2.5%) and consult the area under the normal curve table. In the table, we found that a z score of 1.96 corresponded to 47.5%, so we marked off a range from –1.96 to +1.96 on our curve. We then compared our computed value of z to that range in order to test our hypothesis.

In this case, we have to do something a bit different. Since we have a one-tailed hypothesis but still have an alpha value of .05, we have to mark the entire 5% off on one end of the distribution or the other. Here we have a "greater than" hypothesis so we have to mark the 5% on the positive end of the distribution. You can see this in Figure 5.19.

Now, it is time to determine the critical value of z we are going to use. Using the excerpt from the area under the normal curve table (Table 5.15), we can see that the critical value for z for 45% of the area under the curve lies between 1.64 and 1.65; this leaves us with a critical value of 1.645. That can be seen in Figure 5.20.

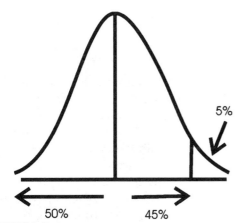

FIGURE 5.19. 5% on the positive side of the mean.

Now, let's compare our computed z to the critical value of z that we have plotted. Is our computed z (i.e., +2) greater than our critical value of z (i.e., 1.645)? Since it is, we will reject the null hypothesis. Your faculty members are making, on average, $74,000 per year. This is significantly more than the national average of $70,000. We can see this in Figure 5.21.

In Figure 5.22, I have shown a couple of other examples using different values of z. When our computed value of z is equal to 1.5, we would not reject the null hypothesis; there is no significant difference between our faculty's salaries and the national average. We do, however, reject the null when z is equal to 1.7. This means our research hypothesis stating that our faculty members are already making more than the national average is supported.

TABLE 5.15. Percentage Value for a z Score of 1.645

z score	0.00	0.01	0.02	0.03	0.04	0.05	0.06	0.07	0.08	0.09
1.6	.4452	.4463	.4474	.4484	.4495	.4505	.4515	.4525	.4535	.4545

■ Testing a One-Tailed "Less Than" Hypothesis

In the preceding example, we had a one-tailed "greater than" hypothesis. Here, let's look at a "less than" case by using an average annual faculty salary of $66,500. Since we already know the national average is $70,000 with a SEM of $2,000, we are concerned that the faculty may have a justifiable complaint. Let's begin to test their claims with this research hypothesis:

■ *Your faculty's average annual salary will be significantly less than the national average salary of faculty members.*

The corresponding null hypothesis would read:

■ *There will be no significant difference between your faculty's average annual salary and the national average salary of faculty members.*

First, let's compute our z score using our faculty member's salary along with our population mean and standard error:

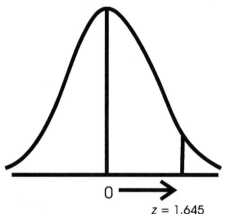

FIGURE 5.20. Critical z for 5% above the mean.

$$z = \frac{(66,500 - 70,000)}{2000}$$

Here the computed z score is –1.75. We can plot this on our distribution to see if a significant difference exists but, before we do, let's consider what our distribution should look like.

As we just saw, when we had a one-tailed hypothesis where we were investigating a "greater than" condition, we had to put the entire 5% alpha value on the right side (i.e., greater than zero) of the distribution. Now, since we are looking at a "less than" relationship, we have to put the 5% on the left (i.e., negative) side of the distribution shown in Figure 5.23.

Since we are using 5%, the critical value of z (i.e., 1.645) will be the same as above; we will just put a negative sign in front of it and use it on the opposite side of the distribution. We can see this in Figure 5.24.

If we compare our computed value of z (i.e., –1.75) to our critical value of z, it is less than –1.645. Because of that, the professor seems to have a case; the faculty's average annual salary is significantly lower than the national average. This can be seen in Figure 5.25.

I have included two more examples in Figure 5.26. In the case where the computed z value is –3.0, you might need to consider giving your faculty a raise. In the other, where z = –1, perhaps the administration should consider giving you a raise for having to deal with employees like this.

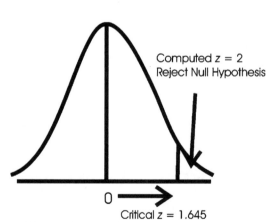

FIGURE 5.21. Rejecting the null hypothesis with a computed z of 2.

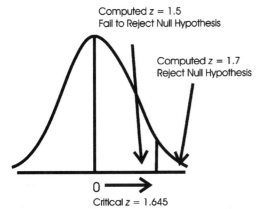

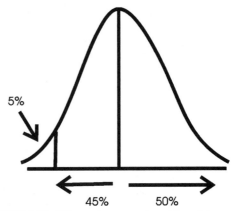

FIGURE 5.22. Examples of rejecting and failing to reject based on computed z scores.

FIGURE 5.23. 5% on the negative side of the mean.

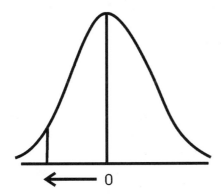

Critical z = -1.645

FIGURE 5.24. Critical z for 5% below the mean.

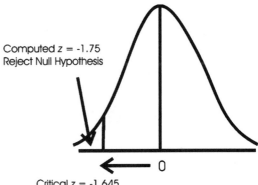

FIGURE 5.25. Rejecting the null hypothesis with a computed z of -1.75.

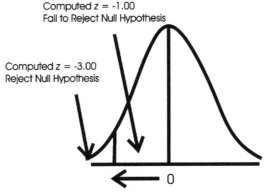

FIGURE 5.26. Examples of rejecting and failing to reject based on computed z scores.

Summarizing What We Just Said

Although we went into a lot of detail to get to this point, we can summarize determining our critical value of z using the following steps:

1. Determine the alpha value we want to use. In this example, let's use $\alpha = .05$.

2. If we have a two-tailed hypothesis (i.e., nondirectional), divide our alpha value by 2; in this case that would leave us with $.05/2 = .025$. If we have a one-tailed (i.e., directional) hypothesis, we will use the entire alpha value.

3. We then subtract the result from Step 2 from .50. In this case, if alpha is equal to .05 and we have a two-tailed hypothesis, we would be left with .475 (i.e., .50 − .025 = .475). If we have a directional hypothesis, we will subtract our entire alpha value from .50, leaving .450 (i.e., .50 − .05 = .45).

4. We then use either of those two values to determine our critical value of z in the area under the normal curve table. In this case, our critical value for a two-tailed hypothesis would be 1.96; our value for a one-tailed hypothesis would be 1.645.

5. We have to determine our range of critical z scores:

   a. If we have a two-tailed hypothesis, our range of z scores would be from −1.96 to +1.96.

   b. If we have a one-tailed "greater than" hypothesis, our range of critical z scores would include everything to the left of zero and everything to the right of zero up to and including +1.645.

   c. If we have a one-tailed "less than" hypothesis, our range of critical z scores would include everything to the right of zero and everything on the left of zero down to and including −1.645.

6. At this point we compare our computed value of z to our critical value of z and either reject or fail to reject our null hypothesis.

Be Careful When Changing Your Alpha Values

We saw this exact same heading in the last chapter—it almost makes you think we need to pay close attention when we're changing alpha values, doesn't it! While we do usually use an alpha value of .05, there are instances where we might want to change. When we do, we have to think back to the idea of Type I and Type II errors. Let's include the same table we used back then but add another column on the right showing the probability of a Type I or Type II error; this is shown in Table 5.16.

Remember, using this table we demonstrated that higher alpha values lead to smaller z scores and confidence intervals. Let's tie those ideas together with the idea of hypothesis testing so we can understand how these same values affect the probability of Type I and Type II errors.

TABLE 5.16. Tying Together the z Score, Alpha Value, Confidence Interval, and Error Probability

Confidence interval	Alpha	z score	Lower end of CI	Mean of CI	Upper end of CI	Width of CI	Hypothesis error probability
90%	.10	1.645	105.50	110	114.50	9.0	Higher probability of Type I error Lower probability of Type II error
95%	.05	1.96	104.63	110	115.37	10.74	Acceptable probability of Type I and Type II errors
99%	.01	2.575	102.94	110	117.06	14.12	Higher probability of Type II error Lower probability of Type I error

When we test a hypothesis and use a larger alpha value, we create a narrower confidence interval. This means we're actually creating a smaller range of values that we would consider not significantly different from the value we're interested in. In this case, if we have a two-tailed hypothesis and use an alpha value of .10, we say that any value from 105.50 to 114.50 is not significantly different from our average of 110 and we would fail to reject the null hypothesis based on that value. Anything outside of that range would be considered significantly different and would cause us to reject our null hypothesis. It's easy to see, then, that a larger alpha gives us a far better chance of rejecting the null hypothesis. Unfortunately, this also means that we're increasing our Type I error rate; we may be rejecting a null hypothesis when we shouldn't.

At the other end of the spectrum, if we decrease our alpha value to .01, we're greatly widening the range of values we would consider not significantly different from our mean score. In this case, we would consider anything from 102.94 to 117.06 to be different from 110 strictly due to chance. This, of course, lowers our probability of rejecting the null hypothesis and greatly increases the probability of a Type II error; we might fail to reject the null hypothesis when we actually should.

As we've said earlier, at this point a good consumer of statistics knows two things. First, we can't count on being 100% right in our statistical decision making because of random error. Second, because of that, we can control our level of risk by using an appropriate alpha value. As we just saw, having too large or too small of an alpha value creates problems; a good consumer of statistics will usually use an alpha value of .05 and its acceptable probability of making either a Type I or Type II error.

The Heart of Inferential Statistics

We have seen, in just a few pages, the very heart of inferential statistics. In this case, using a one-sample z-test, we tested a hypothesis about a population mean using information from both the population and the sample. In upcoming cases we will compare two samples, three samples, and so on. Although the different tests we will use (*t*-tests, ANOVAs, etc.) have different formulas, the basic premise of hypothesis testing remains the same.

As easy as this is, I am the first to admit that calculating all of these values, using the tables, and plotting the results can be somewhat tedious. Given that, let's look at an alternate way of making decisions about our hypotheses. I believe you will find this way a lot easier.

Probability Values

Instead of working through all of those calculations, statisticians use statistical software to compute a *p* value. This *p* value is the probability that a particular outcome is due to chance. In the case we just discussed, another way of describing the *p* value is to call it the probability of our university's salaries being equivalent to the salaries from other universities. These *p* values range from 0.00 (no probability that the sample mean came from the population being considered) to 1.00 (an absolute certainty that the sample mean came from the population being considered). Obviously, if we are trying to reject the null hypothesis, we want as small a *p* value as possible. This will help us ensure that any differences we find are "real" and not due to chance.

Rather than testing a hypothesis by computing an observed value of *z*, finding a critical value of *z*, and then comparing the two, statisticians can just compare their computer-generated *p* value to their predetermined alpha value. For a particular case, if the computed *p* value is less than the alpha value, a researcher will reject the null hypothesis; this means the differences are significant and not attributable to chance. If the *p* value is greater than or equal to the alpha value, the statistician will fail to reject their null hypothesis. This means any differences found are not significant or are attributable only to chance. Table 5.17 shows an example of these ideas.

TABLE 5.17. Tying Together the z Score and the p Value

Computed value of z	Critical value of z	Alpha	Reject null hypothesis?	p value
2.00	1.645	.05	Yes—the computed value of z is greater than the critical value of z.	The p value has to be less than alpha since we are rejecting the null hypothesis.
1.5	1.645	.05	No—the computed value of z is less than the critical value of z.	The p value has to be greater than or equal alpha because we are not rejecting the null hypothesis.
1.645	1.645	.05	No—the computed value of z is equal to the critical value of z.	The p value has to be greater than or equal alpha because we are not rejecting the null hypothesis.

A Few More Examples

Look at the following hypotheses and their alpha and p values. Based on them, let's decide whether we need to reject or fail to reject our null hypothesis.

■ *There will be a significant difference in achievement between students who eat breakfast and students who do not eat breakfast.*

The null hypothesis would read:

■ *There will be no significant difference in achievement between students who eat breakfast and those who do not eat breakfast.*

Suppose we have collected achievement scores for a random sample of students and then asked them to tell us whether or not they eat breakfast. After using a computer to calculate the descriptive statistics, we have found the average score for students eating breakfast is 82 while that of students skipping breakfast is 79. So far, it looks like we might be supporting our research hypothesis but we can't be sure.

Using our standard alpha value of .05, let's assume the output from that procedure shows a p value of .05; what would we do? Looking at the table, we only reject the null hypothesis if the p value is less than the alpha value. In this case, p is equal to alpha, so we will fail to reject the null. Even though the mean scores are different, apparently eating breakfast does not significantly affect achievement in school one way or the other.

Suppose we had the following research hypothesis:

■ *There will be a significant difference in scores on the Beck Depression Inventory between patients receiving behavioral therapy and those receiving cognitive therapy.*

Our null hypothesis is:

■ *There will be no significant difference in scores on the Beck Depression Inventory between patients receiving behavioral therapy and those receiving cognitive therapy.*

At this point, we are going to make up a pretty far-fetched case study. Let's suppose we have a pool of depressed patients. We randomly select one group to receive cognitive therapy and the other to receive behavioral therapy. After an appropriate period of time, say 10 sessions with the psychologist, we ask each of the patients to complete the *Beck Depression Inventory*. After computing the descriptive statistics, we see that the cognitively treated patients have a mean score of 15 while the behaviorally treated patients have a mean score of 30.

Things are looking pretty good for the cognitive therapists, aren't they? Let's use our alpha value of .05, and, using the data we collected, our software computes a p value of .01. Since this is less than our alpha value of .05, obviously the cognitive therapists know something the behaviorists do not; their clients' scores are significantly less than those of the behaviorists' clients.

Let's take a look at one last research hypothesis:

■ *There will be a significant difference in elementary school achievement between students attending public school and those being home-schooled.*

Here's the corresponding null hypothesis:

■ *There will be no significant difference in elementary school achievement between students attending public school and those being home-schooled.*

In order to compare levels of achievement, let's use the results of a standardized test that the students take at the end of each school year. The results show a mean score of 56 for the public school students and 62 for the home-schooled students. This time, let's set alpha equal to .10 and find that p is equal to .11—what would you do? In this case, p is greater than our alpha value, so we fail to reject our null hypothesis. It is apparent that it makes no difference where students attend elementary school—they all do equally as poorly on the standardized tests!

Everything we have talked about concerning the p value is summarized in Table 5.18.

TABLE 5.18. The Relationship between the p Value and a Decision Made about a Hypothesis

Situation	What happens to the null hypothesis?	What happens to the research hypothesis?
Computed p is greater than *alpha*. *or* Computed p is equal to *alpha*.	We fail to reject the null hypothesis. Any observed differences may be due to random sampling error.	We do not support the research hypothesis. Any observed differences may be due to sampling error.
Computed p is less than *alpha*.	We reject the null hypothesis. The observed differences are probably not due to sampling error.	We support the research hypothesis. The observed differences are probably not due to sampling error.

Great News—We Will Always Use Software to Compute Our p Value

Although it is important to understand why and how we perform the manual calculations we have just gone through, we should always rely on a statistical software package to compute our p value. Not only is p difficult to compute by hand, but using the software will eliminate the probability of us making errors in our math. Starting in the next chapter, you'll see how easy it is to use p when testing our hypotheses. We're about to finally learn to identify the appropriate statistical test we need in order to test a given hypothesis but, before we do, let's check our knowledge of this last section.

Look at Tables 5.19, 5.20, and 5.21. In each, you'll see a null and research hypothesis, data values for the population and sample, an alpha value, and the standard error of the mean. Use these values to compute the appropriate z score, obtain the critical z score, and then determine whether or not you should reject the null hypothesis. Finally, based on everything, determine if the p value would be less than .05. Again, you don't have to compute a p value; just state whether or not it would be less than .05 based on whether or not you would reject the null hypothesis based on the other computed values. In the last line of the table, support your decision about the p value. I have supplied the answers at the end of the book so you can check your work.

TABLE 5.19. Intelligence—Is There a Significant Difference?

Null hypothesis:	There will be no significant difference in the intelligence level between our class and the overall university level.
Research hypothesis:	The average intelligence level of our class will be significantly higher than the average intelligence level of all other classes in the university.
Population average:	100
Class average:	105
SEM:	3
Alpha:	.05
Computed z value:	
Critical z value:	
Reject null hypothesis?	
How do you know?	
Is p less than .05?	
How do you know?	

TABLE 5.20. Rainfall—Is There a Significant Difference?

Null hypothesis:	There will be no significant difference in the yearly amount of rainfall between Texas and Florida.
Research hypothesis:	There will be a significant difference in the yearly amount of rainfall between Texas and Florida.
Texas average:	70 inches
Florida average:	62 inches
SEM:	10 inches
Alpha:	.05
Computed z value:	
Critical z value:	
Reject null hypothesis?	
How do you know?	
Is p less than .05?	
How do you know?	

TABLE 5.21. Truancy—Is There a Significant Difference?

Null hypothesis:	There will be no significant difference in the number of truancy cases at our school and the district average number of truancy cases.
Research hypothesis:	There will be a significant difference in the number of truancy cases in our school and the district average number of truancy cases.
District average:	12 cases
Our school's average:	17 cases
SEM:	2 cases
Alpha:	.05
Computed *z* value:	
Critical *z* value:	
Reject null hypothesis?	
How do you know?	
Is *p* less than .05?	
How do you know?	

STEP 5

Choose the Right Statistical Test

It is finally time to learn how to choose the right statistical test for a given situation. Although we are going to go about it in an easily understood manner, it has been my experience that this is one of the most difficult things for the beginning statistician to do. As I said before, one might choose from literally hundreds of different parametric and nonparametric procedures. We will only look at the small number of tests you will probably use.

You Already Know a Few Things

First, you already know a few things about choosing the right statistical procedure for the data you have collected. For instance, we talked about how, if the data you collected for your dependent variable are either nominal or ordinal, we will use nonparametric procedures. If the data are quantitative (interval or ratio level), most of the time we will use a parametric test. In rare instances, if we have quantitative data that are not normally distributed, we will use a nonparametric procedure. We will talk more about this later in the book.

After we have determined the type of data we are collecting, we can use the number of independent variables, the number of levels of each of the independent variables, and the number of dependent variables to select the appropriate statistical test from Table 5.22. (This table is also found in Appendix G.)

A Couple of Notes about the Table

Table 5.22 is very straightforward; it takes information from each of the steps and shows us the appropriate statistical test for a given situation. You'll notice, however, that there are three places marked with asterisks. We need to discuss these before moving forward.

In the second row from the top, you can see that the one-sample *t*-test and the one-sample *z*-test are marked with one asterisk; this indicates that it is not logical to consider either the number of independent variables or their levels. You are simply comparing a sample statistic to a population parameter.

The rightmost column of the table, labeled "Alternate Statistical Test," is marked with two asterisks. As you can see, alternate statistical tests are listed in only a few of the rows; this is because not every parametric test has a corresponding nonparametric test, and, when one does exist, they are used only when a statistician has collected quantitative data that are not normally distributed or in instances where rank-level data were collected. These two things are generally exceptions to the rule. Most consumers of statistics will seldom be asked to use these nonparametric tests. I will explain more about this when we get to the specific tests.

You can see that the third place marked with asterisks is in the row that shows the Pearson and Spearman correlation procedures. You do not have independent and dependent variables in correlations; instead you have predictor variables and criterion variables. While these names are different, they are used in a manner somewhat similar to their independent and dependent cousins. We will discuss this in great detail when we get to that chapter.

TABLE 5.22. Choosing the Right Statistical Test

Number of independent variables	Number of levels in the independent variable	Number of dependent variables	Type of data the dependent variable represents	Statistical test to use	Alternate statistical test**
N/A*	N/A*	1	Quantitative	One-sample *z*-test *or* one-sample *t*-test	N/A
1	2	1	Quantitative	Dependent-sample *t*-test *or* independent-sample *t*-test	Wilcoxon *t*-test *or* Mann–Whitney *U* test
1	3 or more	1	Quantitative	Analysis of variance (ANOVA)	Kruskal–Wallis *H* test
2	2 or more	1	Quantitative	Factorial ANOVA	N/A
0***	0	0***	Quantitative	Pearson correlation	N/A
0***	0	0***	Ordinal	Spearman correlation	N/A
1 or more	2 or more	1 or more	Nominal	Chi-square	N/A

*When you're working with one-sample *z*-tests or one-sample *t*-tests, it is really not logical to point out the independent variable or its levels. As we have just seen with the one-sample *z*-test, we are just comparing a sample statistic to a population parameter. We will see the same holds true, albeit with smaller sample sizes, when we visit the one-sample *t*-test in the next chapter.

**Not all parametric tests have a nonparametric equivalent, or the test mentioned is already a nonparametric test.

***In correlational procedures, you simply want to look at the relationship between two variables; these are called predictor and criterion variables.

Summary of Step 5: Choose the Right Statistical Test

The table we can use to select the statistical test we need is very straightforward and includes the tests you will need to use, most of the time. Notice there was no mention of tests such as the multivariate analysis of variance (MANOVA), multiple regression, or any of the other types of statistical tests that you, the consumer of statistics, will hardly ever use.

Now we need to move forward and discuss the use of the statistical tests identified in the table. We will start each chapter with a brief overview of the statistical tool being considered and then use case studies to help work though each of the six steps in our model.

Do You Understand These Key Words and Phrases?

alpha value	a priori
beta	central limit theorem
confidence interval	level of significance
lower confidence limit	mean of means
p value	sampling distribution of the means
sample standard error of the mean	significantly different
statistic	standard error of the mean (SEM)
Type I error	Type II error
z score	

Do You Understand These Formulas and Symbols?

Confidence interval:

$$\bar{x} \pm (z_{a/2})\left(\frac{\sigma}{\sqrt{n}}\right)$$

Mean of the sampling distribution of means:

$$\mu_{\bar{x}}$$

Population standard error of the mean:

$$\sigma_{\bar{x}} = \frac{\sigma}{\sqrt{n}}$$

Sample standard error of the mean:

$$s_{\bar{x}} = \frac{s}{\sqrt{n}}$$

z score for sampling distribution of the means when the population mean is known:

$$z = \frac{(\bar{x} - \mu)}{\sigma_{\bar{x}}}$$

Remember: In the z score formula, both the population mean and the standard error of the mean can be replaced by the *mean of means* or the *sample standard error of the mean* if needed.

Quiz Time!

Using the decision table, identify the statistical test that should be used to test each of the following hypotheses and then explain why you chose that test.

1. There will be no significant difference in the number of fish caught in lakes, streams, or rivers.

2. Employees assigned to well-lighted offices will have higher levels of productivity than employees assigned to dimly lighted offices.

3. There will be a significant difference in health problems between citizens of Canada, citizens of the United States, and citizens of Mexico.

4. The miles per gallon demonstrated by foreign-manufactured cars is significantly different from 19.

5. There will be a significant difference in the number of children between families labeled as lower, middle, and high socioeconomic status.

6. Audience appreciation for shows appearing on Broadway will be significantly lower than the audience appreciation of the touring version of the same show.

7. There will be a significant correlation between the number of siblings and the annual income of their parents.

8. There is a significant difference in government apathy between voters labeling themselves as Republicans, voters labeling themselves as Democrats, and voters labeling themselves as Liberals.

9. There will be no significant difference in the percentage of males and females in our office building when compared to the national percentage.

10. There will be a significant difference in authors' procrastination levels before being called by their editor and after they've been called by their editor.

CHAPTER 6

The One-Sample *t*-Test

Introduction

We're finally up to the point where we can choose our statistical test, use SPSS to analyze our data, and interpret our hypothesis based on the SPSS report. From here through the end of the book, we will dedicate a chapter to each of the tests in our table. We'll work through a lot of cases, going through the entire six-step process, to ensure we know when and how to use each test.

In the last chapter, we learned to use a statistic from a large sample of data to test a hypothesis about a population parameter. In our case, using a *z*-test, we tested a hypothesis about a population mean using the mean of one sample drawn from that population. We agreed that by "large sample" we mean any sample that has 30 or more data values. What happens, however, when we have a sample with less than 30 items? We cannot use the one-sample *z*-test, so what can we do? The answer is, we can use its counterpart, the one-sample *t*-test. The logic underlying the one-sample *t*-test is very similar to that of the one-sample *z*-test and is easily understood. You'll find that this material is critical to understanding the next two chapters, on the independent-sample *t*-test and the dependent-sample *t*-test. For now, let's take a trip.

Welcome to the Guinness Breweries

All of the sudden we have taken a dramatic turn, haven't we? We have gone from discussing statistics to an apparent trip to a brewery! I hate to burst your bubble, but we are only going to visit the brewery to meet a man named William Gossett.

Gossett, a chemist, worked for Guinness in the early 1900s. His job involved testing samples of ale to ensure quality. He decided it was too costly and time consuming to use large samples (i.e., 30 or more), so he was determined to develop a statistical tool that would allow him to use smaller samples and still be able to test hypotheses about a population parameter (i.e., the batch of ale). We can start understanding his work by looking at the *t* distribution.

The *t* Distribution

We have seen, in prior chapters, when we had large (i.e., greater than 30), mound-shaped quantitative data distributions, the empirical rule showed us how to understand the distribution of the data. We were able to locate the relative position of a data point in terms of the standard deviation by computing *z* scores using the sample mean, the population mean, and the standard error of the mean. If we did not know the standard deviation of the population and therefore could not compute the standard error of the mean, we also saw that we could use the standard deviation of the sample to compute the sample standard error of the mean. This is a very close approximation of the same population parameter and can be used in our formula to compute the *z* score.

Gossett, while experimenting with possible solutions, noticed something interesting about the mound-shaped distribution of data values when the sample size was less than 30. Each time he decreased the sample size by one, and plotted the means of repeated samples of the same size, the shape of the distribution flattened out. You can see the plots for 30 sample means, 25 sample means, and 20 sample means in Figure 6.1.

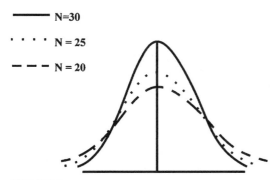

By looking at the picture we can see that, when the means of individual samples with 30 sample means are plotted, the distribution looks like a normal curve. When we use only 25 sample means, the sampling distribution gets flatter and more spread out; this is even more evident when we have only 20 sample means. In other words, the fewer the number of data values you have, the more spread out on both ends (i.e., platykurtotic) our distribution is.

FIGURE 6.1. Changes in distribution shape based on number of samples plotted.

Using this idea, Gossett further discovered that the empirical rule applied to this distribution and, if you compensate for the number of data values less than 30, you can test a hypothesis by comparing a computed value of *t* to a critical value of *t*. This is exactly what we did when we compared a computed value of *z* to a critical value of *z* in the *z*-test. This was the key, he determined, to testing a population parameter using only a relatively few data values.

Putting This Together

As I just said, when we have samples with less than 30 members, we can test a hypothesis using a computed value of *t* just like we used the computed value of *z*. In order to better understand this, let's state a hypothesis and then use a dataset with 15 values to test it:

> ■ *Persons living in urban environments have anxiety levels significantly different from the national average.*

Let's say we have a test that is used to measure anxiety and the scores from it range from 0 (no anxiety) to 80 (high anxiety). Let's use the 15 scores in Table 6.1 as an example so we can continue our discussion. Since we already know how to use SPSS to compute descriptive statistics we'll skip that part; Figure 6.2 shows the output.

TABLE 6.1. Anxiety Scores

34	32	28	32	35
32	28	30	39	32
29	33	34	32	30

Descriptive Statistics

	N	Mean		Std. Deviation	Variance
	Statistic	Statistic	Std. Error	Statistic	Statistic
Anxiety	15	32.0000	.74322	2.87849	8.286
Valid N (listwise)	15				

FIGURE 6.2. Descriptive statistics.

The formula for the computed value of t is similar to the formula for the computed value of z; we just need to include different values:

$$t = \frac{\bar{x} - \mu}{s_{\bar{x}}}$$

As always, the symbol $\bar{x}$ stands for the mean of the dataset—in this case 32. From the descriptive statistics in Figure 6.2, we see the standard error of the sample (i.e., $s_{\bar{x}}$) is .74322. In this case, let's assume the mean of our population (i.e., μ) is 30; this is the value that we want to compare our sample mean against. We can insert these values into our equation and calculate a t value of 2.691 using the following three steps:

1. $t = \dfrac{32 - 30}{.74322}$

2. $t = \dfrac{2}{.74322}$

3. $t = 2.691$

Let's use SPSS to verify our calculations. We've entered our data in Figure 6.3, selected Analyze, Compare Means, and One-Sample T Test. We then select the Anxiety value and set the Test Value to 30, shown in Figure 6.4; this represents the mean value of our population. Clicking on OK causes SPSS to run the one-sample *t*-test, comparing the mean score of our data to the population mean of 30, and will produce Figure 6.5.

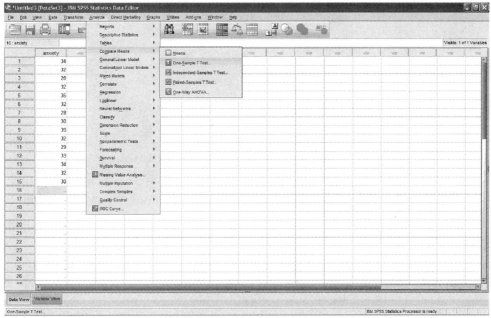

FIGURE 6.3. Selecting the Compare Means and One-Sample T-Test command.

We are familiar with all of these terms, so let's not go into detail just yet; we will refer back to our *p* value, shown as Sig. 2-tailed by SPSS, and the mean difference later. For now, the value that we're most interested in is 2.691, our computed value of *t*. This is the value we will compare to a critical value of *t* in order to test our hypothesis.

Determining the Critical Value of t

Just as we did with the *z* score, we will determine our critical value of *t* using a table developed especially for that purpose. The table showing the area under the curve for *t* values (i.e., a *t* table) will look somewhat different from that for *z* values since it has to compensate for the fewer number of items in the sample. We compensate for the smaller sample sizes by using a value called *degrees of freedom*.

Degrees of Freedom

Throughout the remainder of the book, most of the inferential statistics we compute will include *degrees of freedom*; this is the number of values in the dataset we've col-

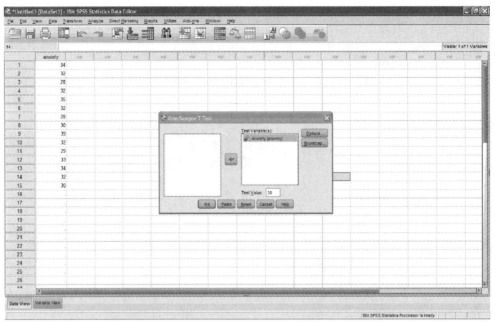

FIGURE 6.4. Identifying the Test Variable and Test Value for the One Sample T-Test.

lected that are "free to vary" when you try to estimate any given value. For example, suppose we have a one-sample *t*-test where we know the population mean is 10. I might ask you to tell me five numbers that, when summed, would give me an average of 10. You might start out by saying "6, 8, 10, 12 … " but then I would have to interrupt you by saying, "That is enough, if I know the first 4 numbers, I can calculate the fifth."

Here's how I know. In order to compute a mean of 10 with a dataset of five values, the sum of the values in the dataset would have to equal 50 (i.e., 50/5 = 10). We could set up an equation to show the relationship between the values we know and the sum we need; *x* is used in the equation to represent the value we do not know:

$$6 + 8 + 10 + 12 + x = 50$$

A little simple math would show the sum of the four values we know is 36:

$$6 + 8 + 10 + 12 = 36$$

One-Sample Test

	t	df	Sig. (2-tailed)	Mean Difference	95% Confidence Interval of the Difference Lower	95% Confidence Interval of the Difference Upper
			Test Value = 30			
Anxiety	2.691	14	.018	2.00000	.4059	3.5941

FIGURE 6.5. Inferential statistics from a One-Sample Test.

We can subtract that value from 50 and see the value we need to complete our equation is 14:

$$50 - 36 = 14$$

In order to double-check our work, we can then put these numbers together and do the required math:

$$6 + 8 + 10 + 12 + 14 = 50$$

And then:

$$50/5 = 10$$

In short, we've shown that, if we know the mean and all of the values in the dataset except one, we can easily determine the missing value. In this case, four of the values in the dataset can vary, but the fifth one cannot. By definition then, our degrees of freedom are 4, that is, $(n - 1)$.

 Be Careful Computing Degrees of Freedom

Although it is just that easy to compute the degrees of freedom for a one-sample *t*-test, you have to be careful. As we will see in later chapters, different statistical tests compute degrees of freedom in different ways; it depends on the number of items in your dataset and the number of levels of the independent variable. For now, let's use our degrees of freedom to look at the critical values of *t* table (Table 6.2). An example of a complete critical values of *t* is shown in Appendix B.

TABLE 6.2. Critical Values of *t*

df	α = .05	α = .025	df	α = .05	α = .025	df	α = .05	α = .025
1	6.314	12.70	11	1.796	2.201	21	1.721	2.080
2	2.920	4.303	12	1.782	2.179	22	1.717	2.074
3	2.353	3.182	13	1.771	2.160	23	1.714	2.069
4	2.132	2.776	14	1.761	2.145	24	1.711	2.064
5	2.015	2.571	15	1.753	2.131	25	1.708	2.060
6	1.943	2.447	16	1.746	2.120	26	1.706	2.056
7	1.895	2.365	17	1.740	2.110	27	1.703	2.052
8	1.860	2.306	18	1.734	2.101	28	1.701	2.048
9	1.833	2.262	19	1.729	2.093	29	1.699	2.045
10	1.812	2.228	20	1.725	2.086	30	1.697	2.042

Although the *t* table looks a lot like the *z* table, there are a few differences. First, you can see degrees of freedom abbreviated as *df*. Next, there are three highlighted columns. The leftmost highlighted column shows degrees of freedom from 1 through 10, the middle highlighted column shows degrees of freedom from 11 through 20, and the rightmost column shows degrees of freedom from 21 through 30. To the right of each of these values, you can see the critical value of *t* for that given degree of freedom; this value is shown for both alpha = .05 and alpha = .025.

Before we move forward, one thing you may have noticed is there is no column for α = .01; I did that on purpose simply to get the idea across. This is just an example of the critical values of the *t* table. You'll find that others have all the values of *t* in one column; others include different alpha values, and so on. For now, we'll just use what we have.

In order to use the *t* table, let us suppose we have a scenario with 12 degrees of freedom and want the critical value of *t* for alpha = .05. All we need to do is find 12 in the table and look directly to the right in the column labeled alpha = .05. As you can see, the critical value of *t* is 1.782. In another case, we might have 29 degrees of freedom and want the critical value of *t* when alpha is .025. Again, we would find the appropriate row and see that the critical value of *t* is 2.045.

Let's Get Back to Our Anxiety Hypothesis

Remember we are dealing with a two-tailed (i.e., nondirectional) hypothesis. Just as we did with the *z*-test, this means we have to divide our alpha value by 2 (i.e., .05/2 = .025). Our degrees of freedom are 14 (i.e., our sample size of 15 minus 1). Using these values, the table shows we need to use a critical value of *t* equal to 2.145; let's plot that value on our curve and test our hypothesis.

Plotting Our Critical Value of t

In our earlier discussion, we agreed that the fewer the number of items in a dataset, the flatter the curve will be. In Figure 6.6, we will have a fairly flat curve like the one below; notice that the critical values of *t* are already marked on either end.

Since our computed value of *t* (i.e., 2.691) is not within the range of the critical values, we can reject the null hypothesis. By looking at our descriptive statistics, it appears that the anxiety levels of people living in urban areas may be significantly higher than the national average. Let's check that, however, with our output from SPSS that was shown above.

We can see here that our *p* value, labeled "Sig. (2-tailed)" is .018; this is less that our alpha value of .025. This verifies the decision we made when we compared the critical value of *t* to the computed value of *t*; we must reject our null hypothesis and support our research hypothesis.

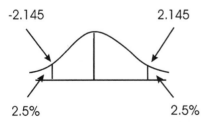

FIGURE 6.6. The critical value of *t* for a two-tailed alpha value of .05.

The Statistical Effect Size of Our Example

Although we have rejected the null hypothesis, some statisticians feel this isn't enough. At this point, we know the groups being compared are significantly different, but we know nothing about the degree to which they're different. In order to address this issue, statisticians have developed tools, called *effect size indices*, which help us better understand the magnitude of any significant difference we've uncovered. An easy way to think about the effect size is that it helps us better understand the extent to which the independent variable affects the dependent variable; this is sometimes called the *practical significance*. In this case, the effect size for the one-sample *t*-test, called Cohen's delta (i.e., *d*), is computed by subtracting the mean of our sample from the mean of the population; we then divide the remainder by the standard deviation of our sample.

$$d = \frac{\overline{x} - \mu}{s}$$

Let's replace the symbols in our formula using the values from above:

1. $d = \dfrac{32 - 30}{2.87849}$

2. $d = \dfrac{2}{2.87849}$

3. $d = .695$

This leaves us with an effect size of .695, but what does that mean?

First, it's plain to see that the computed effect size is nothing more than the percentage of the standard deviation that the difference in the mean scores represents. In discussing this difference, Cohen defined effect sizes as being "small" (i.e., .2 or smaller), "medium" (i.e., between .2 and .5), or "large" (i.e., greater than .5). In this case, we have a large effect size; the groups are significantly different, and the levels of the independent variable had a dramatic effect on the dependent variable. In short, there is a strong relationship between where persons live and their level of anxiety.

As you'll see in the following chapters, different inferential tests require effect sizes to be computed using different values and different formulas. Even so, interpreting them will be about the same. Because they contribute so much to our goal of becoming a good consumer of statistics, we will make computing the effect size a normal part of our descriptive statistics from now on.

Let's Look at a Directional Hypothesis

Let's move forward with our discussion by looking at the following one-tailed hypothesis:

- *Men who use anabolic steroids will have significantly shorter life expectancies than men who do not.*

In this case, suppose we know the average life expectancy for men is 70 years but we only have access to information on 12 men who used steroids of this type. Their age of death is shown in Table 6.3. Using this data, SPSS would compute the descriptive statistics shown in Figure 6.7. Since SPSS does not compute the effect size for a one-sample *t*-test, we can use these statistics, along with the formula, and compute an effect size of –.535:

1. $d = \dfrac{68 - 70}{3.742}$

2. $d = \dfrac{-2}{3.742}$

3. $d = .535$

TABLE 6.3. Steroid Users' Ages at Death

64	66	77	67
66	66	70	66
68	72	64	70

We are interested only in the absolute value, so we drop the negative sign and wind up with an effect size of .535. SPSS would then compute the inferential statistics in Figure 6.8.

Descriptive Statistics

	N	Mean		Std. Deviation
	Statistic	Statistic	Std. Error	Statistic
Age	12	68.0000	1.08012	3.74166
Valid N (listwise)	12			

FIGURE 6.7. Descriptive statistics from a One-Sample Test.

One-Sample Test

	Test Value = 70					
					95% Confidence Interval of the Difference	
	t	df	Sig. (2-tailed)	Mean Difference	Lower	Upper
Age	-1.852	11	.091	-2.00000	-4.3773	.3773

FIGURE 6.8. Inferential statistics from a One-Sample Test.

Here our computed value of *t* is –1.852. If we checked our critical *t* value, we would find it is 1.796 for alpha = .05. Remember, though, the table doesn't show negative values, so, if we are going to test a one-tailed "less than" hypothesis, we need to put a negative sign in front of our value; this means we wind up with –1.796. Since we have a one-tailed hypothesis, as seen in Figure 6.9, we would then plot our critical value on the "less than" side of the distribution.

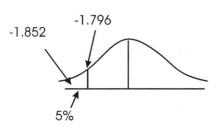

Our computed *t* value, –1.852, is less than the critical value of *t* (i.e., –1.796). Because of that, we reject our null hypothesis and support our research hypothesis. This is supported by a large effect size of .535; the independent variable does have a large effect on the dependent variable. Apparently, men who use anabolic steroids live significantly fewer years than their peers who do not.

FIGURE 6.9. Comparing the computed and critical values of *t* for a one-tailed alpha value of .05.

Just as was the case with the *z*-test, this is not very difficult at all to comprehend. To make things even better, just as we did with the large samples, we can use our *p* value to help us test our null hypothesis.

Using the p Value

As we saw earlier, the *p* value for our anxiety hypothesis was .018, thereby supporting our decision to reject the null hypothesis and support the research hypothesis. Since we were dealing with a nondirectional hypothesis, that was as far as we needed to take it. In this case, we're dealing with a directional hypothesis; because of that, we need to stop and talk about issues related to *p* values and one-tailed tests.

We know that when we have a two-tailed hypothesis, we have to allow for error on both sides of the mean; in order to do that, we have to divide our alpha value by 2. When we have a one-tailed (i.e., directional) hypothesis, we use the entire alpha value on one side of the distribution or the other, depending on the direction stated in the hypothesis.

A directional hypothesis also affects how we interpret our *p* value. SPSS only computes the value for a nondirectional hypothesis. That means our Sig. (2-tailed) value of .091 represents the *p* value for a two-tailed test. Since we are testing a directional hypothesis, we have to divide our *p* value by 2, leaving us with a one-tailed *p* value of .0455. This means we now have a *p* value less than our alpha value of .05 and supports our decision to reject the null hypothesis. It is interesting to note, however, that had we stated a nondirectional hypothesis and used the corresponding two-tailed value (i.e., .091), we would not have rejected the null hypothesis since it is greater than our alpha value of .05. Be aware, however, that the way the *p* value is presented differs between software packages; some compute only the two-tailed *p* value, some only the one-tailed value, and others compute both. In order to accurately test your hypotheses, make sure you know which value you're dealing with.

Check Your Mean Scores!

When we reject a null hypothesis, we always need to ensure the mean scores are in the order hypothesized. In this case, the average life span of males using steroids was

68, two years less than the average life span of 70 and in the order we hypothesized. If, however, the average life span of males using steroids was 72, we would still have a difference of two years and we would reject the null hypothesis. In that case, however, since the values are not in the order hypothesized, we would not support the research hypothesis. We can see this in Table 6.4.

TABLE 6.4. Using the *p* Value to Test the Steroid Use Hypothesis

Average lifespan	Steroids	Difference	*p*	Decision
70	68	−2	.0455	Reject null hypothesis. Support research hypothesis.
70	70	+2	.0455	Reject null hypothesis. Fail to support research hypothesis.

One More Time

In the last chapter, when we talked about using the z score to test hypotheses, we compared certification exam scores from graduate students at ABC University to the national average of 800. In that case, we were concerned whether our students' scores were significantly less than or significantly greater than the national average. Given that, we used a two-tailed, or nondirectional, hypothesis. Let's use the same scenario but imagine we're making the argument that the scores of our students are significantly greater than the national average. Because of that, we would use a one-tailed "greater than" hypothesis:

■ *Graduate students at ABC University will have significantly higher scores on the national certification exam than graduate psychology students at other universities.*

Let's use the data in Table 6.5 to test our hypothesis. Figure 6.10 shows the descriptive statistics from SPSS. Based on these descriptive statistics, we would compute a very large effect size of 1.4 and a t value of 7.426. To test our hypothesis, all we need to do is compare our computed value of t, shown in Figure 6.11, to the critical value of t.

TABLE 6.5. Certification Exam Scores

810	800	810	810	810
810	815	815	815	815
810	815	820	830	830
840	835	840	835	850
848	842	844	811	865

Unlike the nondirectional hypothesis, where we had to divide our alpha value by 2, here we have a one-tailed hypothesis, so we will determine our critical value

Descriptive Statistics

	N	Mean		Std. Deviation
	Statistic	Statistic	Std. Error	Statistic
Certification	25	825.0000	3.36650	16.83251
Valid N (listwise)	25			

FIGURE 6.10. Descriptive statistics from a One-Sample Test.

One-Sample Test

	Test Value = 800					
					95% Confidence Interval of the Difference	
	t	df	Sig. (2-tailed)	Mean Difference	Lower	Upper
VAR00002	7.426	24	.000	25.00000	18.0519	31.9481

FIGURE 6.11. Inferential statistics from a One-Sample Test.

of *t* using the entire alpha value of .05. We can determine our degrees of freedom by subtracting 1 from the sample size and then, by looking at Table 6.2, see that our critical value of *t* is 1.711. Since our critical value of *t* is much less than our computed value of *t* (i.e., 7.426), we must reject our null hypothesis and support our research hypothesis. Based on this, it is apparent that our students do score significantly higher on the national certification exams than those students at other universities. This, again, is supported by the large effect size. The relationship between the computed and critical values of *t* is shown in Figure 6.12.

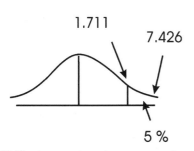

FIGURE 6.12. Comparing the computed and critical values of *t* for a one-tailed alpha value of .05.

Just as was the case earlier, we need to ensure our mean scores are in the direction hypothesized. In this case, the average score of our students is 825, a difference from the mean of +25. Let's compare our current results against the results had our students averaged 775 (Table 6.6).

Important Note about Software Packages

As we saw in our example output, SPSS only gives us the two-tailed *p* value. When testing a directional hypothesis, that means we had to divide our *p* value by 2 prior to comparing it to our alpha value. Remember, this is not the case with all software packages; some will supply both the one-tailed and two-tailed values. These ideas are summarized in Table 6.7.

TABLE 6.6. Using the *p* Value to Test the Certification Exam Score Hypothesis

All universities	ABC	Difference	*p*	Decision
800	825	+25	.0000	Reject null hypothesis. Support research hypothesis.
800	775	−25	.0000	Reject null hypothesis. Fail to support research hypothesis.

TABLE 6.7. The Relationship between Hypothesis Type, Alpha, and *p*

Type of hypothesis	Alpha	*p*
Nondirectional (two-tailed)	Divide by 2.	Use entire value.
Directional (one-tailed)	Use entire value.	Divide by 2.

▓ Let's Use the Six-Step Model!

Now that we know how to test a hypothesis by both comparing a computed value of *t* to a critical value of *t* and comparing a computed *p* value to a preestablished alpha value, let's put this all together and use our six-step model. We'll look at three separate cases to make sure we've got the hang of it; let's get started with ambulance drivers.

The Case of Slow Response Time

Residents in a lower socioeconomic section of town recently complained to local health officials that ambulance response times to their neighborhood were not as fast as average response times throughout town. This, of course, worried the officials, and they decided to investigate. Their records showed the average response time throughout the city to be 90 seconds after an emergency was phoned in. They decided to monitor the next 20 response times to the concerned neighborhood to help determine if the complaint had any merit.

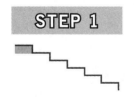

Identify the Problem

In this case, we can clearly see a problem; ambulance response times are perceived to be slower in the lower socioeconomic sections of town. This could lead to undesirable outcomes while waiting for medics at the scene of an accident or illness.

Let's see if this meets our criteria for a good problem:

1. Yes—city officials are interested in this problem.

2. Yes—the scope of the problem is manageable by city officials; they would simply need to collect response times for each section of town.

3. Yes—city officials have the knowledge, time, and resources needed to investigate the problem.

4. Yes—the problem can be researched through the collection and analysis of numeric data.

5. Yes—investigating the problem has theoretical or practical significance.

6. Yes—it is ethical to investigate the problem.

Given that, we'll use this problem statement as we move forward:

◾ *This study will investigate whether ambulance response times vary between different sections of the city.*

Our problem statement is clear and concise, it includes all of the variables we want to investigate, and we've not interjected personal bias.

STEP 2   State a Hypothesis

◾ *Ambulance response times to the lower socioeconomic community are significantly greater than 90 seconds.*

Does this seem odd to you? Were you expecting to say that response times to the lower socioeconomic sections of town were greater than those to the more affluent parts of town? If you were, you are essentially right. In this case, however, we know the exact population value, 90 seconds, we want to compare; therefore the null hypothesis would read:

◾ *Ambulance response times to the lower socioeconomic community are not significantly greater than 90 seconds.*

STEP 3   Identify the Independent Variable

In this case we are interested in looking at times for 20 ambulance calls to the affected neighborhood (i.e., our sample) and compare them to a known population parameter (i.e., the average time for the total number of calls in the town). Because of that, our independent variable is "calls for an ambulance," but we only have one level of that variable: calls to the specific part of town. While it seems that calls to the town overall are another variable, that is not the case. They represent the overall population and cannot be considered as a level of an independent variable; this will become clearer as we move through the next steps.

STEP 4   Identify and Describe the Dependent Variable

The dependent variable, in this case, is the average response time for the 20 ambulance response times in the sample. Just as was the case with the *z*-test, we will compare it to the population average (i.e., the overall average response time for the community). In this case, we know the average response time is 90 seconds.

In order to find the average response time for the 20 calls the officials are going to monitor, let's use the data in Table 6.8. Figure 6.13 shows the descriptive statistics for the sample.

TABLE 6.8. Response Time Data

90	90	101	106	80
82	90	95	95	94
80	89	100	99	99
97	104	102	99	88

One-Sample Statistics

	N	Mean	Std. Deviation	Std. Error Mean
Response Time	20	94.0000	7.69142	1.71985

FIGURE 6.13. Descriptive statistics from a One-Sample Test.

Just as the residents believed, the average response time for calls in their neighborhood is greater than response times for the city in general. Let's move on to the next step to see if their complaints are justified by answering this question, "Is the response time to their neighborhood higher or is it significantly higher?"

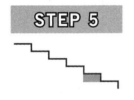

Choose the Right Statistical Test

We have already established, in Step 2, that we are comparing the mean of a small sample (i.e., less than 30 values) to that of the population. Given that, we have one independent variable with one level; we also have one dependent variable representing quantitative data. As a result, we are going to use the one-sample *t*-test.

Use Data Analysis Software to Test the Hypothesis

Figure 6.14 shows us the output we'll use to test our hypothesis. Our one-tailed *p* value is much smaller than our alpha value (i.e., .05), so we reject the null hypothesis. This significant difference is supported by a somewhat large effect size of .520. This is further substantiated by noting that our computed value of *t*, 2.326, is much larger than our critical value of *t*, 1.729.

One-Sample Test

	Test Value = 90					
					95% Confidence Interval of the Difference	
				Mean		
	T	Df	Sig. (2-tailed)	Difference	Lower	Upper
Response Time	2.326	19	.031	4.00000	.4003	7.5997

FIGURE 6.14. Inferential statistics from a One-Sample Test.

Interpreting Our Research Hypothesis

Again, be careful when you interpret the results of these tests. We rejected the null hypothesis, but can we support the research hypothesis? In the case, the mean of our sample is 94 and the population mean is 90. Because of that, the citizens have a legitimate complaint. Remember, however, that if the difference of 4 seconds had been in the other direction (i.e., the mean score of the sample was 86), the same *p* value would have been computed. Like we have said before, the computer does not know our hypothesis, it only processes the data. It is up to us to make sense of what it tells us.

The Case of Stopping Sneezing

Physicians at a premiere hospital are constantly working to lower the amount of time it takes a patient to stop sneezing after being exposed to an allergen. Their experience has shown, on average, that it takes about 2 minutes for standard drugs to be effective. Today, however, they are listening to a representative from a pharmaceutical company who is introducing a drug the company claims will end sneezing in significantly less time. The physicians skeptically agreed to try the drug and decided to use it on a sample of 10 patients. The results of that sample, they believe, will help them decide whether to use the new "miracle drug."

STEP 1

Identify the Problem

The problem here is very clear; patients are exposed to allergens and want to stop sneezing as quickly as possible. While there are drugs that work, the physicians are interested in the problem and are always looking for something better for their patients. Physicians do have the time and expertise and can collect numeric data showing the time it takes patients to stop sneezing. Finally, investigating the problem is ethical and very practically significant (at least to those of us who suffer from allergies!).

> *Persons suffering from allergies desire to be treated with medications that relieve their symptoms as quickly as possible. Physicians will evaluate the new medication to determine if it works as well as claimed by the pharmaceutical representative.*

State a Hypothesis

We have a one-tailed "less than" directional hypothesis:

▪ *Sneezing treated with the new medication will end in significantly less than 2 minutes.*

Our null hypothesis will read:

▪ *Sneezing treated with the new medication will not end in significantly less than 2 minutes.*

STEP 3

Identify the Independent Variable

Here our independent variable is "sneezing," but we only have one level: sneezing in the sample we are interested in. We will be comparing the average time it takes to end sneezing from our sample (i.e., the sample statistic) to the average time for ending sneezing in the overall population (i.e., the population parameter). Because of that, we have an independent variable with only one level.

STEP 4

Identify and Describe the Dependent Variable

The dependent variable is the time it takes sneezing to stop. We know the population average is 2 minutes (i.e., 120 seconds), and we are going to compare it to the average of our 10 cases. We will use data in Table 6.9 with each time shown in seconds; the following Figure 6.15 shows the descriptive statistics for the sample. The pharmaceutical salesperson may be on to something: the new drug does stop sneezing in less time. What we want to know, however, is whether 117 seconds is significantly less than 120 seconds.

TABLE 6.9. Sneezing Data

95	123	120	110	125	113	114	122	122	126

One-Sample Statistics

	N	Mean	Std. Deviation	Std. Error Mean
Time Until Sneezing Stops	10	117.00	9.416	2.978

FIGURE 6.15. Descriptive statistics from a One-Sample Test.

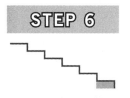

STEP 5

Choose the Right Statistical Test

Since we have one independent variable with one level and one dependent variable representing quantitative data, obviously we are going to use the one-sample *t*-test. Just like before, we are comparing the mean of our small sample (i.e., 10 cases treated with the new drug) to the average time the physicians have experienced with other treatments.

STEP 6

Use Data Analysis Software to Test the Hypothesis

When we use the one-sample *t*-test, Figure 6.16 shows what the SPSS output would look like. It seems the salesperson may be out of luck; the mean time for the new drug was lower, but it was not significantly lower. Our one-tailed *p* value is greater than our alpha value, so we fail to reject the null hypothesis. We could also compute an effect size of .319 to verify that our independent variable had only a moderate effect on the dependent variable. Obviously, our computed value of *t*, –1.007, is much smaller than our critical value of *t*, –1.833.

One-Sample Test

	Test Value = 120					
					\multicolumn{2}{c}{95% Confidence Interval of the Difference}
	t	df	Sig. (2-tailed)	Mean Difference	Lower	Upper
Time Until Sneezing Stops	-1.007	9	.340	-3.00000	-9.7360	3.7360

FIGURE 6.16. Inferential statistics from a One-Sample Test.

There is an interesting issue here, however, that comes up from time to time. In this case, the new drug has an average response time of 117 seconds, while the established drug's average time is 120 seconds. Even though the difference is not significant, the new drug's time is lower. The question to the physicians then remains, "Do we use the new drug and save 3 seconds or, is that so insignificant in the overall scope of things that we continue with what we are used to?"

The Case of Growing Tomatoes

I have lived in big cities most of my life and recently decided to move out into the country. One of the pleasures I've discovered is the ability to grow your own garden and "live off the fruit of the land." In talking to some of the other folks around here, I've discovered that many people consider growing tomatoes a science somewhat akin to nuclear physics. They talk about the amount of rain needed, the type of fertilizer to use, how far plants should be spaced apart, and other really exciting topics. One thing

I found interesting is that they consider a 10-ounce tomato to be about average; most folks can't grow them any bigger than that. After hearing that, I figured a good way for a city boy to fit in would be to grow the biggest, juiciest tomatoes in town!

In order to set my plan in action, I went to the local "feed and seed" store and inquired as to which fertilizer I should use. The guy running the place said he had basically two types. The cheaper variety worked well enough, but he told me, by spending a few extra bucks, I could grow tomatoes I would be proud of! Given that, I bought the costlier of the two, drove home, and put my plan into action!

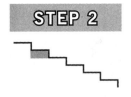

Identify the Problem

Without going into all of the detail, it's apparent I have an interesting, manageable problem: I want to grow the biggest tomatoes I can. I have the time and expertise, the weight of my tomatoes is the numeric data I need, and growing tomatoes is certainly practical and very ethical. Here is my problem statement:

■ *Steve wants to demonstrate to his neighbors that he can grow tomatoes that are equal to or larger than the community's historical average.*

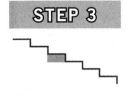

State a Hypothesis

In this case I have a one-tailed "greater than" directional hypothesis:

■ *Steve's tomatoes will weigh significantly more than 10 ounces.*

My null hypothesis is

■ *Steve's tomatoes will not weigh significantly more than 10 ounces.*

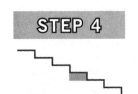

Identify the Independent Variable

In this case, the independent variable is "tomato," but I only have one level; those that I have grown. I'll be comparing the weight of my tomatoes (i.e., the sample statistic) to the average weight of 10 ounces (i.e., the population parameter).

Identify and Describe the Dependent Variable

I show the weights of each tomato in Table 6.10. In this case, the dependent variable is the weight of the tomatoes I have grown. The population average is 10 ounces; the weights of the first 20 tomatoes I picked and the statistics that describe them are shown in Figure 6.17.

TABLE 6.10. Tomato Data

10	11	12	11	10
11	11	12	12	11
8	9	7	8	8
8.5	9	8	7.5	9

One-Sample Statistics

	N	Mean	Std. Deviation	Std. Error Mean
Tomato Weight	20	9.6500	1.63916	.36653

FIGURE 6.17. Descriptive statistics from a One-Sample Test.

Things aren't looking too good, are they? I was determined to grow bigger tomatoes, but mine are actually a little smaller than the average weight reported by my neighbors. Maybe I can salvage some of my dignity, though; I can still hold my head up high if they're not significantly smaller.

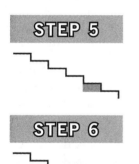

STEP 5

Choose the Right Statistical Test

As we know, I have one independent variable with one level and my dependent variable, the weight of my tomatoes, is quantitative. Because of that, I will use a one-sample *t*-test.

STEP 6

Use Data Analysis Software to Test the Hypothesis

Based on our data, Figure 6.18 shows what SPSS will compute for us. Things might not be as bad as I had feared. My *p* value is .352 but remember, I'm dealing with a one-tailed test so I have to divide that by 2. Even when I do that, my *p* value of .176 is still much larger than the alpha value; this is reflected by a moderate effect size of .213. Because of that, I fail to reject my null hypothesis; this means the research hypothesis is not supported. What does this mean for me? Easy—I get to sit around talking to my neighbors and tell them that my tomatoes are just about average. What does this mean for the guy who sold me the fertilizer? Easy—he may have missed his calling; there are a lot of used cars he could be selling!

One-Sample Test

	Test Value = 10					
					95% Confidence Interval of the Difference	
	t	df	Sig. (2-tailed)	Mean Difference	Lower	Upper
Tomato Weight	-.955	19	.352	-.35000	-1.1172	.4172

FIGURE 6.18. Inferential statistics from a One-Sample Test.

Summary

At the beginning of the chapter I told you that, because of the similarities to the one-sample z-test, it would not take long to discuss the one-sample *t*-test. The key thing to remember from this chapter is the *t* distribution; it is created when multiple samples, each with less than 30 values, are collected and their means plotted. We then used the concept of degrees of freedom to help understand the additional random error introduced by smaller sample sizes. Again, just like the z-test, hypotheses can be tested by computing a *t* value and comparing it to a critical value of *t* or by using a computed *p* value and comparing it to alpha. Regardless of whether we reject or fail to reject the null hypothesis, we can look at the strength of the relationship between our independent and dependent variables using a computed effect size. In the next two chapters we will look at two other tests that use the *t* distribution; the independent- and dependent-sample *t*-tests. While many of the underlying features are the same, we will see that these tests are used to specifically control for the relationship between levels of the independent variable.

Before we move forward with our quiz and into the next chapter, let me tell you a short story about Mr. Gossett. When he was working for Guinness, Gossett was not allowed to publish his findings because the brewery considered his work their intellectual property. After all, they reasoned, the ideas were developed by their employee while he was working in their brewery.

Gossett, realizing the importance of his work, felt it was imperative that he publish his findings. Rather than give his employer a reason to fire him, he published the work under the pseudonym "A. Student." Because of that, many of the older textbooks refer to this distribution as "Student's *t* distribution" and to the test as "Student's *t*-test." While you may not think this is important, just think how much money you could win on the television show *Jeopardy* if one of the categories was "Famous Statisticians"!

Do You Understand These Key Words and Phrases?

computed value of *t*	critical value of *t*
critical value of *t* table	degrees of freedom
effect size	*t* distribution

Do You Know These Formulas?

Computed value of *t*:

$$t = \frac{\bar{x} - \mu}{s_{\bar{x}}}$$

Effect size:

$$d = \frac{\bar{x} - \mu}{s}$$

Quiz Time!

Using the SPSS output provided, what decision would you make for each of the hypotheses below? Manually compute the effect size for each hypothesis; what does that tell you in relationship to the decision made?

1. The average number of cavities for elementary school children in areas without fluoride in the water will be significantly higher than the national average of 2.0 (use Figures 6.19 and 6.20 to test this hypothesis).

One-Sample Statistics

	N	Mean	Std. Deviation	Std. Error Mean
Number of Cavities	15	2.33	1.589	.410

FIGURE 6.19. Descriptive statistics from a One-Sample Test.

One-Sample Test

	Test Value = 2				95% Confidence Interval of the Difference	
	t	Df	Sig. (2-tailed)	Mean Difference	Lower	Upper
Number of Cavities	.813	14	.430	.333	-.55	1.21

FIGURE 6.20. Inferential statistics from a One-Sample Test.

2. Over the course of their careers, physicians trained in Ivy League schools will have significantly fewer malpractice suits filed against them than the national average of 11 (use Figures 6.21 and 6.22 to test this hypothesis).

One-Sample Statistics

	N	Mean	Std. Deviation	Std. Error Mean
Malpractice Suits	15	5.20	2.426	.626

FIGURE 6.21. Descriptive statistics from a One-Sample Test.

One-Sample Test

	Test Value = 11				95% Confidence Interval of the Difference	
	T	df	Sig. (2-tailed)	Mean Difference	Lower	Upper
Malpractice Suits	-9.259	14	.000	-5.800	-7.14	-4.46

FIGURE 6.22. Inferential statistics from a One-Sample Test.

3. The graduation rate of universities that require students to spend their first two years at a community college will be significantly greater than 82% (use Figures 6.23 and 6.24 to test this hypothesis).

One-Sample Statistics

	N	Mean	Std. Deviation	Std. Error Mean
Graduation Rate	15	79.60	4.188	1.081

FIGURE 6.23. Descriptive statistics from a One-Sample Test.

One-Sample Test

	Test Value = 82					
					95% Confidence Interval of the Difference	
	T	df	Sig. (2-tailed)	Mean Difference	Lower	Upper
Graduation Rate	-2.219	14	.043	-2.400	-4.72	-.08

FIGURE 6.24. Inferential statistics from a One-Sample Test.

4. Adult turkeys in the wild will weigh significantly less than 12 pounds (use Figures 6.25 and 6.26 to test this hypothesis).

One-Sample Statistics

	N	Mean	Std. Deviation	Std. Error Mean
Weight of Turkey	15	10.53	2.066	.533

FIGURE 6.25. Descriptive statistics from a One-Sample Test.

One-Sample Test

	Test Value = 12					
					95% Confidence Interval of the Difference	
	T	Df	Sig. (2-tailed)	Mean Difference	Lower	Upper
Weight of Turkey	-2.750	14	.016	-1.467	-2.61	-.32

FIGURE 6.26. Inferential statistics from a One-Sample Test.

CHAPTER 7

The Independent-Sample t-Test

Introduction

We just learned how Gossett successfully used the t distribution, with a one-sample t-test, to compare a sample of ale to the particular batch (i.e., the population) of ale it came from. Gossett soon realized, however, that a problem arose if you have small samples from two different populations and need to determine if they are significantly different from one another.

For example, suppose we are interested in looking at the difference in the number of disciplinary referrals between children from single-parent homes and children who live with both parents. Our independent variable would be "Home Environment," and there would be two levels, "Single Parent" and "Both Parents." Our dependent variable would be the number of times children were disciplined. In another case, we might be interested in comparing the time it takes pigeons to learn via operant or classical conditioning. Again, we would have one independent variable, type of reinforcement, with two levels—operant conditioning and classical conditioning. Our dependent variable would represent the amount of time it took pigeons to learn to do a particular task.

Both of these scenarios meet all of the criteria for an independent- or dependent-sample t-test. Both have one independent variable with two levels, one dependent variable, and the data being collected is quantitative. Our only problem is, how do we decide whether to use the independent-sample t-test or the dependent-sample t-test?

We can answer that question by looking closely at the relationship between the two levels of each of the independent variables. Do you notice anything remarkable? In the first case, you're measuring the number of referrals of two groups of students—those from single-parent homes and those from homes where there are two parents. These two groups are independent of one another; a child who is in one group cannot be in the other. The same goes for the groups of pigeons—one group is being trained with operant conditioning, and the other with classical conditioning. Being in one group excludes the possibility of being in the other.

In both instances the researcher would collect data from two groups, with each group representing a unique population. In Chapter 8, we will see that samples are not always independent, so we will have to use the dependent-sample t-test. In the meantime, in Figure 7.1 we can use our example of students from different home environments to help explain what we have covered up to this point.

■ If We Have Samples from Two Independent Populations, How Do We Know If They Are Significantly Different from One Another?

With the one-sample t-test, we were able to test hypotheses using computed t values, critical T-scores, and p values; we were also able to use Cohen's delta to allow us to measure the effect of the independent variable on the dependent variable. When we have samples from two unrelated populations, we can do the same thing by using the independent-sample t-test. The general idea of hypothesis testing is the same but we need to use a slightly different sampling distribution. Instead of using the *sampling distribution of the means*, which we talked about earlier, we will use the *sampling distribution of mean differences*.

The Sampling Distribution of Mean Differences

Before we get into detail about this distribution, let's conceptually understand where we are going. First, back when we were comparing means from a sample to the population mean, it made sense to take single random samples from a population, calculate

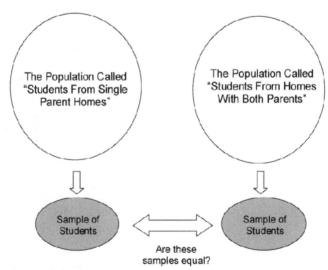

FIGURE 7.1. Comparing samples from two populations.

the mean of the sample and then plot it. By doing this, we created the *sampling distribution of the means*. Given both large enough sample sizes and number of samples, the sample mean was equal to the population mean, and the main measure of dispersion was called the standard error of the mean. We were then able to compare our particular sample mean to the sampling distribution of the means to determine any significant difference.

In this case, however, we are not interested in determining if a single sample is different from the population from which it was drawn. Instead we are interested in determining if the mean difference between samples taken from two like, but independent, populations are significantly different from the mean difference of any other two samples drawn from the same populations. This can be done by creating a *sampling distribution of mean differences* (SDMD) using the following steps.

1. Take repeated samples from each of the two populations we are dealing with.
2. Calculate the mean of each sample.
3. Compute the difference between the two means.
4. Plot the difference between the two means on a histogram.

This is shown in Figure 7.2.

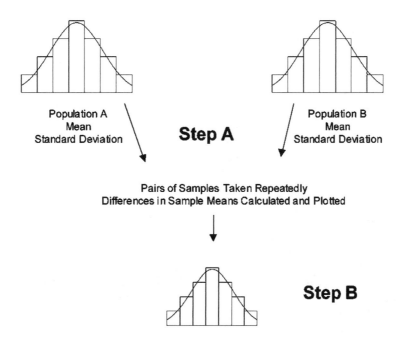

FIGURE 7.2. Creating a sampling distribution of mean differences.

If we continued doing this and plotted each difference on a histogram, we would wind up with a bell-shaped distribution; the larger the number of differences plotted, the more normal the distribution would become. Unlike the sampling distribution of the means, where the mean of the new distribution equals the population mean, the mean of the new distribution will be zero.

In order to understand this, remember that our hypothesis concerning disciplinary referrals involves samples from two populations—students from single-parent homes and students who live with both parents. As we saw with the central limit theorem, most of the values in each sample will cluster around the mean—a mean that we have hypothesized is the same for both populations. When we take samples from each population and then subtract the mean of one from the mean of the other, it naturally follows that the majority of the means of the repeated samplings are eventually going to cancel one another out, leaving a distribution mean of zero. Those means that do not cancel out will leave negative and positive values that create the tails of the distribution.

In order to demonstrate this, let's use the hypothesis about the number of parents and disciplinary referrals. Table 7.1 shows data for 10 samples representing the average number of referrals for each level of the independent variable (i.e., single parent or both parents). After that, I've subtracted the mean number of referrals from two-parent homes from the mean number of referrals from a single-parent home and have included that value in the fourth column:

TABLE 7.1. Referral Data

Sample	Single-parent home	Both parents home	Difference (single – both)
1	3	2	1
2	2	6	–4
3	2	1	1
4	6	5	1
5	1	2	–1
6	0	1	–1
7	3	6	–3
8	5	7	–2
9	5	2	3
10	3	8	–5
Average	3	4	–1

If we use a histogram to plot the difference values from the rightmost column, SPSS will create Figure 7.3. If we continued calculating an infinite number of these mean differences and plotted them on a histogram, we would wind up with the same qualities of the t distribution discussed in Chapter 6.

1. The shape of the distribution will be different for different sample sizes. It will generally be bell-shaped but will be "flatter" with smaller sample sizes.

2. As the number of samples plotted grows, the distribution will be symmetrical around the mean.

3. The empirical rule applies.

4. The measure of dispersion, the standard error of mean differences (SEMD), is conceptually the same as the standard error of the mean.

5. The percentage of values under the curve depends on the alpha value and the degrees of freedom (*df*).

6. The *t* distribution table is used to determine critical *t* values for hypothesis testing.

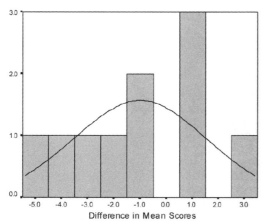

FIGURE 7.3. Bar chart of frequency distribution.

Calculating the t Value for the Independent-Sample t-Test

Looking back, our data show that the students from single-parent homes average three trips per year to see the principal. Their classmates with two parents at home have a higher average: four dreaded trips down the hallway! Although there is a difference between these two groups, is this difference significant or is it due to chance? We can state our question in the form of a null hypothesis and then use the computed *t* value and our table showing the critical values of *t* to help us answer the question.

■ *There will be no significant difference in the number of referrals between students from single-parent homes and students who live with both parents.*

Obviously our research hypothesis would read:

■ *There will be a significant difference in the number of referrals between students from single-parent homes and students who live with both parents.*

Pay Attention Here

Just as I have for the other values throughout the book, I want to show you how to compute the *t* value for the independent-sample *t*-test. While you will probably never be asked to compute a *t* value "by hand," conceptually understanding the process will definitely make you a better consumer of statistics. At first glance, this formula may look complicated, but remember what I said way back in the Introduction; we are only using basic math—addition, subtraction, multiplication, and division.

$$t = \frac{\overline{x}_1 - \overline{x}_2}{\sqrt{\left(\dfrac{SS_1 + SS_2}{n_1 + n_2 - 2}\right)\left(\dfrac{1}{n_1} + \dfrac{1}{n_2}\right)}}$$

Some of this may look like Greek (no pun intended), so let's use Table 7.2 to refresh our memory of the symbols we've already covered as well as add a few new symbols to our gray matter.

TABLE 7.2. Symbols Related to the Independent-Sample *t*-Test

Symbol	Value
x_1	The individual values in Group 1.
x_2	The individual values in Group 2.
x_1^2	The squared values in Group 1.
x_2^2	The squared values in Group 2.
Σx_1	The sum of values in Group 1.
Σx_1^2	The sum of the squared values in Group 1.
Σx_2	The sum of the values in Group 2.
Σx_2^2	The sum of squared values in Group 2.
$\overline{x}_1$	The mean of Group 1.
$\overline{x}_2$	The mean of Group 2.
n_1	The number of values in Group 1.
n_2	The number of values in Group 2.

Now, let's start putting these together to solve the equation. The first thing we see in the denominator (the bottom of the fraction) is that we need to compute the sum of squares for both groups. We've already talked about the idea of the sum of squares earlier, but let's compute it again using the following formulas:

1. $SS_1 = \sum x_1^2 - \dfrac{(\Sigma x_1)^2}{n_1}$

2. $SS_2 = \sum x_2^2 - \dfrac{(\Sigma x_2)^2}{n_2}$

First, we'll modify Table 7.1 by squaring each of the values in our dataset; this is shown in the second and fourth columns of Table 7.3 and is labeled Step 1. In Step 2, we then total the values for each column. Finally, in Step 3, we'll compute the mean for our original two sets of values, shown in the bottom row of Table 7.3.

TABLE 7.3. Computing the Sum of Squares

	Original data from Group 1 x_1	Step 1 x_1^2	Original data from Group 2 x_2	Step 1 x_2^2
	3	9	2	4
	2	4	6	36
	2	4	1	1
	6	36	5	25
	1	1	2	4
	0	0	1	1
	3	9	6	36
	5	25	7	49
	5	25	2	4
	3	9	8	64
Step 2	$\Sigma x_1 = 30$	$\Sigma x_1^2 = 122$	$\Sigma x_2 = 40$	$\Sigma x_2^2 = 224$
Step 3	$\overline{x}_1 = 3$		$\overline{x}_2 = 4$	

We can then compute the sum of squares by using the data from the table.

1. $SS_1 = 122 - \left(\dfrac{30^2}{10}\right)$ $\qquad SS_2 = 224 - \left(\dfrac{40^2}{10}\right)$

2. $SS_1 = 122 - \left(\dfrac{900}{10}\right)$ $\qquad SS_2 = 224 - \left(\dfrac{1600}{10}\right)$

3. $SS_1 = 122 - 90$ $\qquad SS_2 = 224 - 160$

4. $SS_1 = 32$ $\qquad SS_2 = 64$

We can now include our SS_1 and SS_2 values, along with the values we already knew, into our equation and compute our actual *t* value.

1. $t = \dfrac{3-4}{\sqrt{\left(\dfrac{32+64}{10+10-2}\right)\left(\dfrac{1}{10}+\dfrac{1}{10}\right)}}$

2. $t = \dfrac{-1}{\sqrt{\left(\dfrac{96}{18}\right)\left(\dfrac{2}{10}\right)}}$

3. $t = \dfrac{-1}{\sqrt{\left(\dfrac{96}{18}\right)\left(\dfrac{2}{10}\right)}}$

4. $t = \dfrac{-1}{\sqrt{(5.33)(.2)}}$

5. $t = \dfrac{-1}{\sqrt{1.066}}$

6. $t = \dfrac{-1}{1.0325}$

7. $t = -.968$

This time, in order to use SPSS, we have to do something a bit different. On the data screen in Figure 7.4, you see two columns, parents and referrals. In the parents field, I have included a value of 1 if the student has one parent at home and a value of 2 if the student has two parents living at home. These two values will represent the two levels, that is, the number of parents, and the independent variable parents. In the referrals field is the actual number of referrals for that child:

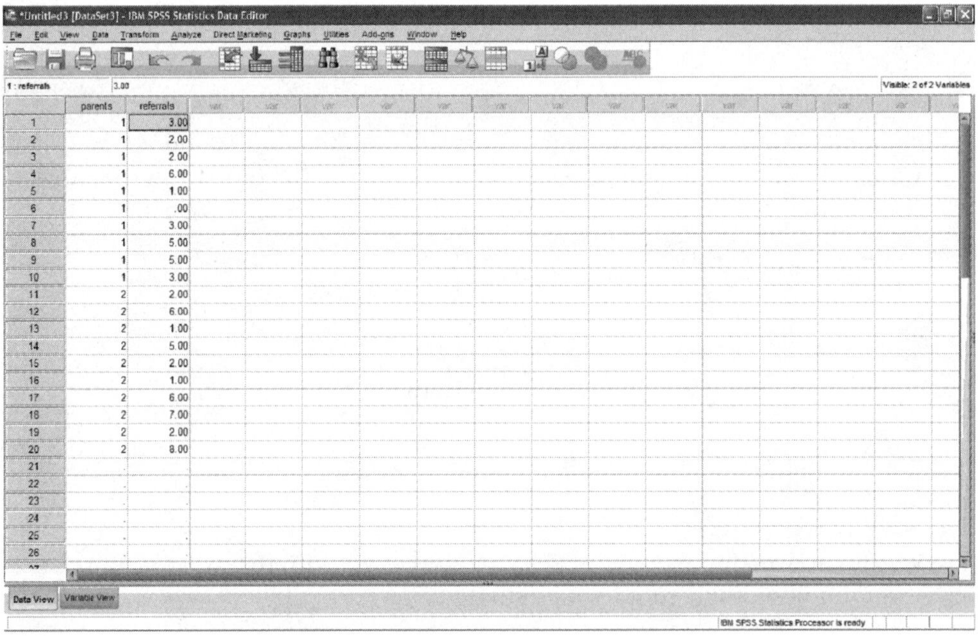

	parents	referrals
1	1	3.00
2	1	2.00
3	1	2.00
4	1	6.00
5	1	1.00
6	1	.00
7	1	3.00
8	1	5.00
9	1	5.00
10	1	3.00
11	2	2.00
12	2	6.00
13	2	1.00
14	2	5.00
15	2	2.00
16	2	1.00
17	2	6.00
18	2	7.00
19	2	2.00
20	2	8.00

FIGURE 7.4. Data View spreadsheet including number of parents and referrals.

At this point, as shown in Figure 7.5, we have to select Analyze, Compare Means, and Independent-Samples T test. The labels used by SPSS diverge from those we normally use. In Figure 7.6 we're asked to identify the Test Variable (i.e., the dependent variable is referrals), the Grouping Variable (i.e., the independent variable is parents), and the Groups (i.e., the levels of the independent variable are one parent or two parents). Clicking on OK will first give us the descriptive statistics shown in Figure 7.7.

Following that, SPSS provides us with the appropriate inferential statistics; we will use these values to actually test our hypothesis. Notice, in the output, that there are several values we've not yet talked about. We'll get to them shortly, but for now let's

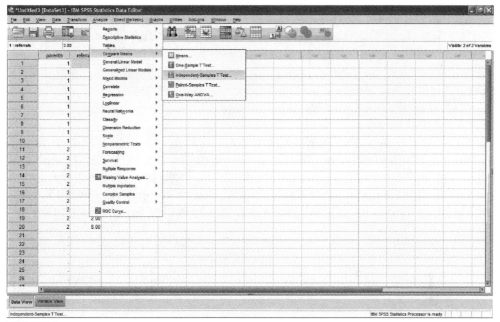

FIGURE 7.5. Using the Compare Means and Independent-Samples T Test commands.

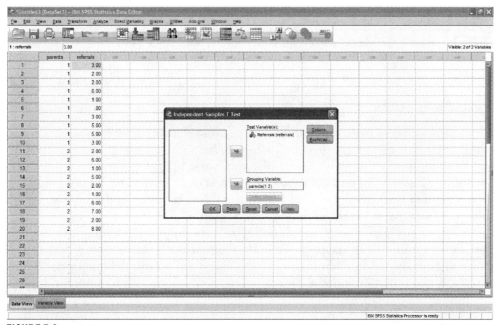

FIGURE 7.6. Defining the Test and Grouping Variables for the Independent-Samples T Test.

Group Statistics

	Parents	N	Mean	Std. Deviation	Std. Error Mean
Referrals	One Parent	10	3.0000	1.88562	.59628
	Both Parents	10	4.0000	2.66667	.84327

FIGURE 7.7. Descriptive statistics from the Independent-Samples T Test.

focus on what we need in Figure 7.8. The column to the far right is labeled "Referrals" and represents the inferential statistics for our dependent variable. Beneath that is a column titled "Equal Variances Assumed." If you look down that column to the row labeled "T," you'll see the value –.968, exactly what we computed by hand. We can now use this information to help us make a decision about the hypothesis we stated.

Independent Samples Test

			Referrals	
			Equal variances assumed	Equal variances not assumed
Levene's Test for Equality of Variances	F		4.787	
	Sig.		.042	
t-test for Equality of Means	T		-.968	-.968
	Df		18	16.200
	Sig. (2-tailed)		.346	.347
	Mean Difference		-1.00000	-1.00000
	Std. Error Difference		1.03280	1.03280
	95% Confidence Interval of the Difference	Lower	-3.16982	-3.18723
		Upper	1.16982	1.18723

FIGURE 7.8. Inferential statistics from the Independent-Samples T Test.

Testing Our Hypothesis

As we saw, students from one-parent homes average three referrals, while their classmates average four. Again, they are different, but are they significantly different? We can determine that by using essentially the same hypothesis-testing procedures we used in the last two chapters. We have computed our t value, and we are going to compare it against the critical value of t from the table. In order to determine the critical value of t, we have to go through a couple of steps.

First, we need to determine which alpha value to use. Let's stick with .05 for now but take a minute to think back. Based on our hypothesis, are we going to use the entire alpha value to determine our critical value? Of course the answer is "No"; since we have a two-tailed hypothesis, we are going to divide our alpha value by 2, giving us an alpha value of .025.

Next we need to determine the necessary degrees of freedom. We do that by using the following formula; in it you can see we add the two sample sizes and then subtract the number of levels in our independent variable, in this case 2:

$$df = n_1 + n_2 - 2$$

Since both of our samples sizes are 10, our degrees of freedom value is 18 (i.e., 10 + 10 – 2 = 18). We can check ourselves by looking at Figure 7.8, shown above.

With an alpha value of .025 and 18 degrees of freedom, we can use Table 7.4, an exact copy of Table 6.2, and find our critical value of *t* is 2.101.

TABLE 7.4. Critical Values of *t*

df	α = .05	α = .025	df	α = .05	α = .025	df	α = .05	α = .025
1	6.314	12.706	11	1.796	2.201	21	1.721	2.080
2	2.920	4.303	12	1.782	2.179	22	1.717	2.074
3	2.353	3.182	13	1.771	2.160	23	1.714	2.069
4	2.132	2.776	14	1.761	2.145	24	1.711	2.064
5	2.015	2.571	15	1.753	2.131	25	1.708	2.060
6	1.943	2.447	16	1.746	2.120	26	1.706	2.056
7	1.895	2.365	17	1.740	2.110	27	1.703	2.052
8	1.860	2.306	18	1.734	2.101	28	1.701	2.048
9	1.833	2.262	19	1.729	2.093	29	1.699	2.045
10	1.812	2.228	20	1.725	2.086	30	1.697	2.042

We can then plot both the critical and computed value and *t* and test our hypothesis. Remember, in Figure 7.9, we have to plot 2.101 as both a positive and a negative value since we are testing a two-tailed hypothesis.

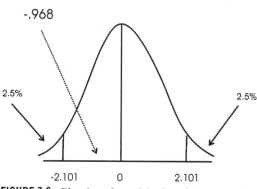

FIGURE 7.9. Plotting the critical and computed values of *t*.

Should we reject or fail to reject our null hypothesis? In this case, we would fail to reject the null since our computed value of *t* (i.e., –.968) is within the range created by our critical value of *t*. That means, in this case, although the students with both parents at home have a higher number of referrals, the difference isn't significant.

The *p* Value

We can also look at our *p* value to help us make a decision. Since we are testing a two-tailed hypothesis, we need to use the two-tailed *p* value. If we look in the column labeled "Equal Variances Assumed" in Figure 7.8, we can see that *p* (i.e., Sig. 2-tailed) of .346 is much greater than our alpha value, and we support our decision to not reject the null hypothesis.

Note on Variance and the t-Test

Although there are columns for both "Equal Variances Assumed" and "Equal Variances Not Assumed," we're only going to use the "Equal Variances Assumed" at this point. We will go into detail regarding the use of these values after we have worked through a couple of examples and have a thorough understanding of the basic ideas of the independent-sample *t*-test.

The Statistical Effect Size of Our Example

As was the case with the one-sample *t*-test, we need to compute an effect size to better understand the strength of the relationship between the mean scores from each of the levels of our independent variable. In this case, we will again use Cohen's *d*, but we will compute it in a slightly different manner:

$$d = \frac{\overline{x}_1 - \overline{x}_2}{S_{\text{pooled}}}$$

You can see, this time, that we're subtracting the mean of one group from the mean of the other and then dividing that value by the *pooled standard deviation* (i.e., S_{pooled}). The pooled standard deviation is calculated by:

1. Adding the variance of the two groups.
2. Dividing that value by 2.
3. Computing the square root of that value.

This is shown in the following formula:

$$S_{\text{pooled}} = \sqrt{\frac{S_1^2 + S_2^2}{2}}$$

Using the values from above, we can compute the effect size in three steps. Notice, SPSS does not include the variance, so we had to square our standard deviation values to include them in the formula:

1. $S_{\text{pooled}} = \sqrt{\dfrac{3.56 + 7.11}{2}}$

2. $S_{\text{pooled}} = \sqrt{5.34}$

3. $S_{\text{pooled}} = 2.31$

Using this value, we can then compute Cohen's delta:

$$d = \frac{3 - 4}{2.31}$$

This gives us an effect size of –.433. As we said earlier, we always use the absolute value of an effect size; that means we just need to drop the negative sign and wind up with an effect size of .433. According to Cohen, this is a medium effect size and, as might be expected given the relatively large *p* value, indicates that the number of parents in a home does not have a large effect on the number of referrals a given student receives.

Let's Try Another Example

Let's say we are interested in determining if there is a significant difference in the frequency of nightmares between men and women. In order to do so, we will test the following null hypothesis:

◼ *There will be no significant difference in the number of nightmares between males and females.*

Our two-tailed research hypothesis would read:

◼ *There will be a significant difference in the number of nightmares between males and females.*

In order to test this hypothesis, we might ask a representative sample of males and females to report how often they have dreams of this type during one month. Obviously, we have one independent variable, gender, with two levels, male and female. We have one dependent variable, frequency of nightmares, and the data collected is quantitative. Since a given participant can fall into only one category, male or female, it appears we will use an independent-sample *t*-test.

Upon administering the survey we might find, in a given month's time, that males had an average of 12 nightmares and females had 7 nightmares. The mean difference between these groups is 5, meaning that males have an average of 5 more nightmares per month than females. Let's use Table 7.5 to represent the 10 males and 10 females who filled out our survey. We can fill out Table 7.6 with the square of the number of dreams for each student.

TABLE 7.5. Nightmare Data

Males	Females
12	2
13	3
17	4
10	9
10	8
10	9
9	10
12	8
15	8
12	9

TABLE 7.6. Computing the Sum of Squares

x_1	x_1^2	x_2	x_2^2
12	144	2	4
13	169	3	9
17	289	4	16
10	100	9	81
10	100	8	64
10	100	9	81
9	81	10	100
12	144	8	64
15	225	8	64
12	144	9	81
$\Sigma x_1 = 120$	$\Sigma x_1^2 = 1496$	$\Sigma x_2 = 70$	$\Sigma x_2^2 = 564$
$\overline{x}_1 = 12.0$		$\overline{x}_2 = 7.0$	

Let's compute the t value by using the values in Table 7.6; I have already computed the sum of squares for you.

$$t = \frac{12 - 7}{\sqrt{\left(\dfrac{56 + 74}{18}\right)\left(\dfrac{1}{10} + \dfrac{1}{10}\right)}}$$

If we worked through the formula, we would end up with a computed t value of 4.16. In Figures 7.10 and 7.11, we can see this exactly matches what SPSS would gener-

Group Statistics

	Gender	N	Mean	Std. Deviation	Std. Error Mean
Nightmares	Male	10	12.00	2.494	.789
	Female	10	7.00	2.867	.907

FIGURE 7.10. Descriptive statistics from the Independent-Samples T Test.

ate for us.

Since we have a two-tailed hypothesis, we must divide our alpha value of .05 by 2; we can then use the new alpha value of .025, along with 18 degrees of freedom, to determine that our critical value of t is 2.101. Since our computed value of t is much greater than our critical value of t, we will reject our null hypothesis (Figure 7.12).

On the SPSS output, our decision is supported, in the column labeled "Equal Variances Assumed," by a two-tailed p value of .001. In short, males do have a significantly greater number of nightmares than females.

Independent Samples Test

		Nightmares	
		Equal variances assumed	Equal variances not assumed
Levene's Test for Equality of Variances	F	.810	
	Sig.	.380	
t-test for Equality of Means	T	4.160	4.160
	Df	18	17.661
	Sig. (2-tailed)	.001	.001
	Mean Difference	5.000	5.000
	Std. Error Difference	1.202	1.202
95% Confidence Interval of the Difference	Lower	2.475	2.472
	Upper	7.525	7.528

FIGURE 7.11. Inferential statistics from the Independent-Samples T Test.

Remember the Effect Size

Let's not forget about our effect size. We know we have a significant difference, but what effect did the levels of the independent variable have on the dependent variable? Again, we could use our effect size formula. Let's compute the pooled variance first,

$$S_{\text{pooled}} = \sqrt{\frac{6.22 + 8.22}{2}}$$

Followed by:

$$S_{\text{pooled}} = \sqrt{\frac{14.44}{2}}$$

Then:

$$S_{\text{pooled}} = \sqrt{7.22}$$

This gives us a pooled standard deviation of 2.69. We can enter that into our effect size formula, along with the values of the two means.

$$d = \frac{12 - 7}{2.69}$$

This results in a very large effect size of 1.86. Obviously, the independent variable has a really large effect on the dependent variable.

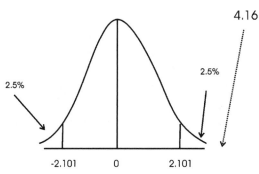

FIGURE 7.12. Plotting the critical and computed values of *t* for a two-tailed test.

What about a Directional Hypothesis?

Using this same example let's suppose we felt, from the outset, that men would have more nightmares than women. That means we would test the same null hypothesis:

■ *There will be no significant difference in the number of nightmares between males and females.*

In this case, however, we would have to change our research hypothesis to be one-tailed, so it would read:

■ *Males will have significantly more nightmares than females.*

This means, of course, that we are going to have the same descriptive statistics and we will compute the same value for t. We do need a different critical value of t, however, because we are testing a directional hypothesis. Because we are dealing with a one-tailed hypothesis, however, we need to refresh our memories on using our alpha value to determine the correct critical value of t we need to use.

As you know, when you state a directional hypothesis, you're saying one of the mean values you're comparing is going to be significantly higher *OR* significantly lower than the other. Since you're not looking at the probability of it being either, this means you're going to have to use your entire alpha value on one end of the curve or the other. Here, we are hypothesizing that the mean of one group is going to be significantly higher than the other. Because of that, we would not divide alpha by 2 as we did with the two-tailed test; instead we would use the t table to look at the one-tailed t value for an alpha level of .05 and 18 degrees of freedom. We would find this gives us a critical value of 1.734; let's plot this in Figure 7.13 and see what happens.

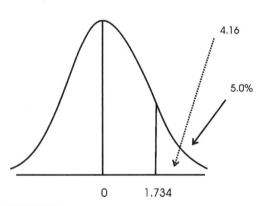

In this case, our computed value of t (i.e., 4.16) is greater than our critical value of t (i.e., 1.734). There seems to be a significant difference in the values, and we can use our p value to help support our decision. We have to be careful here, however. Remember, we are looking at a one-tailed hypothesis, and we used our entire alpha value to determine the critical value of t. Having changed those things, our p value is also affected; it has dropped to .0005. When we compare it to our alpha value of .05, we do support the decision we made when we compared the computed and critical values of t. In Table 7.7, we can see the logical pattern of when to divide either alpha or p by 2 depending on the type of hypothesis.

FIGURE 7.13. Plotting the critical and computed values of t for a one-tailed test.

TABLE 7.7. The Relationship between the Direction of the Hypothesis, Alpha, and *p*

Type of hypothesis	Alpha	*p*
Directional (one-tailed)	Use entire alpha value	Divide *p* by 2
Nondirectional (two-tailed)	Divide alpha by 2	Use entire *p* value

Reminder—Always Pay Attention to the Direction of the Means!

We can see that our computed value of *t* is greater than our critical value of *t*, and our *p* value is less than our alpha value but we always have to be careful! Keep in mind, SPSS doesn't know our hypothesis; it simply works with the data we supply. In cases where the *p* value is less than alpha, you must make your mean values match the direction you hypothesized before you decide to support your research hypothesis.

For example, here we hypothesized that males would have a significantly higher number of nightmares than females. Our mean scores of 12 for males and 7 for females enable us to reject the null hypothesis. Obviously, that means we can support the research hypothesis. Suppose, however, the mean values had been reversed. We may have collected our data and found a mean score of 7 for males and 12 for females. If that's the case, SPSS would compute the same *p* value. This means you would reject the null hypothesis, but you would not support the research hypothesis because the mean difference is not in the order you hypothesized. Be very, very careful with this.

■ Putting the Independent-Sample *t*-Test to Work

Having said all of this and having gained a good understanding of how the independent-sample *t*-test works, it is now time to put it to work for us. Just as we've done in the prior chapters, we will use case studies to help us. After reading the case we will again use the six-step process to help us understand what we need to do. Let's get started with a case about an age-old complaint, "I could do better if I only had better surroundings, equipment, and so forth . . . "

The Case of the Cavernous Lab

A university lab manager has computers set up in two locations on campus. One of these locations, the Old Library, is a dark, cavernous old building; the other is the new brightly lighted, well-decorated Arts and Sciences building. The manager has noted that the equipment layout is about the same in each building and both labs receive the same amount of use.

Each term approximately 50 students use these labs for an "Introduction to Computers" course, with about half in each lab. The faculty member who teaches the classes has noticed that the two groups of students are about the same in that they both have the same demographic and achievement characteristics. Despite this, the teacher has noticed that the students using the lab in the Old Library do not seem to perform as well as students working in the new laboratory. The only thing the professor can figure is that "students always perform better in better surroundings." The teacher has asked the lab manager to investigate and report the findings.

The manager decides the best thing to do is to collect final examination scores from all 50 students and compare the results of the students in the Library to those in the Arts and Sciences building. After collecting the data, the manager realizes he will probably need to use some type of statistical tool to analyze the data, but isn't sure which one. Because of that, the manager asks us to get involved; let's see what we can do to help him.

Identify the Problem

The problem here is very straightforward. The university lab manager has installed new computer labs, but their use seems to have affected student achievement. Not wanting his labs to play a part in students' grades, he decides to investigate. He is certainly interested in a very manageable problem. Since he installed the lab and is working with the teachers, he has the knowledge, time, and resources necessary. He can easily collect numeric data to help ethically address a problem with important practical significance. That allows us to make a very simple problem statement:

■ *This study will investigate whether the location of a computer lab affects student achievement.*

State a Hypothesis

The first thing we tell the manager is that we need to determine if differences in achievement, if they do exist, are significant. Knowing that, the first thing we do is state the hypothesis we will be investigating:

■ *Students taking Introduction to Computers in the new Arts and Sciences laboratory will score significantly higher on their final examination than students taking the same course in the Old Library.*

As we can see, this is a directional hypothesis since we believe one group will do significantly better than the other group. We stated it that way because that's the scenario the manager wants to investigate. We could have just as easily stated a non-directional hypothesis by saying there would be a significant difference between the groups with no direction stated. That isn't, however, what the scenario calls for. The null hypothesis would be:

■ *There will be no significant difference in levels of achievement between students taking Introduction to Computers in the new Arts and Sciences lab and students taking the same course in the Old Library.*

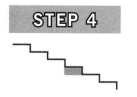

STEP 3

Identify the Independent Variable

Now we need to look at the cause and effect we are trying to investigate. The teacher is suggesting that the location of the laboratory (cause) affects how students perform (effect) in the Introduction to Computers class. This means our independent variable is location and there are two levels: students in the Old Library and students in the new Arts and Sciences building.

STEP 4

Identify and Describe the Dependent Variable

The manager has decided to collect final examination scores from all of the students to help make a decision; this will be our dependent variable. Although it isn't explicitly stated, we can assume the data will be quantitative, with scores ranging from 0 to 100. For our example, let's use the data in Table 7.8. Let's look at the SPSS results in Figure 7.14.

TABLE 7.8. Achievement Data

Students in the Arts and Sciences building					Students in the Old Library				
70	73	73	76	76	65	69	69	72	72
72	79	79	79	79	71	75	75	75	75
82	82	86	82	82	77	78	78	78	78
85	85	85	85	88	81	81	81	82	84
88	88	91	91	94	84	84	87	88	90

Group Statistics

	Building	N	Mean	Std. Deviation	Std. Error Mean
Achievement	Arts & Sciences	25	82.0000	6.39010	1.27802
	Old Library	25	77.9600	6.34087	1.26817

FIGURE 7.14. Descriptive statistics from the Independent-Samples T Test.

We can see that the students in the Arts and Science building do have a higher average score than their counterparts in the Old Library. While it is apparent that the independent variable does have some effect on the dependent variable, that's not enough. We have to determine if the difference between the two groups is significant or if it is due to chance.

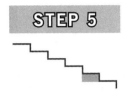

Choose the Right Statistical Test

In this case, we have one independent variable we have named "Building" with two levels—students in the Old Library and students in the new Arts and Sciences building. We are collecting quantitative data, and our data are fairly normally distributed. If we consider all of this information, it appears that the independent-sample *t*-test is exactly what we need to use to test our hypothesis.

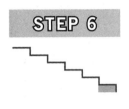

Use Data Analysis Software to Test the Hypothesis

By looking at the prior table, we have already seen that the mean score of the students in the Arts and Sciences building is higher than the mean score of the students in the Old Library, but we also know the difference might be due to chance. We can use the statistics in Figure 7.15 to help us make a decision about our hypothesis.

Independent Samples Test

		Achievement	
		Equal variances assumed	Equal variances not assumed
Levene's Test for Equality of Variances	F	.012	
	Sig.	.913	
t-test for Equality of Means	T	2.244	2.244
	df	48	47.997
	Sig. (2-tailed)	.029	.029
	Mean Difference	4.04000	4.04000
	Std. Error Difference	1.80044	1.80044
	95% Confidence Interval of the Difference Lower	.41996	.41996
	Upper	7.66004	7.66004

FIGURE 7.15. Inferential statistics from the Independent-Samples T Test.

Up to this point, we've used only the "Equal Variances Assumed" column to test our hypotheses. As I said earlier, now that we have a good feeling for how the independent-sample *t*-test works, it is important to understand the meaning of these two values.

Testing Equal Variance in the Groups

Although it rarely affects beginning statisticians, the amount of variance within each group affects the manner in which the *t* value is computed. Again, although SPSS doesn't show the variance as part of the *t*-test results, we could easily square the standard deviation to determine the variance of each group. In this case, these values are very close: 40.833 and 40.207. SPSS uses these values to test the null hypothesis comparing the variance between the groups:

■ *There will be no significant difference between the variance of the Arts and Sciences group and the variance of the Old Library group.*

Notice this is a two-tailed hypothesis and it is always stated in this manner. To test it, SPSS computes an *analysis of variance* (this is often called ANOVA, but more on that later). The results of this ANOVA are shown in the top two rows of Figure 7.15 labeled "Levene's Test for Equality of Variance."

To interpret the results, we first need to set our alpha level at .05. We then compare our alpha value to the computed value of *p* for the Levene test. Since, in this case, the computed value of *p* is .913 and our alpha value is .05, we fail to reject the null hypothesis (be careful here, we are still talking about the values in the rows for the Levene's test, not the *p* values farther down the table). This means there is no significant difference in the variance between the two groups. Given that, we will use the column reading "Equal Variances Assumed" for our independent-sample *t*-test. If the computed *p* value for the Levene test was less than .05, we would reject the null hypothesis and then have to use the column reading "Equal Variances Not Assumed." This idea is summarized in Table 7.9.

TABLE 7.9. Interpreting the Levene Test

Levene test *p* value	Resultant *t*-test decision
p value less than .05	Equal variances not assumed
p value greater than or equal to .05	Equal variances assumed

Testing Our Hypothesis

We are finally ready to test our hypothesis. We have a one-tailed hypothesis; this means we have to divide our Sig. (2-tailed) value of .029 by 2; this leaves us with a one-tailed *p* value of .0145. This is less than our alpha value, .05, and means we reject our null hypothesis. This decision is supported by a fairly large effect size of .635.

Determining the Critical Value of *t* When You Have More Than 30 Degrees of Freedom

We could also check ourselves by comparing the computed value of *t* to the critical value of *t*. In this case, in order to do that, we need to look at an issue that arises when our degrees of freedom value is greater than 30.

Up to this point, we've been dealing with cases where we have had 30 or fewer degrees of freedom and each of those values has been listed in the *t* distribution table. In this case, we have 48 degrees of freedom, but if you look at the complete *t* distribution table in the back of the book, you'll see the values skip from 30 to 40, 40 to 60, 60 to 120, and then 120 to infinity (that's what the symbol ∞ means). You might ask, "What happened to all of the other values?" The short answer is, once we get to a certain number of degrees of freedom, the table can be abbreviated because the area under the curve for t changes so little between the differing degrees of freedom, there is no need to have a value that is 100% correct.

For example, if we have a problem with 40 degrees of freedom, our critical value for t would be 1.684; if we had 60 degrees of freedom, it would be 1.671. Since the difference between these two values is so small (i.e., .013) the table does not break the critical values down any further. Instead, when we're manually computing statistics, we use the next highest df value in the table that is greater than our actual df. In this case, we have 48 degrees of freedom so we would use the value for $df = 60$ (or 1.671). Since our computed value of t is 2.211, we again know we should reject our null.

Make Sure You Check the Direction of the Mean Scores

As we said earlier, if you've stated a directional hypothesis and you have a one-tailed p value less than .05, you have to ensure your mean values match the manner in which you hypothesized in order to support your research hypothesis. For example, here we hypothesized the students in the Arts and Sciences building would do better than the students in the Old Library. Our mean scores of 82 for the Arts and Sciences students and 77.96 for the Old Library students, combined with a p value of .0145, enabled us to reject the null hypothesis. Be careful, though; we might have collected the data and found a mean score of 82 for students in the Old Library and a score of 77.96 for students in the Arts and Sciences building. In that case, even though SPSS would have computed exactly the same p value and we would have rejected the null hypothesis, we still couldn't support the research hypothesis since the means aren't in the order hypothesized. SPSS software doesn't know the hypothesis you're testing, it just works with the numbers you put in. Given that, you have to be careful to interpret the output correctly.

The Case of the Report Cards

Let's continue by looking at a situation in which instructors at a high school are interested in determining if the frequency of report cards has any effect on students' intrinsic motivation. After researching the literature, they are amazed to find many studies have shown mixed results. Some authors contend that immediate feedback contributes to higher levels of motivation, while others believe students might consider it controlling and rebel against it. To investigate this, the teachers agree to give half of their classes weekly report cards and the other half will receive report cards on the normal 9-week schedule. After doing so they plan to analyze their data and come to a conclusion. In order to help them, let's go through the six-step process.

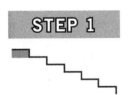 **STEP 1**

Identify the Problem

In this case it seems all of the criteria are met, but let's take a look at the significance of the problem; is it practically or theoretically important? Actually, one could argue both are true. The report cards are being used to address a practical problem; higher achievement is linked to higher levels of motivation. At the same time, the researchers are addressing a problem for which the literature has shown mixed results. Given that, one could definitely argue that the study also addresses current motivational theory and contributes back to the literature about what is known about feedback and intrinsic motivation.

■ *Research has shown that the frequency of report cards may affect student motivation. In order to be able to provide feedback to students in the most meaningful way possible, teachers will investigate the effect on motivation of weekly report cards versus report cards given once per 9 weeks.*

STEP 2   State a Hypothesis

■ *Students receiving report cards each week will have significantly different levels of intrinsic motivation than students receiving report cards every 9 weeks.*

In this case, the research hypothesis is nondirectional because the researchers are stating a difference will exist, but they are not sure which group will be higher. It is based on prior observations and research, the results of which have been mixed.

The null hypothesis would read:

■ *There will be no significant difference in intrinsic motivation between students receiving weekly report cards and students receiving report cards every 9 weeks.*

STEP 3   Identify the Independent Variable

The independent variable is "Report Card Frequency," and there are two levels; one group of students will receive report cards every 9 weeks, and the other group will receive them each week. The two levels are independent of one another. A student in one group cannot be in the other group.

STEP 4   Identify and Describe the Dependent Variable

The dependent variable in this case is the level of a student's intrinsic motivation; we will assume our instrument has 20 questions with a range of 1 to 5 on each question. That means, of course, that scores can range from 20 (a low level of intrinsic motivation) to 100 (a high level of intrinsic motivation) for general levels of intrinsic motivation. Knowing that, let's use the data in Table 7.10 for 10 students in each group. SPSS would create the descriptive statistics shown in Figure 7.16.

We can see the weekly group has a mean score considerably larger than do those students getting the report cards every 9 weeks. Remember, though, the teachers are interested in a significant difference, not just a difference. Knowing that, they need to look at the second table. Our job now is to select the appropriate statistic to help us interpret our data.

TABLE 7.10. Report Card Data

Weekly report cards		9-week report cards	
75	80	69	68
68	82	68	64
87	75	69	68
80	75	72	67
65	55	69	68

Group Statistics

	Report	N	Mean	Std. Deviation	Std. Error Mean
Motivation	Weekly	10	74.2000	9.34285	2.95447
	Nine-week	10	68.2000	1.98886	.62893

FIGURE 7.16. Descriptive statistics from the Independent-Samples T Test.

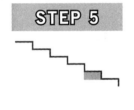

Choose the Right Statistical Test

In this case we again have one independent variable with two groups. The groups are independent of one another, and we have one dependent variable that represents quantitative data. Based on the kurtosis and skewness values, the data appear to be fairly normally distributed so we are going to use the independent-sample *t*-test.

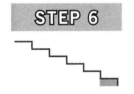

Use Data Analysis Software to Test the Hypothesis

Using Figure 7.17, what should the teachers conclude regarding their null hypothesis? First, the teachers must use Levene's test to decide whether there is a significant difference in the variance between the two groups. In this case, since the *p* value is well below .05, they must use the column labeled "Equal Variances Not Assumed" to test their hypothesis. Notice, when this line is used, the degrees of freedom field isn't calculated in the same manner it is when we use the line that says "Equal Variances Assumed." This is done to control for error introduced by the differences in variance; SPSS will always calculate this for us.

As we said earlier, since the teachers are looking only for a significant difference in their hypothesis and no direction is implied, the hypothesis is two-tailed; that means they should use the Sig. (2-tailed) value shown in the table. In this case, the *p* value is .076, larger than the alpha value of .05. Because of this, the null hypothesis is not rejected; there is not a significant difference in levels of motivation between students who receive report cards every week and those who receive them only every

Independent Samples Test

		Motivation	
		Equal variances assumed	Equal variances not assumed
Levene's Test for Equality of Variances	F	8.881	
	Sig.	.008	
t-test for Equality of Means	T	1.986	1.986
	Df	18	9.814
	Sig. (2-tailed)	.062	.076
	Mean Difference	6.00000	6.00000
	Std. Error Difference	3.02067	3.02067
	95% Confidence Interval of the Difference	Lower	-.34619	-.74780
		Upper	12.34619	12.74780

FIGURE 7.17. Inferential statistics from the Independent-Samples T Test.

9 weeks. We could verify that by checking the computed value of *t* (1.986) against our critical value of *t* (2.101); again, this would show that the teachers would not reject the null hypothesis.

Notice the Effect Size

In this case, although we are failing to reject the null hypothesis, we would compute an effect size of .889. Even though this is very large, remember that the *p* value is a function of both the effect size and the sample size. In this case, although the independent variable had a large effect on the dependent variable, the *p* value was somewhat higher than you might expect because of the small sample size.

A Point of Interest

We have already agreed that a good hypothesis is consistent with prior research or observations. In this case, based on the literature, the teachers hypothesized there would be a significant difference in motivation. They didn't say significantly higher or significantly lower, they just said "different." Interestingly, had they opted for a one-tailed hypothesis such as "Students receiving weekly report cards will have higher levels of intrinsic motivation than those receiving report cards each 9 weeks," they would have rejected the null hypothesis. We know that because, had we divided our two-tailed *p* value by 2, the resultant one-tailed *p* value of .038 would be less than our alpha value. Again, based on the literature, the hypothesis is two-tailed so the teachers have to live with it the way it is. Let's look at another scenario and see what happens.

The Case of the Anxious Athletes

Researchers at a major university are interested in determining if participation in organized sports has any relationship to levels of anxiety. Their initial feelings are that persons involved in physical activity are, for a number of reasons, less likely to report levels of anxiety so high they affect their quality of life. Wanting to test this, the researchers decide to use a well-known anxiety inventory to measure levels of anxiety between two groups—university students involved in organized sports and those not involved in some type of sport. In order to collect their data, professors propose to identify 15 students in each group, ask them to complete the inventory, and then calculate the results.

STEP 1

Identify the Problem

Here, again, we have a problem that is worth investigating. Without going into all of the details, since higher levels of anxiety may lead to lower grades, it is in the university's best interest to determine if this is true. Again, much like the problem investigating motivation, this could be considered both a practical and a theoretical problem.

■ *High levels of anxiety have been shown to be detrimental to students' well-being and quality of life. Since research has shown that involvement in physical activity may lead to lower levels of anxiety, researchers at the university investigate that by comparing anxiety levels of students involved in organized sports versus those who are not.*

STEP 2

State a Hypothesis

In this case, the researchers feel athletes will have significantly lower levels of anxiety than their less active classmates. Given that, they decide to use the following hypothesis:

■ *University students involved in organized sports programs will have significantly lower levels of anxiety than university students not involved in organized sports programs.*

The null hypothesis would read:

■ *There will be no significant difference in levels of anxiety between university students involved in organized sports programs and university students not involved in organized sports programs.*

STEP 3

Identify the Independent Variable

The independent variable in this case is, for lack of a better term, "university student." The two levels are students involved in organized sports and students not involved in organized sports.

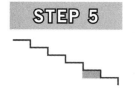

STEP 4

Identify and Describe the Dependent Variable

Again, the professors are going to measure anxiety using an anxiety inventory. The scores reported by the instrument are quantitative in nature and range from 20 (a low level of anxiety) to 80 (a high level of anxiety). Let's use the data in Table 7.11 to represent 15 students in each group. SPSS would generate the descriptive statistics shown in Figure 7.18.

TABLE 7.11. Anxiety Data

Students in organized sports			Students not in organized sports		
35	32	31	24	42	31
33	41	35	25	41	32
25	31	41	31	31	33
40	29	61	31	24	20
44	31	25	41	25	25

Group Statistics

	Sports	N	Mean	Std. Deviation	Std. Error Mean
Anxiety	Plays Sports	15	35.6000	9.04591	2.33565
	Does Not Play Sports	15	30.4000	6.80126	1.75608

FIGURE 7.18. Descriptive statistics from the Independent-Samples T Test.

So far it seems we are doing fine. We have two groups of quantitative data that are fairly normally distributed. The means are 5.2 points apart, and we have a rather large effect size of .650 so we may be on to something. Let's move on to the next step and see what happens.

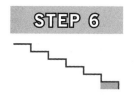

STEP 5

Choose the Right Statistical Test

Here again we have one independent variable with two levels— one group that participates in sports and one that doesn't participate in sports. The dependent variable, anxiety scores, shows two groups with nice bell-shaped distributions. This means we should use an independent-sample *t*-test.

STEP 6

Use Data Analysis Software to Test the Hypothesis

In Figure 7.19, the *p* value for Levene's test is greater than .05, so we will use the scores in the "Equal Variances Assumed" column. Thinking back to our hypothesis, it was one-tailed, so we need to divide the Sig. (2-tailed) value by 2; this leaves us with .043. This means we reject the null hypothesis; there are significant differences in anxiety between students involved in sports and students not involved in sports.

Independent Samples Test			Anxiety	
			Equal variances assumed	Equal variances not assumed
Levene's Test for Equality of Variances	F		.466	
	Sig.		.501	
t-test for Equality of Means	T		1.780	1.780
	Df		28	25.995
	Sig. (2-tailed)		.086	.087
	Mean Difference		5.20000	5.20000
	Std. Error Difference		2.92216	2.92216
	95% Confidence Interval of the Difference	Lower	-.78578	-.80665
		Upper	11.18578	11.20665

FIGURE 7.19. Inferential statistics from the Independent-Samples T Test.

Wait a minute; are we jumping the gun here? We are right; we do have a p value less than our alpha value, but look at the mean scores from the two groups. The students involved in sports have a higher mean score (35.6) than their less active classmates (30.4). In this case, the researchers would technically be able to reject the null hypothesis, but since the results weren't in the hypothesized direction, there is no support for the research hypothesis. For us, this shows we need to pay attention to the relationship between our *printout* and the hypothesis we have stated. Many a good student has been led astray by not paying strict attention to what she was doing!

Just in Case—A Nonparametric Alternative

As we said earlier, the use of the *t*-test mandates we have quantitative data that are normally (or nearly normally) distributed. Although the *t*-test is robust enough to handle minor variances in the normality of the distribution, if the problem is bad enough, we are forced to use the nonparametric *Mann–Whitney U test*. Unlike the independent-sample *t*-test, the U test uses the median of the dataset, rather than the mean, as the midpoint. Because of this, the U test can be used for both quantitative data that is not normally distributed and in cases where the researcher collects ordinal (rank)-level data. Generally speaking, the software packages create output for the U test that is very similar to the *t*-test; it is also just as easy to interpret.

Summary

Using the independent-sample *t*-test is very straightforward. Once we state our hypothesis and identify the statistical tool we need, interpreting the results of a statistical software package is easy. Remember, the key things to look for are Levene's variance test and the computed value of *p* for the data you've entered. Always keep in mind, however, that if you reject the null hypothesis, ensure the means are in the order you hypothesized prior to supporting the research hypothesis. Table 7.12 can act as a guide to help you understand and use the *t*-test.

TABLE 7.12. Decision Making Using the *t*-Test

Step	Decisions to be made
Decide if a *t*-test is needed	One independent variable with two levels? One dependent variable with quantitative data that is normally distributed?
Decide which *t*-test is needed	If levels of independent variable are related, use the dependent sample *t*-test. If levels of independent variable are not related, use the independent sample *t*-test.
Once the test is run, check Levene's *p* value	If *p* is less than .05, equal variances are not assumed. If *p* is greater than or equals .05, equal variances are assumed.
Determine if one-tailed or two-tailed *p* value is needed	If the hypothesis is directional (one-tailed), use the one-tailed *p* value. If you only have the two-tailed value, divide it by 2. If the hypothesis is nondirectional (two-tailed), use the two-tailed *p* value.
Use appropriate *p* value to compare to alpha value	If *p* is less than alpha, reject null hypothesis. Be sure to check if the mean values are in the order you've hypothesized prior to making any decisions. If *p* is greater than or equals alpha, fail to reject null hypothesis.

Do You Understand These Key Words and Phrases?

computed value of *t*	effect size
equal variances assumed	equal variances not assumed
independent-sample *t*-test	Levene's test
Mann–Whitney U test	pooled standard deviation
sampling distribution of mean differences	standard error of mean difference
sum of squares	

Do You Understand These Formulas?

Degrees of freedom:

$$df = n_1 + n_2 - 2$$

Effect size:

$$d = \frac{\bar{x}_1 - \bar{x}_2}{S_{pooled}}$$

Pooled standard deviation:

$$S_{pooled} = \sqrt{\frac{s_1^2 + s_2^2}{2}}$$

Sum of squares for group 1:

$$SS_1 = \sum x_1^2 - \frac{(\sum x_1)^2}{n_1}$$

Sum of squares for group 2:

$$SS_2 = \sum x_2^2 - \frac{(\sum x_2)^2}{n_2}$$

t-score:

$$t = \frac{\bar{x}_1 - \bar{x}_2}{\sqrt{\left(\frac{SS_1 + SS_2}{n_1 + n_2 - 2}\right)\left(\frac{1}{n_1} + \frac{1}{n_2}\right)}}$$

Quiz Time!

The Case of the Homesick Blues

Psychologists are interested in determining whether males or females make more phone calls home during military basic training.

1. What is the hypothesis being tested?

2. What is the independent variable and its levels?

3. What is the dependent variable?

4. Based on Figures 7.20 and 7.21, what decision should they make?

Group Statistics

	Freshman Gender	N	Mean	Std. Deviation	Std. Error Mean
Phone Calls	Males	15	93.2667	9.80136	2.53070
	Females	15	69.8000	22.68165	5.85638

FIGURE 7.20. Descriptive statistics from the Independent-Samples T Test.

Independent Samples Test

		Phone Calls	
		Equal variances assumed	Equal variances not assumed
Levene's Test for Equality of Variances	F	10.936	
	Sig.	.003	
t-test for Equality of Means	T	3.678	3.678
	Df	28	19.052
	Sig. (2-tailed)	.001	.002
	Mean Difference	23.46667	23.46667
	Std. Error Difference	6.37978	6.37978
	95% Confidence Interval of the Difference Lower	10.39828	10.11612
	Upper	36.53505	36.81721

FIGURE 7.21. Inferential statistics from the Independent-Samples T Test.

The Case of the Cold Call

Employees of a telemarketing firm traditionally worked in cubicles where it was easy to hear their coworkers around them. After a recent expansion, employees started complaining that the increased noise level in the room was beginning to interfere with their productivity. Fearing this could lead to fewer sales, the owners of the company decided to put half of their employees into private offices to see if it would lead to more calls per hour.

1. What is the hypothesis being tested?

2. What is the independent variable and its levels?

3. What is the dependent variable?

4. Based on Figures 7.22 and 7.23, what decision should they make?

Group Statistics

	Office Location	N	Mean	Std. Deviation	Std. Error Mean
Marketing Calls	Cubicle	8	15.8750	3.35676	1.18679
	Enclosed Office	8	7.5000	3.66450	1.29560

FIGURE 7.22. Descriptive statistics from the Independent-Samples T Test.

Independent Samples Test

			Marketing Calls	
			Equal variances assumed	Equal variances not assumed
Levene's Test for Equality of Variances	F		.122	
	Sig.		.732	
t-test for Equality of Means	T		4.767	4.767
	Df		14	13.894
	Sig. (2-tailed)		.000	.000
	Mean Difference		8.37500	8.37500
	Std. Error Difference		1.75700	1.75700
	95% Confidence Interval of the Difference	Lower	4.60661	4.60390
		Upper	12.14339	12.14610

FIGURE 7.23. Inferential statistics from the Independent-Samples T Test.

The Case of the Prima Donnas

Female actresses on Broadway are convinced that putting their names on the marquee in front of their male counterparts leads to a larger audience. The owners of a theater are interested in finding out if this is true, so they track tickets sales while changing the order of the names on the marquee periodically.

1. What is the hypothesis being tested?

2. What is the independent variable and its levels?

3. What is the dependent variable?

4. Based on Figures 7.24 and 7.25, what decision should they make?

Group Statistics

	Gender	N	Mean	Std. Deviation	Std. Error Mean
Ticket Sales	Prima-donna	15	955.0667	266.98434	68.93506
	Males	15	879.8667	257.97145	66.60794

FIGURE 7.24. Descriptive statistics from the Independent-Samples T Test.

Independent Samples Test

		Ticket Sales	
		Equal variances assumed	Equal variances not assumed
Levene's Test for Equality of Variances	F	.006	
	Sig.	.939	
t-test for Equality of Means	T	.784	.784
	Df	28	27.967
	Sig. (2-tailed)	.439	.439
	Mean Difference	75.20000	75.20000
	Std. Error Difference	95.85750	95.85750
	95% Confidence Interval of the Difference Lower	-121.15519	-121.16561
	Upper	271.55519	271.56561

FIGURE 7.25. Inferential statistics from the Independent-Samples T Test.

The Case of the Wrong Side of the Road

Some believe that gasoline stations going into a city charge less per gallon than stations do on roads leading out of a city. Their logic is that drivers need to fill up as they're leaving on a trip; since that's the case, the owners of gasoline stations feel they can charge more and get away with it. I'm not sure if I believe this, so I'm going out and collecting gasoline prices from stations going in both directions.

1. What is the hypothesis being tested?

2. What is the independent variable and its levels?

3. What is the dependent variable?

4. Based on Figures 7.26 and 7.27, what decision should I make?

Group Statistics

Gas Station Location		N	Mean	Std. Deviation	Std. Error Mean
Price per Gallon	Into Town	10	3.0180	.16430	.05196
	Out of Town	10	3.0480	.20286	.06415

FIGURE 7.26. Descriptive statistics from the Independent-Samples T Test.

Independent Samples Test

			Price per Gallon	
			Equal variances assumed	Equal variances not assumed
Levene's Test for Equality of Variances	F		1.818	
	Sig.		.194	
t-test for Equality of Means	T		-.363	-.363
	Df		18	17.255
	Sig. (2-tailed)		.721	.721
	Mean Difference		-.03000	-.03000
	Std. Error Difference		.08255	.08255
	95% Confidence Interval of the Difference	Lower	-.20343	-.20397
		Upper	.14343	.14397

FIGURE 7.27. Inferential statistics from the Independent-Samples T Test.

CHAPTER 8

The Dependent-Sample *t*-Test

Introduction

In the last chapter, we saw that when we have one independent variable with two levels that are independent of one another, we used the independent-sample *t*-test. What happens, however, when the levels of the independent variable represent two different measurements of the same thing? For example, imagine comparing pretest and posttest scores for one group of students. If we did, we would have two measurements of the same group. The independent variable would be "Student Test Scores" and there would be two levels, "Pretest Scores" and "Posttest Scores." In this case, a given student's posttest score would be directly related to his or her pretest score. Because of that, we would say the levels are dependent upon, or influence, one another.

As another example, suppose we are interested in determining if a particular drug has an effect on blood pressure. We would measure a person's blood pressure prior to the study, give medicine for a period of time, and then measure the blood pressure again. In this case, for a given set of measurements, the "Before Blood Pressure" and the "After Blood Pressure" are the two levels of the independent variable named "Blood Pressure." Again, these two levels are related to one another, so we would use the dependent-sample *t*-test (sometimes called the *paired-samples t-test*) to check for significant differences.

▨ That's Great, But How Do We Test Our Hypotheses?

Just like in the prior chapter, here we have an independent variable with two levels and one dependent variable where we have collected quantitative data. Given that, you might be asking, "Why not just use the independent-sample *t*-test?" While that's a logical question, unfortunately, we cannot do that. As we'll soon see, the relationship between the levels of the independent variable creates a problem if we try to do it that way.

Independence versus Dependence

When we talked about the independent-sample t-test, the key word was "independent"; scores from one group did not influence scores in the second group. That's not the case when we have two related groups; let's use an example to help us understand what we are getting at.

Using the idea of pretests and posttests we alluded to earlier, let's use the data in Table 8.1 to help us better understand where we are going with this idea.

TABLE 8.1. Student Test Data

Student	Pretest	Posttest
1	52	89
2	48	77
3	51	81
4	45	69
5	50	80
6	60	90
Average	51	81

In this case, the key to understanding what is meant by dependence is based on two things. First, obviously each individual pretest score was made by one given student. Given that, when a student takes the posttest, his score will be "related" to his pretest score. Second, the fact that two sets of scores were collected from one set of people makes them "dependent" on each other; the posttest score should generally be higher than the pretest score for a given student. This might sound somewhat confusing, but we will see more examples that make this concept very clear. The bottom line, for now, is that we have two levels: student pretest scores and student posttest scores.

Computing the t Value for a Dependent-Sample t-Test

Because of the relationship between the two levels of the independent variable, we cannot compute the t value using the same formula as the independent-sample t-test; if we did, we would dramatically increase the probability of making a Type I error (i.e., rejecting the null hypothesis when we shouldn't). In order to avoid that, we have a different formula to compute the t value:

$$t = \frac{\overline{D}}{\sqrt{\frac{\sum D^2 - \frac{(\sum D)^2}{n}}{n(n-1)}}}$$

Testing a One-Tailed "Greater Than" Hypothesis

To help understand this formula, let's look at my wife's job. She's a media specialist and firmly believes that it doesn't matter what children read (obviously within limits); it is getting them to read anything that makes them better readers.

In order to test her theory, she would start by asking her students to tell her how many hours each week they spend reading outside of school. She could then introduce them to the *Harry Potter* series, books that appeal to children at that age level. After a few weeks she would again ask the children how many hours they read per week. Obviously, she hopes the second number is greater than the first. Her research hypothesis would be:

■ *The average number of hours spent reading per week will be significantly greater after allowing students to read books that appeal to them.*

In Table 8.2, we have data for five students with the number of books they read before the new books were introduced, as well as the number of books they read after the new books were introduced. In order to compute the *t* value, we need to include two extra columns to create Table 8.3. The first new column, labeled *D*, shows us the difference between the Before and After scores. The second new column, labeled D^2, shows the squared value of *D*. At the bottom of each of those columns, you can see we've summed these values.

TABLE 8.2. Number of Books Read

Student	Before	After
1	4	6
2	4	6
3	6	6
4	7	9
5	8	12
Average	5.8	7.8

TABLE 8.3. Computing the Difference Values for Books Read

Student	Before	After	D	D^2
1	4	6	+2	4
2	4	6	+2	4
3	6	6	0	0
4	7	9	+2	4
5	8	12	+4	16
Average	5.8	7.8	$\Sigma D = 10$	$\Sigma D^2 = 28$

Looking back at the equation, you can see the first value we need to compute is $\bar{D}$, the average of the difference between scores in a given dataset. We can see the sum of the differences (i.e., ΣD) is 10; to compute the average, we need to divide that by 5, the number of values in our dataset. This leaves us with an average of 2; we can put that into our formula before moving on to the next step.

$$t = \frac{2}{\sqrt{\left(\dfrac{\sum D^2 - \dfrac{(\sum D)^2}{n}}{n(n-1)}\right)}}$$

We can now deal with the denominator of the equation. Again, let's go through step by step. First, in the third column of our table we have calculated the difference between both values and summed them (i.e., 10). In the rightmost column, we have taken each of the difference values and squared it. Adding these values gives us the sum of the differences squared (i.e., $\sum D^2$). We can insert these values, along with n (i.e., 5) into the equation. We can finish computing the equation using the following steps.

1. $t = \dfrac{2}{\sqrt{\left(\dfrac{28 - \dfrac{10^2}{5}}{n(n-1)}\right)}}$

2. $t = \dfrac{2}{\sqrt{\left(\dfrac{28 - \dfrac{100}{5}}{n(n-1)}\right)}}$

3. $t = \dfrac{2}{\sqrt{\left(\dfrac{28 - 20}{5(5-1)}\right)}}$

4. $t = \dfrac{2}{\sqrt{\left(\dfrac{8}{20}\right)}}$

5. $t = \dfrac{2}{\sqrt{(.4)}}$

6. $t = \dfrac{2}{.6325}$

This leaves us with a computed t value of 3.16. Before we can test the hypothesis, we have to determine the critical value of t from the same table we used with the independent-sample *t*-test. This time, however, we will compute our degrees of freedom by subtracting one from the total pairs of data; when we subtract 1 from 5, we're left with 4 degrees of freedom. Using the traditional alpha value of .05, we would refer to our table and find that the critical value of *t* is 2.132.

We can then plot that on our *t* distribution shown in Figure 8.1; remember, we have a one-tailed hypothesis, so the entire *t* value goes on one end. Here our computed value of *t* is greater than our critical value of *t*. Obviously we have rejected the null hypothesis and supported my wife's research hypothesis: children do read more when they are actually interested in what they are reading.

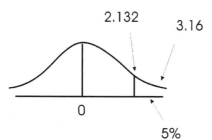

2.132   3.16

0

5%

FIGURE 8.1. Comparing the computed and critical values of *t*.

In order to check our work using SPSS, we need to set up our spreadsheet, shown in Figure 8.2, to include two variables, Before and After; we would then include the data for each. Following that, in Figure 8.3, we select Analyze, Compare Means, and Paired-Samples T Test. As shown in Figure 8.4, we would then identify the pairs of data we want to compare and click on OK. SPSS would first provide us with Figure 8.5; it verifies what we computed earlier in Table 8.3.

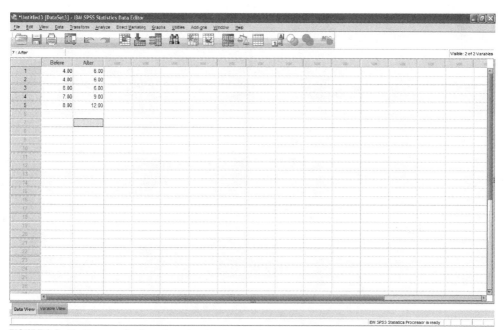

FIGURE 8.2. Before and after data in the Data View spreadsheet.

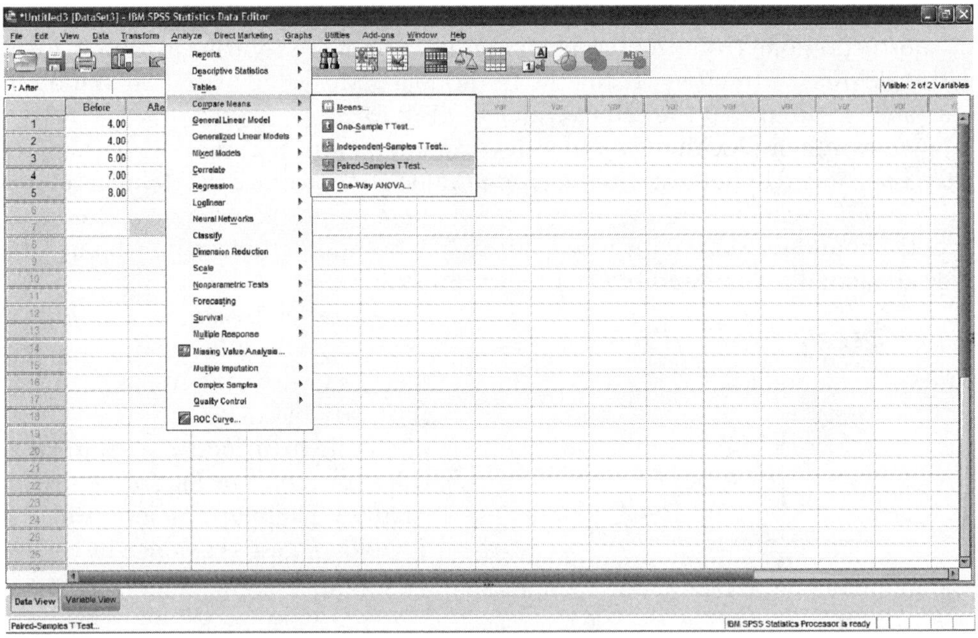

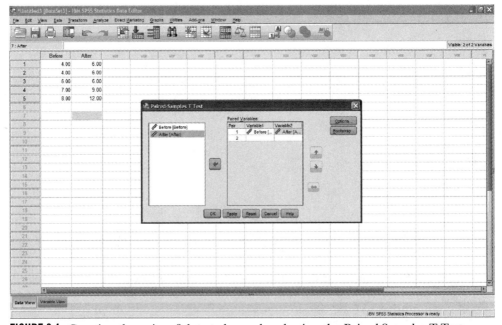

FIGURE 8.3. Using the Compare Means and Paired-Samples T Test command.

FIGURE 8.4. Creating the pairs of data to be analyzed using the Paired-Samples T Test.

		Mean	N	Std. Deviation	Std. Error Mean
Pair 1	Before	5.8000	5	1.78885	.80000
	After	7.8000	5	2.68328	1.20000

FIGURE 8.5. Descriptive statistics from the Paired-Samples test.

As we can see in Figure 8.6, the *t* value of 3.16 is exactly what we computed earlier and our *p* value is less than .05. This means we can support the research hypothesis; kids reading books they enjoy actually did spend significantly more time reading per week than they did when they had no choice in their reading material. What's the bottom line? My wife is happy!

Paired Samples Test

			Pair 1
			After - Before
Paired Differences	Mean		2.00000
	Std. Deviation		1.41421
	Std. Error Mean		.63246
	95% Confidence Interval of	Lower	.24402
	the Difference	Upper	3.75598
T			3.162
Df			4
Sig. (2-tailed)			.034

FIGURE 8.6. Inferential statistics from the Paired-Samples test.

The Effect Size for a Dependent-Sample t-Test

Just as was the case with the independent-sample *t*-test, we can compute an effect size for the dependent-sample *t*-test. Interpreting it is the same as in our earlier example, but the formula is a lot easier to compute; all you do is divide the average mean difference by the standard deviation of the difference.

$$d = \frac{\overline{x}_{\text{difference}}}{s_{\text{difference}}}$$

Using the values from above and the following two steps, we can compute an effect size of 1.42:

1. $d = \dfrac{2}{1.41}$

2. $d = 1.42$

According to Cohen's standards, this is very large and indicates that the treatment had quite an effect on the dependent variable. Just as we did in the preceding chapter, we will always include the effect size as part of our descriptive statistics.

Testing a One-Tailed "Less Than" Hypothesis

We have seen how well this works with a "greater than" one-tailed hypothesis; now let's look at an example where we are hypothesizing that one average will be significantly less than another. Let's suppose we are working with a track coach at our local high school who is trying to improve his team's times in the 400-meter run. The coach has heard that a diet high in protein leads to more muscle mass and figures this should contribute to his athletes running faster. He decides to test the new diet for 6 weeks and measure the results:

> ■ *A track athlete's time in the 800 meter run will be significantly less after following a high-protein diet for 6 weeks.*

During the course of the diet regimen, the coach collected data shown in Table 8.4; each of the times is measured in seconds.

TABLE 8.4. Race Time Data

Athlete	Before	After
Al	105	98
Brandon	110	105
Charlie	107	100
De'Andre	112	107
Eduardo	101	91
Frank	108	103

Just by looking, it seems that the "after" scores are lower, but let's go ahead and compute our t value. First, since we are looking at a one-tailed "less than" hypothesis, we need to subtract the "Before" from the "After" value. This can be seen in Table 8.5.

TABLE 8.5. Computing the Difference Values for Race Times

Athlete	Before	After	D	D^2
Al	105	98	–7	49
Brandon	110	105	–5	25
Charlie	107	100	–7	49
De'Andre	112	107	–5	25
Eduardo	101	91	–10	100
Frank	108	103	–5	25
Sum			$\Sigma D^2 = -39$	$\Sigma D^2 = 273$

To compute $\bar{D}$, we are again going to divide the sum of the differences (–39) by the number of values in the dataset (6); this gives us an average difference of –6.5. Let's go ahead and enter that into our formula.

$$t = \frac{-6.5}{\sqrt{\left(\dfrac{\sum D^2 - \dfrac{(\sum D)^2}{n}}{n(n-1)}\right)}}$$

We continue by inserting the sum of the differences squared (i.e., $\Sigma D^2 = 273$), the sum of the differences (i.e., $\Sigma D = -39$), and n (i.e., 6), into our equation:

$$t = \frac{-6.5}{\sqrt{\left(\dfrac{273 - \dfrac{(-39)^2}{6}}{6(6-1)}\right)}}$$

If we simplify that, we come up with the following formulas:

1. $t = \dfrac{-6.5}{\sqrt{\left(\dfrac{273 - \dfrac{1521}{6}}{30}\right)}}$

2. $t = \dfrac{-6.5}{\sqrt{\left(\dfrac{19.5}{30}\right)}}$

3. $t = \dfrac{-6.5}{.806}$

4. $t = -8.06$

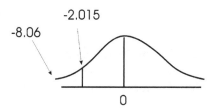

FIGURE 8.7. Using the computed and critical values of *t* to test the hypothesis.

We can now help the coach test his hypothesis. First, using our alpha value of .05 and our degrees of freedom of 5, we find that we have a critical *t* value of 2.015. In order to plot this, keep in mind that we have a one-tailed hypothesis, so the entire critical *t* value goes on one end of the distribution as shown in Figure 8.7. In this case, since we have a one-tailed "less than" hypothesis, it needs to go on the left tail of the distribution. Since our computed value of t is also negative, it needs to go on the left side of the distribution as well.

We can clearly see our computed value of t is far less than our critical value of t; this means we have to reject the null hypothesis. It appears that athletes really do run faster if they follow a high-protein diet.

In Figures 8.8 and 8.9, we can confirm what we have done using SPSS. We are further assured of our decision since the p value of .000 is less than our alpha value; we can also compute an effect size by dividing the mean difference of our scores by the standard deviation.

Paired Samples Statistics

	Mean	N	Std. Deviation	Std. Error Mean
Pair 1 Before the Diet	107.1667	6	3.86868	1.57938
After the Diet	100.6667	6	5.75036	2.34758

FIGURE 8.8. Descriptive statistics from the Paired-Samples test.

Paired Samples Test

		Pair 1 After the Diet - Before the Diet
Paired Differences	Mean	-6.50000
	Std. Deviation	1.97484
	Std. Error Mean	.80623
	95% Confidence Interval of the Difference Lower	-8.57247
	Upper	-4.42753
T		-8.062
Df		5
Sig. (2-tailed)		.000

FIGURE 8.9. Inferential statistics from the Paired-Samples test.

$$d = \frac{-6.5}{1.97}$$

This yields an effect size of –3.30, but as we did with the independent-sample t-test, we have to drop the negative sign and use the absolute value. When we do, we see that our effect size means our intervention had a definite effect on our dependent variable.

■ Testing a Two-Tailed Hypothesis

In order to help us thoroughly understand what we are doing, let's consider a case where we have a two-tailed hypothesis. In this case, we are investigating a drug designed to stop migraine headaches. While we have every indication that it should work, we have discovered that the drug may negatively affect a person's systolic (i.e., the upper number) blood pressure. In some instances, the drug might cause the person's blood pressure to rise; in other cases, it might drop significantly.

Let's create a dataset by taking a patient's blood pressure at the start of our study, administer the migraine drug for 2 weeks and then measure the blood pressure again. After we have completed our study, we wind up with the data in Table 8.6; notice that I've already computed the difference and the squared difference for you.

TABLE 8.6. Computing the Difference Values for Blood Pressure

Patient	Prior	After	D	D^2
1	120	135	15	225
2	117	118	1	1
3	119	131	12	144
4	130	128	−2	4
5	121	121	0	0
6	105	115	10	100
7	128	124	−4	16
8	114	111	−3	9
9	109	117	8	64
10	120	120	0	0
			$\Sigma D = 37$	$\Sigma D^2 = 563$

Just like before, we compute $\overline{D}$ by taking the average of all of the difference scores (i.e., $37/10 = 3.7$) and put it into our formula.

$$ t = \frac{3.7}{\sqrt{\left(\dfrac{\sum D^2 - \dfrac{(\sum D)^2}{n}}{n(n-1)}\right)}} $$

We can insert the values for ΣD^2, ΣD, and n into our equation and use the following steps to compute t:

1. $t = \dfrac{3.7}{\sqrt{\left(\dfrac{563 - \dfrac{1369}{10}}{10(10-1)}\right)}}$

2. $t = \dfrac{3.7}{\sqrt{\left(\dfrac{563 - 136.9}{90}\right)}}$

3. $t = \dfrac{3.7}{\sqrt{\left(\dfrac{426.1}{90}\right)}}$

4. $t = \dfrac{3.7}{\sqrt{(4.73)}}$

5. $t = \dfrac{3.7}{2.18}$

Finally, we are left with a t value of 1.70; this matches Figures 8.10 and 8.11 that would be created by SPSS.

Paired Samples Statistics

		Mean	N	Std. Deviation	Std. Error Mean
Pair 1	Before the Medication	118.3000	10	7.66014	2.42235
	After the Medication	122.0000	10	7.49815	2.37112

FIGURE 8.10. Descriptive statistics from the Paired-Samples test.

Paired Samples Test

			Pair 1 — After the Medication – Before the Medication
Paired Differences	Mean		3.70000
	Std. Deviation		6.88073
	Std. Error Mean		2.17588
	95% Confidence Interval of the Difference	Lower	-1.22218
		Upper	8.62218
T			1.700
Df			9
Sig. (2-tailed)			.123

FIGURE 8.11. Inferential statistics from the Paired-Samples test.

We can see from the descriptive statistics that the average blood pressure before the medication is 118.3, with a slightly higher blood pressure of 122 after the medication has been taken. If we compute our effect size of .538 based on these values, we can see that the intervention had a moderate influence on our dependent variable.

We can now plot our computed *t* value and the appropriate critical value. Remember, since we are dealing with a two-tailed hypothesis, we have to divide alpha by 2 and find the critical value for alpha = .025. Using the table, along with 9 degrees of freedom, our critical value of *t* is 2.262. As shown in Figure 8.12, we would then mark that off on both ends of our normal curve.

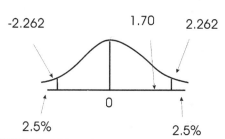

FIGURE 8.12. Using the computed and critical values of *t* to test the hypothesis.

Since our computed value is within the range of the positive and negative critical values, we do not reject the null hypothesis; even though the average "After" blood pressure rose slightly, it wasn't a significant difference. This is verified by the two-tailed *p* value of .123. In other words, after all of this, we have shown that the drug manufacturers have nothing to worry about. Apparently, their new migraine medicine doesn't significantly affect the average systolic blood pressure.

▨ Let's Move Forward and Use Our Six-Step Model

Since we are now proficient in doing the calculations by hand, let's use our six-step approach to work with a middle school math teacher. In this case, the teacher has noticed that many of her students are very apathetic toward math. This, she knows, can contribute to low levels of achievement. The teacher, realizing that something must be done, begins to investigate ways to get her students more interested in their studies.

During her investigation, the teacher finds a new software package that seems to be exactly what she needs. The particular package starts by asking students about their personal interests and activities, and then it creates word problems based on that information. By tailoring the math lesson to each individual student, the teacher hopes to foster more interest in her subject. She hopes, of course, this will lead to less apathy and higher achievement.

This really excites the teacher and a plan is immediately put into effect. The teacher plans to measure student interest in math and then use the software for 10 weeks. At the end of the 10 weeks, she plans on using the same instrument to see if there has been any change in their feelings toward math.

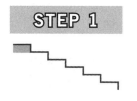
STEP 1

Identify the Problem

There is definitely a straightforward, important problem in the teacher's room and she has the expertise, resources, and time to investigate it. It would be easy for her to collect numeric data to determine levels of apathy toward math, and there appear to be no problems with the ethicality of her investigation. Here is her problem statement:

■ *Math teachers have found many students are apathetic toward their subject matter; this, they feel, may lead to lower achievement. This study will investigate using a software package that tailors the text of word problems to each student's given interests and activities. This, they believe, will lead to lower apathy and higher achievement.*

State a Hypothesis

From the text of the scenario it is easy to develop the hypothesis the teacher will be investigating:

■ *Levels of student apathy will be significantly lower after using the new software system for 10 weeks than they were prior to using the software.*

As we can see, we have a directional hypothesis because we have stated that students will have lower levels of apathy after the intervention than they did before it. Again, we stated it in this manner because it reflects the situation that the teacher wants to investigate. The corresponding null hypothesis would read:

■ *There will be no significant difference in levels of student apathy before using the software and levels of student apathy after using the software.*

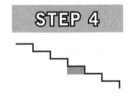

Identify the Independent Variable

When we were using the independent-sample *t*-test, it was easy to identify the cause-and-effect relationship in the hypothesis and, from that, identify the independent variable and its levels. In this case, the independent variable is not as clear but we have two groups we can easily identify—a group of students before using the software and the same students after using the software. Given that, we can call the independent variable "Student" and use those as our levels. When we used the independent-sample *t*-test, we had two separate groups. The fact it was two separate groups is what made it independent. Here we have the same set of students, but we are interested in measuring them at two different points in time (i.e., their pretest scores and their posttest scores). Given that, we will use a dependent-sample *t*-test.

Identify and Describe the Dependent Variable

Our dependent variable is student apathy, and we can see the scores below in Table 8.7. The first column, "Student ID Number," is the unique number identified for each student. The second column shows the student's apathy score at the start of the 10 weeks, and the third column shows the same student's apathy score at the end of the 10 weeks.

TABLE 8.7. Student Apathy Scores

Student ID number	Starting apathy score	Ending apathy score
1	62	50
2	67	55
3	67	55
4	60	68
5	69	60
6	66	60
7	70	65
8	72	65
9	68	70

SPSS would produce Figure 8.13, our descriptive statistics. In this case, we used nine pairs of data representing "Starting Apathy" and "Ending Apathy" scores. In the Mean column, we see two values. The Starting Apathy scores show that students had an average score of 66.78 on the questionnaire administered at the start of the 10 weeks. The Ending Apathy scores, collected at the end of the 10 weeks, show an average of 60.89. For each of the mean values, we also see the standard deviation and the standard error of the mean.

Paired Samples Statistics

		Mean	N	Std. Deviation	Std. Error Mean
Pair 1	Ending Apathy	60.8889	9	6.67915	2.22638
	Starting Apathy	66.7778	9	3.76755	1.25585

FIGURE 8.13. Descriptive statistics from the Paired-Samples test.

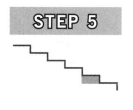

STEP 5

Choose the Right Statistical Test

In this case, we have one independent variable with two levels that are related; we say the levels are related because the quantitative data collected for one level is related to the quantitative data in the other level. It is clear that we need to use a dependent-sample *t*-test.

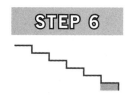

STEP 6

Use Data Analysis Software to Test the Hypothesis

SPSS would compute the results shown in Figure 8.14.

Right off the bat, we can see a mean difference of 5.889, which would allow us to compute an effect size of .856. While these numbers look promising, do not let them take your atten-

Paired Samples Test

			Pair 1
			Ending Apathy - Starting Apathy
Paired Differences	Mean		-5.88889
	Std. Deviation		6.88194
	Std. Error Mean		2.29398
	95% Confidence Interval of the Difference	Lower	-11.17882
		Upper	-.59896
T			-2.567
Df			8
Sig. (2-tailed)			.033

FIGURE 8.14. Inferential statistics from the Paired-Samples test.

tion away from testing our hypothesis using the *p* value. In this case, we have to divide the (Sig. 2-tailed) value of .033 by 2; this gives us a one-tailed *p* value of .0165 so we will reject the null hypothesis; the ending apathy score is significantly lower than apathy at the outset of the study; hopefully this will lead to higher achievement in math!

The Case of the Unexcused Students

In this case, let's try to help one of our friends: the principal of a local high school. It seems, with every year that passes, that students miss more and more days due to unexcused absences. The principal, being a believer in the tenets of behavioral psychology, decides to try to bribe the students with a reward. The students are told that, if they can significantly decrease the number of times they are absent during the term, they will be rewarded with a party. Let's use our six-step approach to see if we can help him investigate whether or not his plan will work.

STEP 1

Identify the Problem

The principal is faced with somewhat of a two-pronged problem. Students are missing more and more days of school; since absences are directly related to lower achievement, it would be in the best interest of the students to be at school as often as possible. At the same time, many states base school funding partially on student attendance rates; it would be in the principal's best interests (and career aspirations) to get students to be in school as often as possible. This is clearly a problem that can be investigated using inferential statistics.

The school in question is experiencing a problem with low student attendance. Since both achievement and school funding are related to student attendance, it is imperative that action be taken to address the problem. The principal will investigate whether an extrinsic reward will help increase attendance.

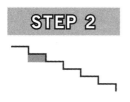

State a Hypothesis

Here's the research hypothesis that corresponds to the principal's plan:

▪ *Students will have significantly fewer absences after the party program than they did before the party program.*

We can see this is a directional hypothesis in that we are suggesting the total number of absences will be less than they were before starting the study. We can state the null hypothesis in the following manner:

▪ *There will be no significant difference in the number of absences prior to the party program and after the party program is announced.*

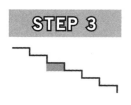

Identify the Independent Variable

We could easily identify the independent variable and its levels by noting our interest in one set of students at two distinct points in time: before the promise of the party and after the party plan was announced.

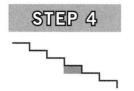

Identify and Describe the Dependent Variable

Our dependent variable is obvious; we know we have to collect the number of absences for each student, both in the semester before the plan and at the end of the semester in which the plan was implemented. In Table 8.8, we can see we have a row that contains the number of absences for students A through H before the start of the study and another row that shows the number of absences for the same students at the end of the term. SPSS would calculate the descriptive statistics shown in Figure 8.15.

TABLE 8.8. Absence Data

	A	B	C	D	E	F	G	H
Before the plan	8	7	6	5	6	8	4	3
After the plan	7	6	7	4	6	7	4	3

Things are looking good; the average number of absences after the implementation of the party program is less than the average number before starting the pro-

Paired Samples Statistics

		Mean	N	Std. Deviation	Std. Error Mean
Pair 1	Ending Absences	5.50	8	1.604	.567
	Starting Absences	5.88	8	1.808	.639

FIGURE 8.15. Descriptive statistics from the Paired-Samples test.

gram. We have to be careful, though; the values are very close. We should run the appropriate statistical test and see what the output tells us.

STEP 5

Choose the Right Statistical Test

Here we have the perfect scenario for a dependent-sample *t*-test because there is one independent variable with two related levels and one dependent variable representing quantitative data.

STEP 6

Use Data Analysis Software to Test the Hypothesis

We can use SPSS to easily compute the inferential statistics we would need to test our hypothesis; these statistics are shown in Figure 8.16.

Paired Samples Test

			Pair 1 Ending Absences - Starting Absences
Paired Differences	Mean		-.375
	Std. Deviation		.744
	Std. Error Mean		.263
	95% Confidence Interval of the Difference	Lower	-.997
		Upper	.247
T			-1.426
Df			7
Sig. (2-tailed)			.197

FIGURE 8.16. Inferential statistics from the Paired-Samples test.

We have a very small standard deviation of the difference (i.e., .744) and an even smaller mean difference score (i.e., .375); this gives us an effect size of .504; things are not looking good for the principal but let's move forward.

Knowing that we have a one-tailed hypothesis, we would divide our Sig. (2-tailed) value and arrive at a one-tailed *p* value of .0985; this means we cannot reject our null

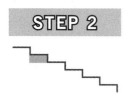

hypothesis. We could verify this by comparing our computed value of *t*, –1.416, to our critical *t* value, 2.365. Although the principal tried, the offer of a party at the end of the term just was not enough of a stimulus to get students to come to school.

The Case of Never Saying Never

Never one to give up, the principal of the high school decides to try again. Thinking back over the original plan, the principal decides the offer of a party just was not enough of a reward to get students to change their behavior. The principal brainstorms with some teachers, and they develop what they feel is a foolproof plan. The literature suggests that the traditional school day, because it starts so early in the morning, is not conducive to getting teenagers to come to school. Knowing that, the principal suggests starting school at noon and running it until 6:00 P.M. This, the principal feels, will allow the students to sleep in and still have time in the evening for all of their activities. Upon announcing the plan to the public, the principal is inundated with criticism as many feel forcing children to stay later in the day will cause even more absenteeism. The principal, vowing at least to test the plan, concedes they may be right but continues to plan for setting up the new schedule.

STEP 1

Identify the Problem

Obviously, the problem is exactly the same as what the principal addressed before. His plan worked to a minor degree, but as we've talked about, that difference may just be caused by error due to the exact students he worked with, the type of year the study was conducted, and so on. Given that, it's time to move forward and try something else. Remember, his students' future, not to mention his job, may hang in the balance! While the problem is the same, the problem statement will change slightly:

> *The school in question is experiencing a problem with low student attendance. Since both achievement and school funding are related to student attendance, it is imperative that action be taken to address the problem. The principal will investigate whether a change in the school's schedule will help increase attendance.*

STEP 2

State a Hypothesis

Based on this story, we can clearly see a hypothesis has formed. Since the principal is not sure if attendance will go up or down, the hypothesis must be stated as two-tailed (Step 1):

> *Students attending high school under the new schedule will have a significantly different number of absences than they did when they were under the old schedule.*

The null hypothesis would read:

■ *There will be no significant difference in absences between the new schedule and the old schedule.*

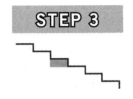

Identify the Independent Variable

Just as in the prior example, the independent variable is the student body. Remember, however, we are measuring them at two different points in time (i.e., before the new schedule and after the new schedule).

Identify and Describe the Dependent Variable

The dependent variable is the number of times the students were absent; let's use the data in Table 8.9 to compute the descriptive statistics shown in Figure 8.17.

TABLE 8.9. Absence Data

	A	B	C	D	E	F	G	H
Old schedule	4	5	6	5	6	7	6	7
New schedule	5	6	7	7	6	8	6	8

Paired Samples Statistics

		Mean	N	Std. Deviation	Std. Error Mean
Pair 1	Old Schedule	5.75	8	1.035	.366
	New Schedule	6.63	8	1.061	.375

FIGURE 8.17. Descriptive statistics from the Paired-Samples test.

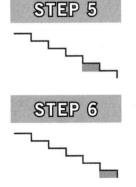

Choose the Right Statistical Test

This scenario calls for a dependent sample *t*-test since the two levels of the independent variable are related, and we use the dependent variable to collect parametric data.

Use Data Analysis Software to Test the Hypothesis

Using the results shown in Figure 8.18, we could compute a fairly large effect size (i.e., 1.37) but that might be the only good news. First, we can clearly see the mean number of absences went up during the trial period. This bad news is compounded when we look at the two-tailed p value, .006, and see that it is clearly less than our alpha value of .05. We have to reject our null hypothesis; there was a significant difference in the number of absences. Unfortunately for the principal, this means the new schedule caused a significant increase in the number of absences!

Paired Samples Test

		Pair 1
		Old Schedule - New Schedule
Paired Differences	Mean	-.875
	Std. Deviation	.641
	Std. Error Mean	.227
	95% Confidence Interval of the Difference Lower	-1.411
	Upper	-.339
T		-3.862
Df		7
Sig. (2-tailed)		.006

FIGURE 8.18. Inferential statistics from the Paired-Samples test.

Just in Case—A Nonparametric Alternative

Just as was the case with the independent-sample *t*-test, there are instances where either the data distribution is not conducive to using parametric statistics or we have collected data that are ordinal (ranked) in nature. In cases where this happens and the levels of the independent variable are dependent on one another, we have to use the nonparametric *Wilcoxon t-test*. Setting up, running, and interpreting the output of the Wilcoxon test is very similar to that of the dependent-sample *t*-test. Again, this is not something that happens often, and, like the Mann–Whitney U test, this test is something you need to keep in the back of your mind for those rare instances.

Summary

The dependent-sample *t*-test, much like its independent counterpart, is easy to understand, both conceptually and from an applied perspective. The key to using both of these inferential techniques is to keep in mind that they can be used only when you have one independent variable with two levels and when the dependent variable measures quantitative data that is fairly normally distributed. Again, the labels "independent" and "dependent" describe the relationship between the two levels of the independent variable that are being compared.

As I said earlier, these two inferential tests are used widely in educational research. What happens, though, when you have more than one independent variable, more than two levels of an independent variable, or even multiple dependent variables? These questions, and more, will be answered in the following chapters.

Do You Understand These Key Words and Phrases?

dependent-sample t-test   Wilcoxon t-test

Do You Understand These Formulas?

Effect size for a dependent-sample t-test:

$$d = \frac{\overline{x}_{\text{difference}}}{S_{\text{difference}}}$$

t score for the dependent-sample t-test:

$$t = \frac{\overline{D}}{\sqrt{\left(\dfrac{\sum D^2 - \dfrac{(\sum D)^2}{n}}{n(n-1)}\right)}}$$

Quiz Time!

As usual, before we wind up this chapter, let's take a look at a couple of case studies. Read through these and answer the questions that follow. If you need to check your work, the answers are at the end of the chapter.

The Case of Technology and Achievement

Proponents of technology in the classroom have suggested that supplying children with laptop computers will significantly increase their grades.

1. What is the hypothesis being tested?

2. What is the independent variable and its levels?

3. What is the dependent variable?

4. Based on Figures 8.19 and 8.20, what decision should they make?

Paired Samples Statistics

		Mean	N	Std. Deviation	Std. Error Mean
Pair 1	Laptops Used	85.2667	15	3.80726	.98303
	No Laptops	86.5333	15	4.79385	1.23777

FIGURE 8.19. Descriptive statistics from the Paired-Samples test.

Paired Samples Test

			Pair 1 Laptops Used - No Laptops
Paired Differences	Mean		-1.26667
	Std. Deviation		6.08824
	Std. Error Mean		1.57198
	95% Confidence Interval of the Difference	Lower	-4.63822
		Upper	2.10489
T			-.806
Df			14
Sig. (2-tailed)			.434

FIGURE 8.20. Inferential statistics from the Paired-Samples test.

The Case of Worrying about Our Neighbors

Citizens are concerned that the annexation of property adjoining their town will decrease the value of their homes. The city government insists there will be no change.

1. What is the hypothesis being investigated?
2. What is the independent variable and its levels?
3. What is the dependent variable?
4. Based on Figures 8.21 and 8.22, are the citizens' concerns warranted?

Paired Samples Statistics

		Mean	N	Std. Deviation	Std. Error Mean
Pair 1	Before Annex	$116,027.73	15	$10,229.535	$2,641.255
	After Annex	$106,555.87	15	$8,521.641	$2,200.278

FIGURE 8.21. Descriptive statistics from the Paired-Samples test.

Paired Samples Test

			Pair 1 After Annex - Before Annex
Paired Differences	Mean		$-9,471.867
	Std. Deviation		$14,517.440
	Std. Error Mean		$3,748.387
	95% Confidence Interval of	Lower	$-17,511.357
	the Difference	Upper	$-1,432.377
t			-2.527
df			14
Sig. (2-tailed)			.024

FIGURE 8.22. Inferential statistics from the Paired-Samples test.

The Case of SPAM

In order to attract new customers, a local entrepreneur is advertising an Internet service that guarantees customers will receive fewer unsolicited e-mails per month than they would using other services. After enrolling for the service, several customers complained that, contrary to the advertisements, they were actually getting more SPAM!

1. What is the hypothesis the entrepreneur is stating?

2. What is the independent variable and its levels?

3. What is the dependent variable?

4. Based on Figures 8.23 and 8.24, should the customers complain?

Paired Samples Statistics

		Mean	N	Std. Deviation	Std. Error Mean
Pair 1	SPAM Before	46.00	16	5.465	1.366
	SPAM After	65.88	16	5.976	1.494

FIGURE 8.23. Descriptive statistics from the Paired-Samples test.

Paired Samples Test

		Pair 1
		SPAM After - SPAM Before
Paired Differences	Mean	19.875
	Std. Deviation	9.444
	Std. Error Mean	2.361
	95% Confidence Interval of the Difference — Lower	14.843
	95% Confidence Interval of the Difference — Upper	24.907
T		8.418
Df		15
Sig. (2-tailed)		.000

FIGURE 8.24. Inferential statistics from the Paired-Samples test.

The Case of "We Can't Get No Satisfaction"

The faculty at a large university recently hired a new dean, and, after a few weeks, the university's administration felt they should survey the faculty to see if they were satisfied with their choice. Based on the results, the faculty seemed fairly happy, leaving the administration feeling they had found exactly the right person for the job. After 3 months, however, the complaints starting piling in—something was going seriously wrong, the faculty seemed very upset! Based on that, the faculty members were asked to complete the same satisfaction survey again. To the administration's dismay, the faculty's satisfaction was lower; the question is, however, was it significantly lower?

1. What is the hypothesis the faculty is stating?

2. What is the independent variable and its levels?

3. What is the dependent variable?

4. Based on Figures 8.25 and 8.26, what should the administration do?

Paired Samples Statistics

		Mean	N	Std. Deviation	Std. Error Mean
Pair 1	Satisfaction Before	60.1333	15	6.63181	1.71233
	Satisfaction After	42.2667	15	5.57375	1.43914

FIGURE 8.25. Descriptive statistics from the Paired-Samples test.

Paired Samples Test

			Pair 1 Satisfaction Before - Satisfaction After
Paired Differences	Mean		17.86667
	Std. Deviation		7.70776
	Std. Error Mean		1.99013
	95% Confidence Interval of the Difference	Lower	13.59825
		Upper	22.13508
T			8.978
df			14
Sig. (2-tailed)			.000

FIGURE 8.26. Inferential statistics from the Paired-Samples test.

CHAPTER 9

The Analysis of Variance

Introduction

In the last two chapters we dealt with situations in which we have one independent variable with two levels. In many instances, however, we may have an independent variable with three or more levels. For example, suppose we are interested in determining if there is a significant difference in the per capita income between residents of Canada, the United States, and Mexico. We could easily state a hypothesis to investigate this question.

> There will be a significant difference in the average per capita income between citizens of Canada, the United States, and Mexico.

Our independent variable is "Country" and the three different countries represent three levels. The dependent variable is the average income for each country's citizens. Given all of this, which statistical test can we use?

At about this point, a lot of beginning statisticians say, "This is easy; we will just use three separate independent-sample t-tests. We will compare Canada to the United States, Canada to Mexico, and then the United States to Mexico. That covers all the bases." They are right. Those three comparisons are inclusive of all of the comparisons that could be made between the three countries, but using three different t-tests creates a problem. Let me show you what I mean.

Let's suppose we actually decided to use the three separate t-tests; by doing so, we would have three separate computed values of t, up to three separate critical values of t (depending on the number in each group), and we would be using alpha in three separate tests. The first two of these things aren't really important, but the three alpha values create quite a problem. In order to understand what might go wrong, think back to a very early chapter where we talked about Type I and Type II errors.

We agreed that Type I errors occur when we reject a null hypothesis when we shouldn't; the larger the alpha value, the greater the probability of making an error of this type. We also said that, if we decrease alpha, we increase the probability of a Type II error where we fail to reject a null hypothesis when we should. We settled on a traditional alpha value of .05 because it is a good balance between the two types of error.

Keeping that in mind, think what might happen if we used three independent-sample t-tests to test our hypothesis. In essence, we would be using an alpha value of .05 three different times in order to test one overall hypothesis. Obviously, this is going to greatly inflate our possibility of making a Type I error. In order to address this problem, an English agronomist, Ronald Fisher, developed the *analysis of variance* or, as it is most always abbreviated, *ANOVA*.

Fisher, originally an astronomer, began working in 1919 as a biologist for an agricultural station in England. During his tenure there, Fisher contributed greatly to the fields of genetics and statistics. Like his friend Gossett and his t-test, Fisher was interested in working with small data samples and eventually developed the analysis of variance to help him in his studies. His list of accomplishments in statistics and genetics is so great and his awards are so impressive that he was even knighted by Queen Elizabeth in 1952!

Understanding the ANOVA

The key to understanding when and how to use the ANOVA is much the same as it is for any statistical test. First, you need to know the number of independent variables and the number of levels within each variable. You also need to know how many dependent variables you have and the type of data collected; let's use an example we can relate to:

- *There will be a significant difference in achievement between students studying math in a lecture-based classroom, students studying math in a technology-based classroom, and students studying math in a classroom where both lecture and technology are used.*

It is easy to see, in this hypothesis, that the independent variable is "instructional type" and there are three levels—lecture, technology, and a combination of both; the dependent variable is achievement. Notice, also, that we have stated a nondirectional hypothesis since it would be difficult to state one null hypothesis that would cover all of the different combinations we could possibly check. Given that, when you're dealing with a situation that calls for an analysis of variance, you always state a nondirectional hypothesis.

Fisher felt that the only way to accurately investigate significant differences between groups was to look at the variance of scores within each of the levels of the independent variable as well as the variance between the different groups. We will discuss the underlying logic and formulas in a bit, but for now let's take a look at the different types of ANOVAs.

▇ The Different Types of ANOVAs

Like the *t*-test, there are several types of ANOVA; the one we use in a particular case is based on the number of independent variables, the number of dependent variables, and the type of data collected. From the outset, let's make one thing clear. If your dependent variables represent nonparametric data, you will be using one of the nonparametric ANOVAs, techniques that are rarely used by beginning statisticians. Because of that, they are beyond the scope of this book. Suffice it to say that nonparametric ANOVAs exist; we just will not be talking about them. Having said that, let's focus on the types of ANOVAs that can be used to analyze a single quantitative dependent variable.

One-Way ANOVA

The most elementary ANOVA is called the one-way, or simple, ANOVA. When we say something is "one way," we mean there is only one independent variable and one dependent variable. The independent variable can have three or more levels, and the dependent variable represents quantitative data.

Factorial ANOVA

When we say an ANOVA is "factorial," we simply mean there is more than one independent variable and only one dependent variable. One of the most common ways we refer to factorial ANOVAs is by stating the number of independent variables. For example, if we have a three-way ANOVA, we are saying we have a factorial ANOVA with three independent variables. Sometimes we take it a step further and tell the reader the number of levels within each of the independent variables. For example, let's say we had a two-way ANOVA where the independent variables were "college class" and "gender" and we are interested in using it to determine if there is a significant difference in achievement. As we know, "college class" has four levels: freshman, sophomore, junior, and senior; obviously, gender has two levels, female and male. In order to be as clear as possible, someone performing research using this particular ANOVA might call it a 4 by 2 (sometimes you see it abbreviated as 4×2) ANOVA rather than just a factorial ANOVA. This tells the reader two things.

First, there are two independent variables. The first independent variable is represented by the "4," which also shows it has four levels (i.e., freshman, sophomore, junior, and senior). The second independent variable is represented by the "2," which means it has two levels (i.e., male and female). The great thing about using this annotation is we can tell exactly how many measurements we are going to make by multiplying the two numbers together. In this case, we would be collecting eight measurements ($4 \times 2 = 8$). This is shown in Table 9.1.

TABLE 9.1. Cells for Gender and Class for Factorial ANOVA

Achievement of male freshmen	Achievement of male sophomores	Achievement of male juniors	Achievement of male seniors
Achievement of female freshmen	Achievement of female sophomores	Achievement of female juniors	Achievement of female seniors

In another case, we might have a three-way ANOVA where we are collecting data about (1) the gender of a child, (2) whether or not the child went to kindergarten, and (3) the number of parents in the home. Using these variables, we are interested in determining if there is a significant difference in achievement in the first grade. Here, we would have a three-way ANOVA because we have three independent variables. (This three-way ANOVA could also be called a $2 \times 2 \times 2$ ANOVA because gender has two levels, male or female; kindergarten attendance has two levels, yes or no; and number of parents in the home has two levels, in this case one or two parents.) We would then have to collect data about eight different groups ($2 \times 2 \times 2 = 8$). Again, we can graphically represent this in Table 9.2.

TABLE 9.2. Cells for Gender and Number of Parents for Factorial ANOVA

Males attending kindergarten—one parent at home	Females attending kindergarten—one parent at home	Males not attending kindergarten—one parent at home	Females not attending kindergarten—one parent at home
Males attending kindergarten—two parents at home	Females attending kindergarten—two parents at home	Males not attending kindergarten—two parents at home	Females not attending kindergarten—two parents at home

The factorial ANOVA's underlying logic, and particularly the computations, are very complicated. We will touch briefly on its use at the end of the chapter, but we will spend most of our effort focusing on the one-way ANOVA.

Multivariate ANOVA

There is one last type of ANOVA that bears mentioning, one where we have more than one dependent variable, regardless of the number of independent variables. For example, suppose we have three groups of students and we want to measure their intrinsic motivation and achievement. In this case, we have one independent variable with three levels and two dependent variables; because of this we would use a *multivariate analysis of variance* or *MANOVA*.

The multivariate analysis of variance, because of all the interactions that multiple independent and dependent variables create, can be a complicated test to compute, analyze, and interpret. Thinking back to our chart, we did not mention it as one of the tests we are going to cover. Instead, we are going to look extensively at the one-way ANOVA, and we will briefly cover the factorial ANOVA. Before we move any further, however, let's talk a little about the assumptions we make when using any ANOVA.

▧ Assumptions of the ANOVA

Unfortunately, even when you have the right number of independent and dependent variables and you are collecting the right type of data, sometimes the ANOVA is not the appropriate test to work with. The ANOVA, in fact, works under four major assumptions:

Random Samples

The first of these assumptions is that, if your dependent variable scores represent a sample of a larger population, then it should be a random sample. Other types of samples, especially where the researcher picks subjects because of ease or availability, can make the results of the ANOVA not generalizable to other samples or situations.

Independence of Scores

The second assumption is that the scores collected are independent of one another. For example, if you are collecting data from school children, this means the score that one child gets is not influenced by, or dependent on, the score received by another child. This is the same idea as was the case with the independent-sample *t*-test.

Normal Distribution of Data

The third assumption is that each sample or population studied needs to be somewhat normally distributed. As we saw in the *t*-test, this does not mean a perfect bell-shaped distribution is required, but there has to be a certain degree of "normality" for the underlying calculations to work correctly.

Homogeneity of Variance

The fourth assumption requires that there be "homogeneity of variance." This means that the degree of variance within each of the samples should be about the same. As we will see later, SPSS will provide us with this information, thereby making our decision-making process far easier.

The first two assumptions do not have anything to do with the actual statistical process; instead they are methodological concerns and should be addressed when the data are being collected. For the third assumption, we've already seen that it's very straightforward to check if data are normally distributed using various numeric and graphical descriptive statistics. The last assumption, homogeneity of variance, is something we address with secondary statistical procedures appropriate for each test. We saw the same thing when we used the Levene test when interpreting the results of the independent-sample *t*-test; we have to do exactly the same thing when we are using the ANOVA. In other words, if we are using the ANOVA and we find the p value of the Levene statistic is less than .05, we have decisions to make. For now, however, let's talk about the math and logic underlying the one-way ANOVA.

▪ Calculating the ANOVA

The computations for the ANOVA are very straightforward. Simply put, Fisher demonstrated that, in order to test hypotheses using a one-way analysis of variance, one needed to understand four things:

1. The *total variance* in the data. This is the sum of the *within-group variance* and the *between-groups variance*.
2. The *total degrees of freedom*. This is the sum of the *within-group degrees of freedom* and the *between-groups degrees of freedom*.
3. The *mean square* values for the *within-group* and *between-groups* variance.
4. The *F* value that can be calculated using the data from the first three steps.

We can better understand these values by using the data in Table 9.3. In this case, we are interested in looking at the effect of different types of medication on headache relief. Here we have three levels: Pill 1, Pill 2, and Pill 3. The dependent variable, which we will call "time to relief," shows the number of minutes that elapse between the time the pill is taken and the time the headache symptoms are relieved. This will allow us to state the following hypothesis:

TABLE 9.3. Amount of Time for Pill Type

Pill 1	Pill 2	Pill 3
8	6	8
8	6	9
7	5	2
3	4	8
5	2	2

▪ *There will be no significant difference in the amount of time needed to relieve headache symptoms between Pill 1, Pill 2, and Pill 3.*

Descriptive Statistics

We can use the actual ANOVA procedure to calculate the descriptive statistics but, before we do that, let's look at how we would set up the data in SPSS. In Figure 9.1, you can see a variable labeled Pill; there are three values representing the three types of pills. There is also a variable labeled Time; this is the amount of time before relief.

Once we have entered our data, we will then tell SPSS to run the ANOVA for us. In Figure 9.2, we begin by selecting Compare Means and then One-Way ANOVA. At this point, as shown in Figure 9.3, we will be asked to do several things. First, we include our variable Time in the Dependent List (i.e., the dependent variable); we follow that by including Pill in the Factor Type (i.e., the independent variable). In this case, unlike the Independent Sample T Test, we do not need to include the exact values for the levels; SPSS simply uses each value entered under Pill as a level. We have also se-

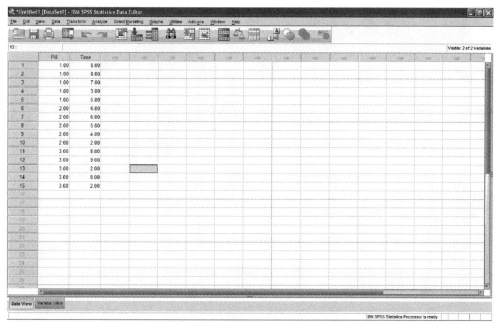

FIGURE 9.1. Data for pill and time in the Data View spreadsheet.

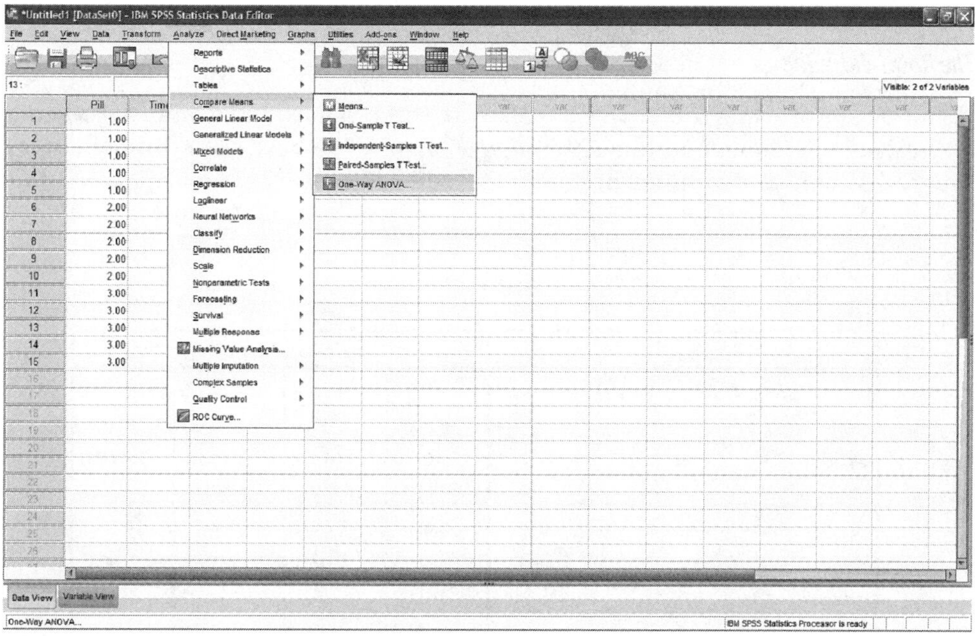

FIGURE 9.2. Using the Compare Means and One-Way ANOVA commands.

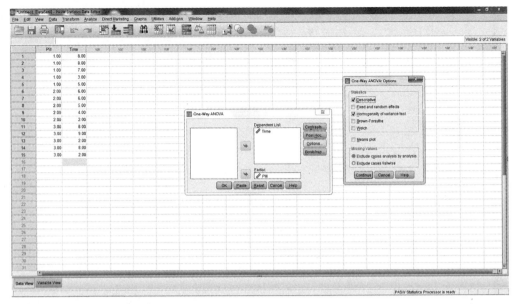

FIGURE 9.3. Identifying the independent and dependent variables and the statistics to be computed.

lected Options and asked for Descriptives and the Homogeneity of variance test. This will generate the output shown in Figure 9.4.

By looking at the mean scores, it appears that the time it takes Pill 2 to work may be significantly less than the other two pills, but, in order to make sure, we need to continue by computing the component parts of the F statistic.

The Total Variance

Fisher began by using these data to calculate three values; the *total sum of squares*, the *between sum of squares*, and the *within sum of squares*. We have seen this idea of sum of squares earlier, so it will not be a lot of trouble. We start by squaring each value in

Descriptives

Time

		Pill One	Pill Two	Pill Three	Total
N		5	5	5	15
Mean		6.2000	4.6000	5.8000	5.5333
Std. Deviation		2.16795	1.67332	3.49285	2.47463
Std. Error		.96954	.74833	1.56205	.63895
95% Confidence Interval for	Lower Bound	3.5081	2.5223	1.4631	4.1629
Mean	Upper Bound	8.8919	6.6777	10.1369	6.9037
Minimum		3.00	2.00	2.00	2.00
Maximum		8.00	6.00	9.00	9.00

FIGURE 9.4. Descriptive statistics from the ANOVA.

Table 9.4; I have substituted x_1 for Pill 1, x_2 for Pill 2, and x_3 for Pill 3 and have gone ahead and totaled each column.

TABLE 9.4. Computation of Sums of Squares

x_1	x_1^2	x_2	x_2^2	x_3	x_3^2
8	64	6	36	8	64
8	64	6	36	9	81
7	49	5	25	2	4
3	9	4	16	8	64
5	25	2	4	2	4
$\Sigma x_1 = 31$	$\Sigma x_1^2 = 211$	$\Sigma x_2 = 23$	$\Sigma x_2^2 = 117$	$\Sigma x_3 = 29$	$\Sigma x_3^2 = 217$

The Total Sum of Squares

At this point, we have all of the information necessary to compute the *total sum of squares* using the following formula; I have abbreviated "sum of squares" as *SS*:

$$SS_{\text{Total}} = \sum x^2 - \frac{(\sum x)^2}{N}$$

First we need to determine the value Σx^2. Since it represents each value in the dataset squared, we sum the second, fourth, and sixth columns:

1. $\Sigma x^2 = 211 + 117 + 217$
2. $\Sigma x^2 = 545$

To calculate the second part of the equation (i.e., $\frac{(\sum x)^2}{N}$), we follow these steps:

1. Compute Σx by adding the totals of the first, third, and fifth columns (i.e., $31 + 23 + 29 = 83$).
2. Compute $(\Sigma x)^2$ by squaring Σx (i.e., $83 * 83 = 6889$).
3. Divide $(\Sigma x)^2$ by the total number of data values collected (i.e., $6889/15 = 459.27$).

After we have completed these steps, we now have everything we need to include in the equation:

1. $SS_{\text{Total}} = \sum x^2 - \frac{(\sum x)^2}{N}$
2. $SS_{\text{Total}} = 545 - 459.27$
3. $SS_{\text{Total}} = 85.73$

The Between Sum of Squares

The *between sum of squares* also has a rather lengthy formula but, again, it is just basic math:

$$SS_{\text{Between}} = \frac{(\sum x_1)^2}{n_1} + \frac{(\sum x_2)^2}{n_2} + \frac{(\sum x_3)^2}{n_3} - \frac{(\sum x)^2}{N}$$

First, we square the totals for columns 1, 3, and 5 and then divide each total by the number of data items in the column. Let's break this down into four steps:

1. $SS_{\text{Between}} = \dfrac{(31)^2}{5} + \dfrac{(23)^2}{5} + \dfrac{(29)^2}{5} - \dfrac{(\sum x)^2}{N}$

2. $SS_{\text{Between}} = \dfrac{961}{5} + \dfrac{529}{5} + \dfrac{841}{5} - \dfrac{(\sum x)^2}{N}$

3. $SS_{\text{Between}} = 192.2 + 105.8 + 168.2 - \dfrac{(\sum x)^2}{N}$

4. $SS_{\text{Between}} = 466.2 - \dfrac{(\sum x)^2}{N}$

Now, in order to determine the value that we have to subtract from 466.2, all we do is complete the following three steps. Notice, this is exactly what we did when we were computing the total sum of squares:

1. Add the totals for the first, third, and fifth columns (i.e., 31 + 23 + 29 = 83).
2. Square the value from Step 1 (i.e., 83 * 83 = 6889).
3. Divide the value from Step 2 by the total number of data values collected (i.e., 6889/15 = 459.27).
4. SS_{Between} = 466.2 – 459.27.

We can now enter that value into our equation:

$$SS_{\text{Between}} = 466.2 - 459.27$$

This leaves us with

$$SS_{\text{Between}} = 6.93$$

The Within Sum of Squares

Just as we did with the others, we will break the formula for the *within sum of squares* down into its component parts:

$$SS_{\text{Within}} = SS_1 + SS_2 + SS_3$$

Computing the within sum of squares requires that we first compute the sum of squares for each level of the independent variable; each takes the same general form:

$$SS_1 = \sum X_1^2 - \frac{(\sum x_1)^2}{N}$$

To complete this equation, we go through the following three general steps:

1. We square the sum of all of the values from the first level of the independent variable shown in column one (i.e., 31 * 31 = 961).

2. We divide the result of Step 1 by the number of data values in that column by N (i.e., 5).

3. We then subtract that value from the sum of the squared values shown in column 2 of the table (i.e., 211).

We can then include our values in the equation and compute SS_1 using the following four steps:

1. $SS_1 = 211 - \dfrac{(31)^2}{5}$

2. $SS_1 = 211 - \dfrac{961}{5}$

3. $SS_1 = 211 - 192.2$

4. $SS_1 = 18.8$

We can then use the same formula to compute SS_2 (i.e., 11.2) and SS_3 (i.e., 48.8) and then insert all three values into our original equation:

$$SS_{\text{Within}} = 18.8 + 11.2 + 48.8$$

This gives us a SS_{Within} value of 78.8.

A SHORTCUT

There is an easier way to compute the total sum of squares; all you need to do is add the between sum of squares to the within sum of squares. This is shown in the following formula:

$$SS_{\text{Total}} = SS_{\text{Between}} + SS_{\text{Within}}$$

That means if we know only two of the values, we can easily compute the third. For example, if we know the total sum of squares and the sum of squares between, we can insert them into the following formula:

$$85.73 = 6.93 + SS_{\text{Within}}$$

We then subtract 6.93 from both sides and arrive at a total sum of square value of 78.8:

$$85.73 - 6.93 = 6.93 + SS_{\text{Within}} - 6.93$$

If you manually compute these values, you might find that computing the between sum of squares is the most tedious. Because of that, it is much easier to compute the total sum of squares and the within sum of squares and use those to compute the between sum of squares value.

Computing the Degrees of Freedom

As has been the case throughout the book, computing the degrees of freedom is very straightforward. For the ANOVA, however, you have to compute three different degrees of freedom values—one for between groups, one for within groups, and a total value.

1. The between-groups degrees of freedom are equal to the number of levels of the independent variable, minus one. In this case, we have three levels, so we have two degrees of freedom.

2. The within-group degrees of freedom are equal to the total number of data items minus the number of levels in the independent variable. In this case, we have 12 degrees of freedom (i.e., $15 - 3 = 12$).

3. The total degrees of freedom are equal to the within-group degrees of freedom plus the between-groups degrees of freedom (i.e., $12 + 2 = 14$ degrees of freedom). You can also compute this value by subtracting 1 from the total number of data items (i.e., $15 - 1 = 14$).

Computing the Mean Square

As we said earlier, the *mean square* is one of the primary components needed to compute the F value. The mean square is the sum of the *between-groups mean square* and the *within-group mean square*. Both are easy to compute by dividing the sum of squares for each group by the number of degrees of freedom for that group. In this case, the mean square values are:

1. Between-groups mean square = 3.47 (i.e., 6.93/2)
2. Within-group mean square = 6.57 (i.e., 78.8/12)

Computing the F value

Finally we can compute the F value, obviously named for Fisher, by dividing the between-groups mean square by the within-group mean square. This gives us an F value of .528 (i.e., 3.47/6.57).

We Have the Four Components—Now What?

Before we go any further, let's look at the output from SPSS, shown in Figure 9.5. As you can see, our computations match exactly.

ANOVA

Pill

	Sum of Squares	df	Mean Square	F	Sig.
Between Groups	6.933	2	3.467	.528	.603
Within Groups	78.800	12	6.567		
Total	85.733	14			

FIGURE 9.5. Inferential statistics from the ANOVA.

The F Distribution

At this point, we have our computed value of *F*. Now we need an *F* distribution so that we can determine the critical value needed to test our hypothesis. This process is similar to how *t* distributions are created, in that an infinite number of *F* values are computed from random samples of data and then plotted. This process is shown in Figure 9.6.

This distribution is certainly a lot different from the normal curves we have been using up to this point. First, there are no plotted values less than zero; this is explained by the fact that we are using the variance which, as you know, can never be less than zero. As for the strange shape of the distribution, that takes a little more explaining.

When Fisher first developed the *F* distribution, he discovered that the particular shape of the distribution depends on the "within" and "between" degrees of freedom. In the following example, you can see a plot of three distributions with different degrees of freedom. As you can see in Figure 9.7, the larger the between-groups degrees of freedom, shown on the horizontal axis, the more the distribution is skewed to the right; the more degrees of freedom you have for within groups, shown on the vertical axis, the more peaked the distribution will be.

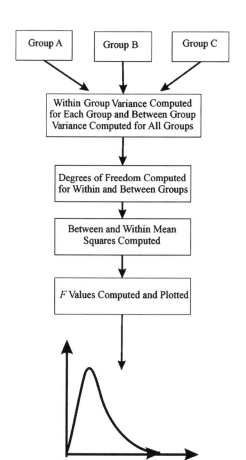

FIGURE 9.6. Creating the *F* distribution.

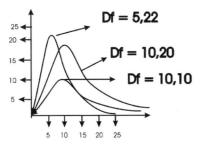

FIGURE 9.7. Changes in the shape of the *F* distribution based on sample sizes.

Determining the Area under the Curve for *F* Distributions

Like the *z* (i.e., normal) distribution and the *t* distribution, we get the critical value of *F* from a table related to its distribution. An example is shown in Table 9.5.

Since it is so small, obviously this isn't a complete *F* table; in fact, it is only one part of the *F* table for alpha = .05. Since, like the other tests, the size of the critical region changes depending on the alpha value, there are completely different *F* tables for alpha = .10 and alpha = .01. This is done because having two separate degrees of freedom requires entries on the top and left sides of the table. In order to avoid confusion, it is far easier to limit the entries on a given table to a particular alpha value. Examples of tables for all alpha values are shown in Appendices C, D, and E.

TABLE 9.5. Critical Values of *F*

Within-group degrees of freedom	Between-groups degrees of freedom									
	1	2	3	4	5	6	7	8	9	10
1	1.61	2.00	2.16	2.25	2.30	2.34	2.37	2.39	2.41	2.42
2	18.5	19.0	19.2	19.3	19.3	19.4	19.4	19.4	19.4	19.4
3	10.1	9.55	9.28	9.12	9.01	8.94	8.89	8.85	8.81	8.79
4	7.71	6.94	6.59	6.39	6.26	6.16	6.09	6.04	6.00	5.96
5	6.61	5.79	5.41	5.19	5.05	4.95	4.88	4.82	4.77	4.74
6	5.99	5.14	4.76	4.53	4.39	4.28	4.21	4.15	4.10	4.06
7	5.59	4.74	4.35	4.12	3.97	3.87	3.79	3.73	3.68	3.64
8	5.32	4.46	4.07	3.84	3.69	3.58	3.5	3.44	3.39	3.35
9	5.12	4.26	3.86	3.63	3.48	3.37	3.29	3.23	3.18	3.14
10	4.96	4.10	3.71	3.48	3.33	3.22	3.14	3.07	3.02	2.98
11	4.84	3.98	3.59	3.36	3.20	3.09	3.01	2.95	2.90	2.85
12	4.75	3.89	3.49	3.26	3.11	3.00	2.91	2.85	2.80	2.75

Having said that, if you're like me, upon first seeing the *F* table, you're probably thinking, "This isn't so bad, it looks a lot like our *t* and *z* tables." If so, you're right; there's only one small difference. First, you can see the two degrees of freedom values we have talked about. The horizontal row across the top shows you the "between-groups degrees of freedom," and the vertical column on the left shows you the "within-group degrees of freedom."

Now we are going to use this table, along with our data about the different headache remedies, to test this hypothesis:

■ *There will be no significant difference in the amount of time needed to relieve headache symptoms between Pill 1, Pill 2, and Pill 3.*

First, we have to locate the critical value of *F* for situations where we have 2 between-groups degrees of freedom and 12 within-group degrees of freedom. To do this, we go across the top row until we get to the column showing 2 degrees of freedom; we then go down the left column until we get to 12 degrees of freedom. The point where that row and that column intersect shows us an *F* value of 3.89; we can mark that on our *F* distribution shown in Figure 9.8. In Figure 9.9, we then plot our computed value of *F*.

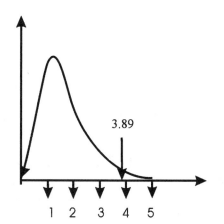

FIGURE 9.8. The critical value of *F* equals 3.89.

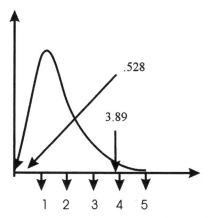

FIGURE 9.9. Comparing a computed value of *F* to a critical value of *F*.

In order to test our hypothesis, we compare these two values. If our computed value of *F* falls to the right of the critical value, we reject the null hypothesis. If our computed value of *F* is equal to or to the left of the critical value of *F*, then we do not reject the null hypothesis. This, of course, is exactly the same logic we used with the *z*- and *t*-tests. In this case it means we cannot reject the null hypothesis; obviously, there is no significant difference in headache relief time between the three types of pills.

The p Value for an ANOVA

The decision to reject the null hypothesis is supported by a *p* value of .603, shown as Sig., in Figure 9.5. This simply means there are no significant differences between any of the combinations of pills: Pill 1 compared to Pill 2, Pill 1 compared to Pill 3 or Pill 2 compared to Pill 3. You'll see that when we reject a null hypothesis, we have to take it a step further. We'll use tools called *post-hoc tests* to help us determine where the difference actually lies. That's all we need to know for now, so let's move forward.

Effect Size for the ANOVA

You can also compute an effect size for the analysis of variance. Again, all we are doing is investigating the effect of the independent variable on the dependent variable. Calculating the effect size, called *eta-squared*, is easy:

$$\eta^2 = \text{between sum of squares/total sum of squares}$$

Substituting our values into the formula, we will compute an effect size of .08.

1. $\eta^2 = 6.933/85.733$
2. $\eta^2 = .08$

This supports our decision not to reject the null hypothesis. Apparently, the type of pill you take does not make a lot of difference. At the same time, a person with a really bad headache might look at the mean scores and choose Pill 2; while it is not significantly faster, it is somewhat faster!

The Case of Multiple Means of Math Mastery

Let's move on by looking at another case. We are going to use our six-step model and, again, we will do all of the calculations manually.

A schoolteacher has a large group of children in her math class. The class is equally distributed in terms of gender, ethnic breakdown, and academic achievement levels. The teacher is a progressive graduate student and has learned in one of her classes that computer-assisted instruction (CAI) may help children realize higher achievement in mathematics.

The teacher decides to divide her class into three equal groups; the first group will receive a lecture, the second will use computer-assisted instruction, and the last group will receive a combination of both. The teacher is interested in finding out, after a semester of instruction, if there is any difference in achievement. She can then use the six-step model in her investigation.

STEP 1   Identify the Problem

In this case, a problem or opportunity for research clearly exists. The teacher wants to know if using technology will help increase achievement in her class. Let's see if this meets our criteria for a good problem:

1. Yes—she is interested in this problem.
2. Yes—the scope of the problem is manageable; she would simply need to collect grades from each of her students.
3. Yes—the teacher definitely has the knowledge, time, and resources needed to investigate the problem.
4. Yes—she can collect numeric data representing her students' grades.
5. Yes—investigating the problem has theoretical or practical significance.
6. Yes—it is ethical to investigate the problem.

It seems that our problem definitely meets all criteria; remember, as we write it, it must be clear and concise, contain all variables to be considered, and not interject the bias of the researcher:

■ *This study will investigate the effect of technology on the achievement, both alone and in combination with lecture, of students in a math class.*

STEP 2

State a Hypothesis

The hypothesis the teacher wants to investigate is clearly formulated. Remember, because of the shape of the distribution, we will only state nondirectional hypotheses.

■ *There will be a significant difference in levels of achievement between students taking math in a lecture group, students taking math in a CAI group, and students getting a combination of CAI and lecture.*

Our null hypothesis would be:

■ *There will be no significant difference in levels of achievement between students taking math in a lecture group, students taking math in a CAI group, and students getting a combination of CAI and lecture.*

STEP 3

Identify the Independent Variable

The independent variable is "type of instruction" or "instructional method," and there are three levels: students in the lecture group, students in the CAI group, and students using a combination of lecture and CAI.

STEP 4

Identify and Describe the Dependent Variable

The dependent variable is achievement. Let's assume she'll be using a final grade based on exams, homework, and quizzes; this is shown in Table 9.6. Figure 9.10 shows our descriptive statistics.

TABLE 9.6. Final Grade Data for Instructional Type

Lecture	CAI	CAI/lecture
80	84	90
82	86	92
78	84	94
76	82	90
80	88	100
72	90	94
82	92	94
88	88	96
74	86	92
68	84	94

Descriptives

Grade

	Lecture	CAI	Combination	Total
N	10	10	10	30
Mean	78.0000	86.4000	93.6000	86.0000
Std. Deviation	5.73488	3.09839	2.95146	7.61124
Std. Error	1.81353	.97980	.93333	1.38962
95% Confidence Interval for Lower Bound	73.8975	84.1835	91.4887	83.1579
Mean Upper Bound	82.1025	88.6165	95.7113	88.8421
Minimum	68.00	82.00	90.00	68.00
Maximum	88.00	92.00	100.00	100.00

FIGURE 9.10. Descriptive statistics from the ANOVA.

In this case, the mean scores between the three groups are different, but the same question has to be asked, "Are they significantly different?"

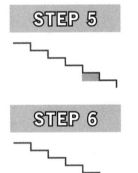

STEP 5

Choose the Right Statistical Test

Here we have one independent variable with three levels and one dependent variable where the data are quantitative. In this case, it appears we need to use an analysis of variance (ANOVA).

STEP 6

Use Data Analysis Software to Test the Hypothesis

Before we commit ourselves to an ANOVA, however, let's go back and think about the assumption we talked about earlier: the *homogeneity of variance*. As we said, we need to have about the same amount of variance within each of the samples. It is important that we consider this because, if the degree of variance is significantly different between the groups, the calculations underlying the ANOVA could be thrown off and would affect the validity of our results. Because manually calculating these values is something we would never do, SPSS creates the table shown in Figure 9.11.

We will use this table to test the null hypothesis: "There is no significant difference in the variance within the three sets of data." Just like with the *t*-test, all we have to do is compare our *p* value (i.e., Sig.) to an alpha value of .05. Obviously, since our *p* value is larger than alpha, there is no significant difference in the degree of variance

Test of Homogeneity of Variances

Grade

Levene Statistic	df1	df2	Sig.
2.557	2	27	.096

FIGURE 9.11. Homogeneity of variance statistics from the ANOVA.

within the three sets of data. If the p value was less than .05, we might need to revert to the ANOVA's nonparametric equivalent; the Kruskal–Wallace H test. Keep in mind, however, that the difference in the variances between the groups has to be very, very large in order for that to ever happen. Given that, we can go ahead and use the SPSS output shown in Figure 9.12.

ANOVA

Grade

	Sum of Squares	Df	Mean Square	F	Sig.
Between Groups	1219.200	2	609.600	35.719	.000
Within Groups	460.800	27	17.067		
Total	1680.000	29			

FIGURE 9.12. Inferential statistics from the ANOVA.

We finally have everything we need to test our hypothesis:

■ *There will be no significant difference in levels of achievement between students taking math in a lecture group, students taking math in a CAI group, and students getting a combination of CAI and lecture.*

First, since this is our first case with an ANOVA, let's plot both our computed value of F and our critical F value. Remember, we determine the critical value of F by using both the within- and between-groups degrees of freedom in combination with our desired alpha value. In Figure 9.13, we would find that these values represent a critical value of 3.35.

Our computed value of F (i.e., 35.719) falls very far to the right of our critical value of F (i.e., 3.35); there is a significant difference between the three groups. SPSS also computed a p value of .000, and we could manually compute an effect size of .726. Both of these values support our decision.

Here we have somewhat of a conundrum; we have rejected the null hypothesis and can feel assured that a significant difference exists between the groups. The questions that arise, however, are "Which groups are different?" Is the CAI group different from Lecture? Is the Lecture group different from the Combination group? Is the Combination group different from the CAI group? How can we figure this out?"

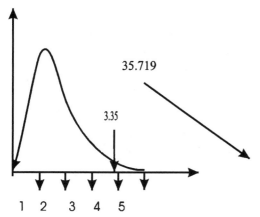

FIGURE 9.13. Comparing critical and computed values of F to test a hypothesis.

The Post-Hoc Comparisons

As we said earlier, in order to determine which groups are significantly different, we could use separate *t*-tests to compare the lecture and CAI groups, the lecture and combination groups, and the CAI and combination groups. Remember, however, that approach is not statistically sound. This would mean that you would have a separate alpha value for each of the *t*-tests, which, of course, would greatly inflate the Type I error rate. Instead of doing that, we can use a set of tools called multiple comparison tests to help us make a more sound decision.

Multiple-Comparison Tests

The multiple-comparison tests do exactly what their name implies. They make comparisons between the levels of an ANOVA and tell us which are significantly different from one another. We won't get into the manual calculations; suffice it to say that they are a modified version of the independent-sample *t*-test that controls for the inflated Type I error rate. We will always use computer software to compute the right value for us.

There are several multiple-comparison tests we can choose from, and most of the time they will provide us with exactly the same results. There are situations that require a specific multiple-comparison test, but these do not arise that often. For our purposes, one of the most commonly used multiple-comparison tests is the Bonferroni test. The procedure is easy to understand, and it gives the researcher a good estimate of the probability that any two groups are different while, at the same time, controlling for Type I and Type II errors. Typical output for the Bonferroni procedure is shown in Figure 9.14.

You can see that there are four rows in the table; the bottom three rows represent the three levels of our independent variable—the Lecture group, the CAI group, and the Combination group. If we look at the lecture row, for example, we can see we are

Multiple Comparisons

Grade

Bonferroni

(I) Method	(J) Method	Mean Difference (I-J)	Std. Error	Sig.	95% Confidence Interval Lower Bound	Upper Bound
Lecture	CAI	-8.40000*	1.84752	.000	-13.1157	-3.6843
	Combination	-15.60000*	1.84752	.000	-20.3157	-10.8843
CAI	Lecture	8.40000*	1.84752	.000	3.6843	13.1157
	Combination	-7.20000*	1.84752	.002	-11.9157	-2.4843
Combination	Lecture	15.60000*	1.84752	.000	10.8843	20.3157
	CAI	7.20000*	1.84752	.002	2.4843	11.9157

*. The mean difference is significant at the 0.05 level.

FIGURE 9.14. Multiple-comparison tests following the ANOVA.

given statistics representing it and its relationship to the other levels of the independent variable. There is one quirk, however: because of the total number of possible comparisons, part of the table is redundant.

For example, in the third row where you compare the CAI group to the Combination group, the mean difference is –7.2 points. In row 4, you compare the same two groups; only, this time, you subtract the average CAI score from the average combination score. This gives you a difference, obviously, of +7.2 points. The order in which you make the subtraction doesn't matter; the point here is that there is a 7.2-point difference between the groups.

We can also see an asterisk next to each of the mean difference values; this indicates that the mean difference is significant when alpha equals .05. For example, we see that the mean difference between the Lecture and the CAI group is –8.4; this difference is statistically significant. In this case, we can see asterisks next to all of our mean difference values; each of the groups is significantly different from one another.

We can verify our results by looking at the p value in the fourth column (i.e., Sig.). Here, we can see the p value between each of these; this means the two groups in that given comparison are significantly different from one another. Again, the p value supports the significant difference shown by the asterisks next to all mean difference comparisons.

Always Observe the Means!

At this point, we have significant differences in scores between each of the three groups. The average score of the combination group is significantly higher than the average score of the CAI group, and it, in turn, is significantly higher than the lecture group. As was the case with the t-tests, it is imperative to remember that we have to look at the mean values from the descriptive statistics before we make any decisions. In this case, educators might want to adopt the Combination approach, but remember, the same low p values would exist if the mean differences were in the opposite direction (e.g., the Combination group was significantly lower than the Lecture group).

So, what do we know from all of this? Apparently, the use of a combination of lecture and CAI results in achievement scores significantly higher than either lecture alone or CAI alone. At the same time, the use of CAI is significantly more effective than the use of lecture alone. The teacher should use this information to plan future class sessions accordingly. Let's move forward and see if we can help another educator.

The Case of Seniors Skipping School

A principal (are we back to this guy again?) has heard a rumor that seniors, because they are getting close to the end of their high school careers, are starting to skip class regularly which, if bad grades follow, could affect their ability to graduate. Obviously the principal doesn't think this is a very good idea, and so he initiates an investigation.

STEP 1

Identify the Problem

Yes, the principal is still interested in this problem and has the authority, time, resources, and knowledge to address it. This problem is practically significant and well within the scope of the principal's authority. It would be unethical if the principal didn't investigate this problem by simply counting the number of absences of each senior. We could state it in the following manner:

- *The administration and faculty in the high school have noticed what they believe is an inordinate number of absences by students in the senior class when compared to students in other grades. Afraid that higher levels of absenteeism could lead to lower grades and fewer graduates, personnel at the school will attempt to determine if seniors are skipping more classes than students in other grades.*

STEP 2

State a Hypothesis

In order for the principal to decide if a significant difference exists in the number of absences by class in the high school, he could use this hypothesis:

- *There will be a significant difference in the number of absences between freshmen, sophomores, juniors, and seniors.*

The corresponding null hypothesis would state that no significant difference would exist.

STEP 3

Identify the Independent Variable

Our independent variable is the class a given student is in and there are four levels: freshman, sophomore, junior, and senior.

STEP 4

Identify and Describe the Dependent Variable

The dependent variable is the number of days absent; we'll use Table 9.7 to show data the principal collected from the school's registrar. Figure 9.15 shows the descriptive statistics that would be computed by SPSS.

Here, the means for the freshman, sophomore, and junior classes are all approximately equal; the average for seniors is much larger. Our question, as always is, "Are the number of absences different, or are they significantly different?" Let's pick the appropriate statistical test and find out.

TABLE 9.7. Number of Absences for Each Class

Freshman		Sophomore		Junior		Senior	
4	1	5	1	3	2	5	6
3	2	4	1	4	4	5	7
2	1	3	2	2	2	5	6
4	5	4	3	3	4	7	7
5	3	5	2	4	4	5	4
2	4	1	3	3	2	4	3
1	4	1	5	1	4	4	5
0	3	2	4	1	2	6	3
2	2	0	3	2	3	6	5

Descriptives

Absences

		Freshman	Sophomore	Junior	Senior	Total
N		18	18	18	18	72
Mean		2.6667	2.7222	2.7778	5.1667	3.3333
Std. Deviation		1.45521	1.56452	1.06027	1.24853	1.69506
Std. Error		.34300	.36876	.24991	.29428	.19977
95% Confidence Interval for Mean	Lower Bound	1.9430	1.9442	2.2505	4.5458	2.9350
	Upper Bound	3.3903	3.5002	3.3050	5.7875	3.7317
Minimum		.00	.00	1.00	3.00	.00
Maximum		5.00	5.00	4.00	7.00	7.00

FIGURE 9.15. Descriptive statistics from the ANOVA.

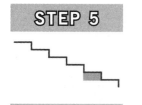

STEP 5

Choose the Right Statistical Test

Since we have one independent variable with four levels and our dependent variable is quantitative data, we are going to use the one-way analysis of variance.

STEP 6

Use Data Analysis Software to Test the Hypothesis

In Figure 9.16, our Levene p value is much greater than .05, so we have no problems with the homogeneity of variance between the different datasets. This means we can continue with the ANOVA table shown in Figure 9.17.

Since our p value is less than our alpha value of .05, we will reject our null hypothesis and support our research hypothesis. We could compute a moderate effect size of .396 that supports our decision. After looking at the descriptive statistics, we are fairly sure where the significant difference lies, but let's look at our multiple comparison tests in Figure 9.18 just to make sure.

Test of Homogeneity of Variances

Absences

Levene Statistic	df1	df2	Sig.
1.318	3	68	.276

FIGURE 9.16. Homogeneity of variance statistics from the ANOVA.

ANOVA

Absences

	Sum of Squares	df	Mean Square	F	Sig.
Between Groups	80.778	3	26.926	14.859	.000
Within Groups	123.222	68	1.812		
Total	204.000	71			

FIGURE 9.17. Inferential statistics from the ANOVA.

Multiple Comparisons

Absences

Bonferroni

(I) Class	(J) Class	Mean Difference (I-J)	Std. Error	Sig.	95% Confidence Interval Lower Bound	Upper Bound
Freshman	Sophomore	-.05556	.44871	1.000	-1.2750	1.1639
	Junior	-.11111	.44871	1.000	-1.3305	1.1083
	Senior	-2.50000*	.44871	.000	-3.7194	-1.2806
Sophomore	Freshman	.05556	.44871	1.000	-1.1639	1.2750
	Junior	-.05556	.44871	1.000	-1.2750	1.1639
	Senior	-2.44444*	.44871	.000	-3.6639	-1.2250
Junior	Freshman	.11111	.44871	1.000	-1.1083	1.3305
	Sophomore	.05556	.44871	1.000	-1.1639	1.2750
	Senior	-2.38889*	.44871	.000	-3.6083	-1.1695
Senior	Freshman	2.50000*	.44871	.000	1.2806	3.7194
	Sophomore	2.44444*	.44871	.000	1.2250	3.6639
	Junior	2.38889*	.44871	.000	1.1695	3.6083

*. The mean difference is significant at the 0.05 level.

FIGURE 9.18. Multiple-comparison tests following the ANOVA.

This table verifies what we suspected. Since there are asterisks beside each of the mean difference values between the seniors and every other class, it is apparent they have a significantly higher number of absences than those in the other classes. There is no significant difference in the mean scores between any of the other classes. As always, be careful; the same statistics would have been computed had the seniors had significantly fewer absences.

The Case of Quality Time

Many social workers are concerned that, as their number of children grows, parents are spending less and less quality time with each newborn child. They are concerned because they believe this lack of parental involvement leads to higher levels of anger, hostility, and instances of antisocial behavior as the child grows older. In order to investigate the social workers' concerns, a group of researchers decides to identify a population, determine the birth order of each person in the population, and then ask each person the number of times they have been arrested. They begin with a clear, well-defined problem statement.

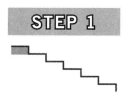

Identify the Problem

This problem clearly meets all six criteria for a good problem statement; let's go ahead and write a problem statement:

- *Social workers are concerned that children with older siblings will receive less parental attention as they grow older. This leads to higher levels of anger, hostility, and antisocial behavior later in life.*

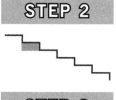

State a Hypothesis

- *There will be a significant difference in the number of arrests and the relative birth order of a child.*

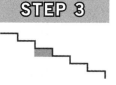

Identify the Independent Variable

In our hypothesis, relative birth order is our independent variable. For our example, we are going to assume there are no more than five levels. A child was either born first, second, third, fourth, or fifth.

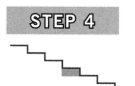

Identify and Describe the Dependent Variable

The dependent variable is the number of arrests; Table 9.8 shows data for 50 people. Figure 9.19 shows our descriptive statistics.

TABLE 9.8. Number of Arrests by Birth Order

Born first	Born second	Born third	Born fourth	Born fifth
4	3	3	4	1
2	2	4	3	2
3	3	2	3	1
3	3	5	5	0
4	4	5	4	2
1	3	3	6	1
4	3	2	3	1
4	2	1	3	0
4	1	5	5	2
1	3	4	3	1

Descriptives

Arrests

		Born First	Born Second	Born Third	Born Fourth	Born Fifth	Total
N		10	10	10	10	10	50
Mean		3.0000	2.7000	3.4000	3.9000	1.1000	2.8200
Std. Deviation		1.24722	.82327	1.42984	1.10050	.73786	1.42414
Std. Error		.39441	.26034	.45216	.34801	.23333	.20140
95% Confidence Interval for Mean	Lower Bound	2.1078	2.1111	2.3772	3.1127	.5722	2.4153
	Upper Bound	3.8922	3.2889	4.4228	4.6873	1.6278	3.2247
Minimum		1.00	1.00	1.00	3.00	.00	.00
Maximum		4.00	4.00	5.00	6.00	2.00	6.00

FIGURE 9.19. Descriptive statistics from the ANOVA.

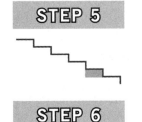

STEP 5

Choose the Right Statistical Test

Here again we have met the criteria for a one-way ANOVA; one independent variable with five levels, and our dependent variable is quantitative.

STEP 6

Use Data Analysis Software to Test the Hypothesis

Again, we assume we are using the one-way ANOVA, but let's check the Levene test in Figure 9.20.

Our p value is fairly low but still above our alpha level of .05; this means we do not have to worry about our assumption that the variance is homogeneous between the five groups. Based on that, Figure 9.21 shows us the ANOVA table.

Here we have a p value less than .05, so we will reject our null hypothesis. This is supported by a moderate effect size of .454. Apparently, the social workers are right; there is a significant difference in the number of arrests when you look at birth order. We do need to look at our post-hoc tests in Figure 9.22 to see where those differences

Test of Homogeneity of Variances

Arrests

Levene Statistic	df1	df2	Sig.
2.218	4	45	.082

FIGURE 9.20. Homogeneity of variance statistics from the ANOVA.

ANOVA

Arrests

	Sum of Squares	df	Mean Square	F	Sig.
Between Groups	45.080	4	11.270	9.340	.000
Within Groups	54.300	45	1.207		
Total	99.380	49			

FIGURE 9.21. Inferential statistics from the ANOVA.

Multiple Comparisons

Arrests

Bonferroni

(I) Birth Order	(J) Birth Order	Mean Difference (I-J)	Std. Error	Sig.	95% Confidence Interval	
					Lower Bound	Upper Bound
First	Second	.30000	.49126	1.000	-1.1502	1.7502
	Third	-.40000	.49126	1.000	-1.8502	1.0502
	Fourth	-.90000	.49126	.736	-2.3502	.5502
	Fifth	1.90000*	.49126	.004	.4498	3.3502
Second	First	-.30000	.49126	1.000	-1.7502	1.1502
	Third	-.70000	.49126	1.000	-2.1502	.7502
	Fourth	-1.20000	.49126	.186	-2.6502	.2502
	Fifth	1.60000*	.49126	.021	.1498	3.0502
Third	First	.40000	.49126	1.000	-1.0502	1.8502
	Second	.70000	.49126	1.000	-.7502	2.1502
	Fourth	-.50000	.49126	1.000	-1.9502	.9502
	Fifth	2.30000*	.49126	.000	.8498	3.7502
Fourth	First	.90000	.49126	.736	-.5502	2.3502
	Second	1.20000	.49126	.186	-.2502	2.6502
	Third	.50000	.49126	1.000	-.9502	1.9502
	Fifth	2.80000*	.49126	.000	1.3498	4.2502
Fifth	First	-1.90000*	.49126	.004	-3.3502	-.4498
	Second	-1.60000*	.49126	.021	-3.0502	-.1498
	Third	-2.30000*	.49126	.000	-3.7502	-.8498
	Fourth	-2.80000*	.49126	.000	-4.2502	-1.3498

*. The mean difference is significant at the 0.05 level.

FIGURE 9.22. Multiple-comparison tests following the ANOVA.

lie. If they are right, the people born last will have a significantly higher number of arrests.

What does this tell us? The first thing it tells us is that, for the social workers, things can go from looking pretty good to looking pretty bad in just a matter of seconds. First, if they had paid attention to the descriptive statistics in Step 3, they would have already known that children born last have fewer arrests; the fact that they are significantly less is supported by the p value. The birth order of children born first through fourth makes no difference in the number of arrests but children born fifth have significantly fewer arrests than their siblings.

The Case of Regional Discrepancies

The president of a manufacturing company with factories in four locations throughout the United States has recently been informed that one of the factories may be lagging the others in production. This bothers the president since each factory supports a specific region of the country. Lower production means fewer products for customers. When stores run out of a specific product, buyers might just turn to the competitor's brand; it's time to take action!

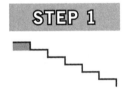

STEP 1

Identify the Problem

The company's president has reason to be alarmed; factories lagging in production could certainly affect the company's bottom line. This, of course, might cause the Board of Directors to consider the president's bottom line. Since he has three kids in college, it behooves him to investigate whether there is actually a problem!

■ *Different areas of the country depend on one of four regional manufacturing factories for needed products. If a given factory falls behind in production, it is possible that consumers will turn to products from another manufacturer to meet their needs.*

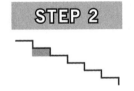

STEP 2

State a Hypothesis

The company's president would begin with the following research hypothesis:

■ *There will be a significant difference in production between the four plants.*

The corresponding null hypothesis would simply state that no significant difference would exist.

STEP 3

Identify the Independent Variable

Our independent variable is plant, and there are four levels. For convenience we'll just call them Plants A through D.

STEP 4

Identify and Describe the Dependent Variable

The dependent variable is production. We'll use the data values in Table 9.9 to represent the number of units manufactured by each plant during the past 12 months. Figure 9.23 shows our descriptive statistics.

TABLE 9.9. Number of Units Manufactured by Each Plant

Plant A	Plant B	Plant C	Plant D
1037	968	983	969
1002	953	953	1013
1003	959	1024	1015
1013	1026	984	1019
1050	968	965	1014
954	1027	1033	1029
1034	1015	1019	970
1036	1032	1019	1001
998	1024	1006	957
980	1031	1028	1044
975	1023	989	1018
957	983	1035	974

Descriptives

Units Produced

	N	Mean	Std. Deviation	Std. Error	95% Confidence Interval for Mean		Minimum	Maximum
					Lower Bound	Upper Bound		
Plant A	12	1003.2500	32.15057	9.28107	982.8225	1023.6775	954.00	1050.00
Plant B	12	1000.7500	31.53678	9.10388	980.7125	1020.7875	953.00	1032.00
Plant C	12	1003.1667	27.65315	7.98278	985.5967	1020.7366	953.00	1035.00
Plant D	12	1001.9167	27.62561	7.97483	984.3642	1019.4691	957.00	1044.00
Total	48	1002.2708	28.86781	4.16671	993.8885	1010.6532	953.00	1050.00

FIGURE 9.23. Descriptive statistics from the ANOVA.

We can see, right away, that there is probably not a significant difference in production rates between the four plants since the mean scores seem to be very, very close. Let's use the appropriate statistical test just to make sure.

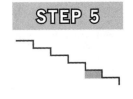

Choose the Right Statistical Test

Here we have one independent variable with four levels: Plant A, Plant B, Plant C, and Plant D, and our dependent variable is quantitative in nature; it appears we will use an ANOVA to analyze our data.

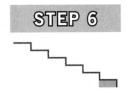

Use Data Analysis Software to Test the Hypothesis

There is no problem with the homogeneity of variance. Given that, we'll use the analysis of variance; we'll start by looking at Figure 9.24 to check our homogeneity of variance and then use the ANOVA table in Figure 9.25 to test our hypothesis.

Test of Homogeneity of Variances

Units Produced

Levene Statistic	df1	df2	Sig.
.428	3	44	.734

FIGURE 9.24. Homogeneity of variance statistics from the ANOVA.

ANOVA

Units Produced

	Sum of Squares	Df	Mean Square	F	Sig.
Between Groups	50.396	3	16.799	.019	.996
Within Groups	39117.083	44	889.025		
Total	39167.479	47			

FIGURE 9.25. Inferential statistics from the ANOVA.

Our p value, .996, is very large; obviously, we will not reject our null hypothesis. We could also compute a very small effect size of .0012 that supports our decision. Although we wouldn't have to, we can check our multiple comparison tests in Figure 9.26 just to see how they look when no significant difference is detected.

In this case, there are no asterisks marking any of the mean differences; this verifies what we saw with the large p value and small effect size. Apparently, the rumor regarding low production was only that. Now all the president has to worry about is if they are all running up to capacity!

■ The Factorial ANOVA

At the beginning of this chapter, we learned that a factorial ANOVA is used when a researcher has more than one independent variable but only one dependent variable

Multiple Comparisons

Units Produced

Bonferroni

(I) Plant	(J) Plant	Mean Difference (I-J)	Std. Error	Sig.	95% Confidence Interval	
					Lower Bound	Upper Bound
Plant A	Plant B	2.500	12.173	1.000	-31.13	36.13
	Plant C	.083	12.173	1.000	-33.55	33.71
	Plant D	1.333	12.173	1.000	-32.30	34.96
Plant B	Plant A	-2.500	12.173	1.000	-36.13	31.13
	Plant C	-2.417	12.173	1.000	-36.05	31.21
	Plant D	-1.167	12.173	1.000	-34.80	32.46
Plant C	Plant A	-.083	12.173	1.000	-33.71	33.55
	Plant B	2.417	12.173	1.000	-31.21	36.05
	Plant D	1.250	12.173	1.000	-32.38	34.88
Plant D	Plant A	-1.333	12.173	1.000	-34.96	32.30
	Plant B	1.167	12.173	1.000	-32.46	34.80
	Plant C	-1.250	12.173	1.000	-34.88	32.38

FIGURE 9.26. Multiple-comparison tests following the ANOVA.

that represents quantitative data. Most entry-level statistics books do not cover this topic, for three reasons.

1. Although it may appear to be quite clear cut, you'll find when you get into scenarios where you have three or more independent variables that things can get quite complicated in a hurry. For example, if you have a $4 \times 2 \times 3$ factorial ANOVA, you're dealing with 24 different comparisons.

2. Second, the math underlying this test is far beyond what we want to cover in this book. Because of that, we will not manually calculate it.

3. If you violate any of the assumptions that underlie using an analysis of variance, there is not a viable nonparametric alternative. The best one can do is run a Kruskal–Wallis H test, the nonparametric alternative for the one-way ANOVA, for each of the independent variables. This, of course, inflates the probability of making a Type I error. If you are really ambitious, there is a test called Friedman's ANOVA that can be modified to work in this scenario.

Having said all of that, let's look at the basics.

The Case of Age Affecting Ability

Recently, a university's accrediting agency mandated that professors, in all disciplines, should use technology more frequently in their classrooms. One of the young assistant professors noted, however, that it was not as easy as that. What, the professor wondered, would happen if technology proved to have a negative effect in the classroom?

To investigate this issue, the young professor arranged to have two classes in statistics taught in different ways. In the first class, students would use hand-held calculators to work through their problems. In the second class, students would use personal computers to help in their work. The professor followed this plan throughout the term and, after collecting final grades, decided to test whether the grades in the two groups were equivalent. Noting there was one independent variable (class) with two levels (the calculator section and the computer section) and one dependent variable (course grade), the professor analyzed the data using an independent-sample t-test. Upon running the analysis, the results showed a p value greater than the predetermined alpha value of .05. Given that, rejecting the null hypothesis was not feasible—there was apparently no significant difference between the two groups.

As the young professor was presenting the results in a faculty meeting, one of his colleagues noted that a lot of variables affect whether or not technology adoption is successful. The young professor decided to investigate this and, after some additional research, found that a student's age is an important predictor of success with technology. In fact, he read, students that are older than 30 appear to have more problems when trying to use technology in a classroom. "That's true." the young professor thought. "I did notice some of the older students looking puzzled when we started using the computers. Maybe I should look at this."

STEP 1

Identify the Problem

Here the university has a problem it is interested in addressing; in this case, the accreditation board pointed out the problem. Given that, it's certainly ethical and practically significant; numeric data can be collected by which a hypothesis can be tested. The scope of the problem is well within the authority of the university, and they had better make sure they have the knowledge, time, and resources needed to address it. They begin with a clear statement of the problem:

   ▪ *An accrediting agency has mandated that technology be used at all levels of university instruction. Faculty members are concerned that the achievement of older students may be affected because of this requirement.*

STEP 2

State a Hypothesis

Before we state a hypothesis, we need to take a closer look at what we are doing. First, it is obvious we are going to be hypothesizing using two independent variables, age and technology type used (computer or calculator). If we were using a one-way

ANOVA to look at each of these variables alone, we would state the following two null hypotheses:

■ *There will be no significant difference in achievement between students who use computers in a statistics class and those who use calculators in a statistics class. There will be no significant difference in achievement between students in a statistics class who are older than 30 and those who are 30 or younger.*

If we think about it, however, this is not really what we are interested in; there are actually four groups that we want to look at. To make this clear, look at Table 9.10.

TABLE 9.10. Cells for Age and Technology Type for Factorial ANOVA

	People 30 or younger	People older than 30
Computers	Group 1	Group 2
Calculators	Group 3	Group 4

You might be saying to yourself, "What's the big deal? We have four groups; we still have the two independent variables and have already stated hypotheses about them—what's the problem?" The problem is that we can't just look at the effects of each of the independent variables, called the *main effects*; we also have to look at the interaction between the two variables. In other words, it is important to know if these two variables interact to cause something unexpected to happen. For example, the literature suggests that younger people will perform better than older people when computers are involved; this is called an interaction of technology type and age. It is imperative that we look at the interaction of these variables to see if achievement is different based on the manner in which these independent variables affect one another. Given that, in addition to our two main effect hypotheses that we stated earlier, we also have to state an *interaction effect* hypothesis:

■ *There will be no significant interaction between technology use and age and their effect on achievement in a statistics class.*

Obviously, this is radically different from any hypotheses we have stated up to this point, but it will become clear as we work through the example.

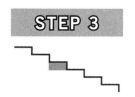

STEP 3

Identify the Independent Variable

Here our independent variables are age and technology type. Age has two levels: over 30 and 30 and below. Technology type also has two levels: computers or calculators.

STEP 4

Identify and Describe the Dependent Variable

The dependent variable is achievement in the statistics class. We'll use the data in Table 9.11 and move forward with our analysis.

We have two groups of 20 students: those who use calculators and those who use computers. Each of these groups includes 10 students age 30 and younger; the remaining 10 students are older than 30. In order to input this into the SPSS spreadsheet, shown in Figure 9.27, we would identify the students using computers as 1 in the Technology column; those with calculators would be identified as 2. Students 30 or younger would be identified as 1; students older than 30 would be identified with a 2.

TABLE 9.11. Score Data for Technology and Age Groups

Technology	Age group	Scores	
Computer	30 or younger	68	69
		67	70
		66	71
		73	73
		75	74
	Over 30	79	75
		80	75
		76	72
		71	71
		72	74
Calculator	30 or younger	73	77
		73	78
		77	69
		77	70
		75	71
	Over 30	70	69
		74	73
		75	75
		72	76
		72	72

Our software would generate the descriptive statistics shown in Figure 9.28. As you can see, the average scores for the computer group (72.55) and the calculator group (73.40) are about the same; so are the average scores for the 30 or younger group (72.30) and the students older than 30 (73.65). We can take this a step further and break the scores down by age group within the technology group; our software would give us the following results:

Again, just by looking at the data, it would seem that our young professor might be right; there doesn't appear to be any great difference between the age of a student

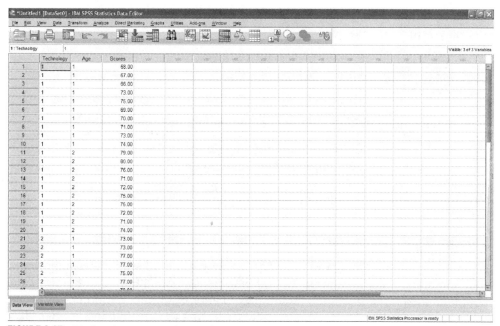

FIGURE 9.27. Technology type, age, and score entered into the Data View spreadsheet.

Descriptive Statistics

Dependent Variable:Scores

Technology	Age	Mean	Std. Deviation	N
Computer	30 or Younger	70.6000	3.09839	10
	Over 30	74.5000	3.17105	10
	Total	72.5500	3.64872	20
Calculator	30 or Younger	74.0000	3.26599	10
	Over 30	72.8000	2.25093	10
	Total	73.4000	2.79850	20
Total	30 or Younger	72.3000	3.55557	20
	Over 30	73.6500	2.81490	20
	Total	72.9750	3.23829	40

FIGURE 9.28. Descriptive statistics for the factorial ANOVA.

and whether or not they are using computers or calculators. Let's move forward with our analysis just to make sure.

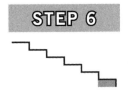

Choose the Right Statistical Test

Here we have two independent variables and one dependent variable that is quantitative in nature. Given that, we are going to use a factorial analysis of variance. Remember, also, we can call this a two-way analysis of variance since we have two independent variables or we can call it a "2 × 2" analysis of variance because there are two levels in the first independent variable (age group) and two levels in the second independent variable (type of technology).

Use Data Analysis Software to Test the Hypothesis

When we look at the averages for all students using calculators, they seem to be fairly equal (i.e., 74.0 and 72.8), regardless of a given student's age. There is a larger difference between students using computers in that persons 30 and younger have an average score of 70.6 while their older classmates have an average of 74.5. Our question, however, is the same as it always is, "Are these differences due to chance, or are the averages from the two groups significantly different?" We would use our computer to compute both the Levene test and the factorial ANOVA. In order to do that, we would select Analyze, General Linear Model, and Univariate, shown in Figure 9.29. This would be followed by identifying the Dependent Variable and Fixed Factors (Figure 9.30).

FIGURE 9.29. Selecting the General Linear Model and Univariate to run a factorial ANOVA.

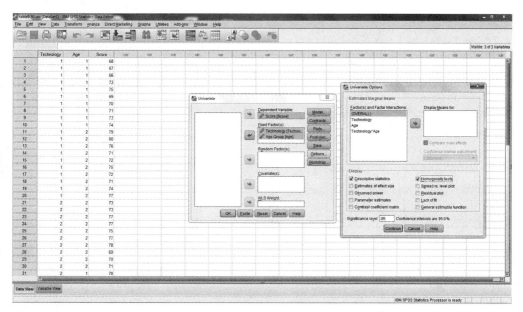

FIGURE 9.30. Identifying the independent variables, dependent variable, and descriptive statistics for the factorial ANOVA.

Figure 9.31 shows that the *p* value from the Levene test is .462, much greater than our alpha value of .05. This means we can proceed with testing our hypotheses with the ANOVA table shown in Figure 9.32.

Levene's Test of Equality of Error Variances

Dependent Variable:Scores

F	df1	df2	Sig.
.878	3	36	.462

FIGURE 9.31. Levene's test for the factorial ANOVA.

The table in Figure 9.32 is very similar to the ANOVA table we used earlier, but there are a few differences. First, there is a lot of technical information given in the first two rows, so let's skip down to the third and fourth rows, labeled "Technology" and "Age." These lines, called the Main Effects, show what the results of a one-way ANOVA would be if it were calculated using only one independent variable at a time. In this case, if we separated the groups by either Age or Technology used, the differences are not significant; our *p* value (i.e., Sig.) is greater than our alpha value of .05. This, of course, validates the observation we made earlier when we noticed that the means for each of these groups were fairly close to one another.

Tests of Between-Subjects Effects

Dependent Variable:Scores

Source	Type III Sum of Squares	df	Mean Square	F	Sig.	Partial Eta Squared
Corrected Model	90.475[a]	3	30.158	3.409	.028	.221
Intercept	213014.025	1	213014.025	24076.938	.000	.999
technology	7.225	1	7.225	.817	.372	.022
age	18.225	1	18.225	2.060	.160	.054
technology * age	65.025	1	65.025	7.350	.010	.170
Error	318.500	36	8.847			
Total	213423.000	40				
Corrected Total	408.975	39				

a. R Squared = .221 (Adjusted R Squared = .156)

FIGURE 9.32. Inferential statistics from a factorial ANOVA.

Interpreting the Interaction p Value

After we have looked at the results of analyzing the independent variables separately, we then need to look at the p value of the interaction between Technology and Age, shown in the fifth row. In this case, we can see the p value is equal to .010, far less than our alpha value of .05. This means we have a significant interaction effect and can reject our null hypothesis.

> ▨ *There will be no significant interaction between technology and age and their effect on achievement in a statistics class.*

We have rejected the null hypothesis, but what does it mean? How do we interpret these results so that they are meaningful to us? To do that, we will rely on that old adage "one picture is worth a thousand words"; many people find it easier to help interpret the results of a factorial ANOVA by plotting the interactions using the average scores from each subgroup. We would do this by selecting the Plots subcommand from the earlier screen and identifying the x-axis and y-axis variables as shown in Figure 9.33. SPSS would then create Figure 9.34, an interaction plot.

To interpret this chart, first look at the legend at the top right of the chart. It tells us a solid line represents the scores for students 30 and younger and a dashed line represents the scores of students older than 30. When we look at the actual graph, we can see that the bottom left of the graph represents those students in the computer group and the bottom right side shows those students in the calculator group.

Given this, you can see that the dashed line on the left side of the graph (computer group) starts at the point on that line which represents the average score of the over-30 group (74.5). It then goes down to the point on the right side of the graph (the calculator side) that is equal to the average score of that group (72.8). The solid line for the 30 and younger group starts on the computer side at 70.6 and goes up to the level of the calculator group (74.00). Based on this, it seems the young professor's

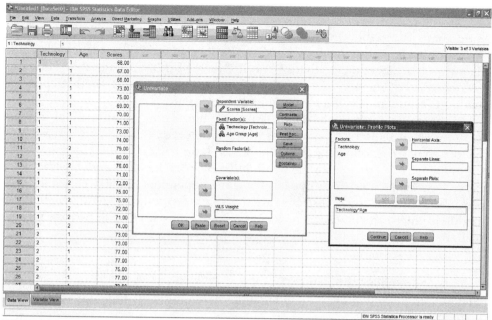

FIGURE 9.33. Selection of the dependent variable and fixed factors in a factorial ANOVA.

colleague was right: age as an independent variable does interact with the type of technology a student uses. Let's look at this in a little more detail.

When looking at the right side of the chart, the first thing we notice is that there is not a big difference in achievement, by age, for those students using calculators. We actually already knew this because, using the descriptive statistics, we saw students under 30 who used a calculator had an average score of 74.0 while students 30 and older using calculators had an average of 72.8. The more striking observation we can make is that students 30 and younger have somewhat lower achievement (i.e., 70.6) than their older classmates when using computers (i.e., 74.5). While we knew this already from looking at the data, plotting them on this graph makes the differences quite clear.

At this point, we can safely state that a significant interaction does exist (remember, however, we already knew that because of the low p value). We can also see that when considered by the type of technology used, the average scores change when we move between levels of age. Before we move forward, let's spend some time talking about the types of interactions that might exist.

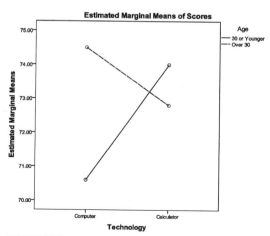

FIGURE 9.34. Disordinal interaction.

Ordinal and Disordinal Interactions

As we saw in our graph, the two lines, when plotted, crossed one another. We call this a *disordinal* interaction, and it exists when one level of an independent variable has an effect dramatically opposite that of another level of the same variable. For example, here we saw that the age of a student affects achievement differently depending on whether the student is using a computer or a calculator.

In other cases, we might have an *ordinal* interaction. In an ordinal interaction, the lines, when plotted, do not cross one another. For example, if our plotted data had looked like the following graph, we would have had an ordinal interaction. If you look at Figure 9.35, the reasoning behind this is clear; the independent variable called Technology Type only affects one level of the independent variable called Age Group. As can be seen, students whose ages are less than or equal to 30 have a higher average than their older classmates when calculators are being used. The two age groups are equal, however, when they are all using computers.

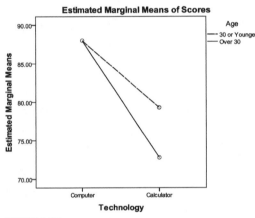

FIGURE 9.35. Ordinal interaction.

Always Use Your *p* Value!

Remember, although it is easy to interpret this graph, you cannot use it alone to decide if your interaction is significant; for that you must rely on the *p* value. The graphs are only used to help demonstrate the degree of the interaction. Having said that, let's move on to the next example.

The Case of the Coach

Let's look at another example in which a high school coach is interested in improving the endurance of his football players. He's heard there are two things that greatly contribute to endurance: the type of conditioning exercise players are involved in and their diet. In order to investigate this issue, he sets up a training regimen where half of his students will work out by running; others will go to an aerobics class. He also assigns half of each group of student athletes to a carbohydrate-rich diet, while the others will eat as much protein as possible. The coach is interested in determining if either of these approaches, or a combination of both, will lead to higher levels of endurance.

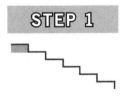

Identify the Problem

This case doesn't specifically state a problem, but one is certainly implied. The coach is interested in winning as many games as possible, so he's looking for a way to increase endurance. I'm sure there is some way the coach can numerically measure endurance. (How should I know? I'm a statistician, not a coach!)

All of the other criteria for a good problem are met, so let's write our problem statement:

- *This study will investigate the effect of combining different diets and exercise regimens on the endurance of high school athletes.*

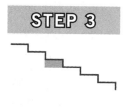

STEP 2

State a Hypothesis

The coach will be investigating the following two main-effect hypotheses:

- *There will be no significant difference in the degree of endurance of high school athletes following a diet rich in protein and athletes eating a diet rich in carbohydrates. There will be no significant difference in the degree of endurance between high school athletes who run and athletes who go to aerobics classes.*

He will also be investigating a hypothesis for the interaction effect:

- *There will be no significant interaction between exercise regimen and diet for high school athletes.*

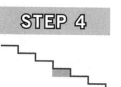

STEP 3

Identify the Independent Variable

There are two independent variables, diet and exercise regimen. Diet has two levels: carbohydrate rich and protein rich. Exercise regimen also has two levels: running and aerobics.

STEP 4

Identify and Describe the Dependent Variable

The dependent variable is endurance and, for the purposes of this example, it will be measured by a physical skill test the coach has used in the past. Results of the test are quantitative and range from 0 to 100.

At this point, it is obvious that we are going to use the factorial ANOVA since we have two independent variables, with one dependent variable used to collect quantitative data. Rather than create an entire dataset, let's just use the results shown in Figure 9.36 to test the coach's hypothesis.

Using the descriptive statistics, we can see something interesting is happening. The protein group has a much higher mean (i.e., 87.1) than the carbohydrate group (i.e., 73.0), but it appears to make no difference if the protein group runs (i.e., 87.4) or takes aerobics (i.e., 86.8). It does seem that athletes eating carbohydrates and taking aerobics have a slightly lower average (i.e., 72.6) than others in the aerobics class who are eating protein (i.e., 73.4). Let's check to see if this difference is significant.

Descriptive Statistics

Dependent Variable:Endurance

Food	Exercise	Mean	Std. Deviation	N
Protein	Aerobic	86.8000	.83666	5
	Running	87.4000	1.81659	5
	Total	87.1000	1.37032	10
Carbohydrate	Aerobic	72.6000	1.14018	5
	Running	73.4000	2.40832	5
	Total	73.0000	1.82574	10
Total	Aerobic	79.7000	7.54321	10
	Running	80.4000	7.64780	10
	Total	80.0500	7.40181	20

FIGURE 9.36. Factorial ANOVA descriptive statistics.

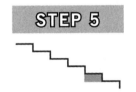

Choose the Right Statistical Test

Since we have two independent variables and our dependent variable is quantitative, we are going to use the factorial ANOVA.

Use Data Analysis Software to Test the Hypothesis

First, let's check the Levene statistic in Figure 9.37 to make sure we are not violating the homogeneity of variance assumption.

Our p value is exactly equal to our alpha of .05. Remember, we only reject the null hypothesis when p is less than .05; there are no problems here. Given that, we can use the following ANOVA table in Figure 9.38 to test our hypothesis.

Immediately we can tell, from our computed p value (i.e., .895), that a significant interaction between diet and type of exercise does not exist. Even though the interaction is not significant, we can still look at the interaction plot in Figure 9.39.

Here we have an ordinal interaction; the lines do not cross one another. In fact, the lines are nearly parallel. Since the interaction is not significant, it's necessary to look at the p values of the main effects.

Levene's Test of Equality of Error Variances

Dependent Variable:Endurance

F	df1	df2	Sig.
3.234	3	16	.050

FIGURE 9.37. Factorial ANOVA Test of Equality.

Tests of Between-Subjects Effects

Dependent Variable:Endurance

Source	Type III Sum of Squares	df	Mean Square	F	Sig.	Partial Eta Squared
Corrected Model	996.550[a]	3	332.183	119.706	.000	.957
Intercept	128160.050	1	128160.050	46183.802	.000	1.000
Food	994.050	1	994.050	358.216	.000	.957
Exercise	2.450	1	2.450	.883	.361	.052
Food * Exercise	.050	1	.050	.018	.895	.001
Error	44.400	16	2.775			
Total	129201.000	20				
Corrected Total	1040.950	19				

a. R Squared = .957 (Adjusted R Squared = .949)

FIGURE 9.38. Inferential statistics for the factorial ANOVA.

In this case, we can see there is not a significant main effect for the type of exercise that a person performs because the p value is equal to .361. Of course, we suspected that before examining the p value because our mean scores were so close to one another for the two groups. A significant main effect does exist for the type of diet an athlete follows. In this case, the p value of .000 shows that athletes eating a diet high in protein can develop higher levels of endurance than athletes eating a diet high in carbohydrates. We can verify this by looking at the interaction plot and noting that the scores on the left side (protein) are dramatically higher than those on the right side (carbohydrates). Given all of this, the only null hypothesis we can reject is our first one; it seems that a diet rich in protein is better for developing superior levels of endurance.

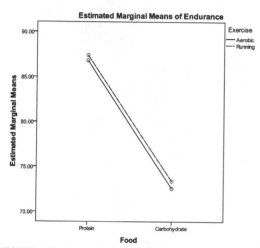

FIGURE 9.39. Ordinal interaction of food and exercise.

Summary

Although we have covered quite a lot of material in this chapter, the analysis of variance is fairly straightforward. We use the simple (one-way) ANOVA when we have one independent variable that has three or more levels and one dependent variable that is represented by quantitative data. When we have more than one independent variable and only one dependent variable, we will use a factorial ANOVA. As I mentioned, there are other types of ANOVAs but, for now, we have talked about as much as we need to; it is time to move forward!

Do You Understand These Key Words and Phrases?

analysis of variance	assumptions
between groups variance	cell
degrees of freedom	disordinal interaction
F value	factorial analysis of variance
interaction effect	main effect
mean square	multiple-comparison test
multivariate ANOVA (MANOVA)	one-way ANOVA
ordinal interaction	post-hoc tests
within-group variance	

Do You Understand These Formulas?

Between sum of squares:

$$SS_{Between} = \frac{(\sum x_1)^2}{n_1} + \frac{(\sum x_2)^2}{n_2} + \frac{(\sum x_3)^2}{n_3} - \frac{(\sum x)^2}{N}$$

Total sum of squares:

$$SS_{Total} = \sum x^2 - \frac{(\sum x)^2}{N}$$

Within sum of squares:

$$SS_W = SS_1 + SS_2 + SS_3$$

Quiz Time!

Let's test our mastery of this topic. Work through these examples and check your work at the end of the book.

The Case of Degree Completion

A university is interested in knowing if age is a determining factor in the number of years it takes to complete a doctoral degree. To aid in their investigation, they've placed each of their graduates into one of four age categories: 20–29, 30–39, 40–49, and 50 and above.

1. What are the null and research hypotheses the university is investigating?

2. What is the independent variable?

3. What are the levels of the independent variable?

4. What is the dependent variable?

5. Based on the information in Figures 9.40, 9.41, and 9.42, what could the university conclude?

Descriptives

Time to Finish

	N	Mean	Std. Deviation	Std. Error	95% Confidence Interval for Mean		Minimum	Maximum
					Lower Bound	Upper Bound		
20 thru 29	5	4.4000	1.08397	.48477	3.0541	5.7459	2.50	5.00
30 thru 39	5	3.7000	.83666	.37417	2.6611	4.7389	3.00	5.00
40 thru 49	5	3.6000	.89443	.40000	2.4894	4.7106	3.00	5.00
50 and older	5	4.8000	.83666	.37417	3.7611	5.8389	4.00	6.00
Total	20	4.1250	.98509	.22027	3.6640	4.5860	2.50	6.00

FIGURE 9.40. ANOVA descriptive statistics.

Test of Homogeneity of Variances

Time to Finish

Levene Statistic	df1	df2	Sig.
.073	3	16	.973

FIGURE 9.41. ANOVA tests of homogeneity of variance.

ANOVA

Time to Finish

	Sum of Squares	df	Mean Square	F	Sig.
Between Groups	4.937	3	1.646	1.951	.162
Within Groups	13.500	16	.844		
Total	18.438	19			

FIGURE 9.42. Inferential statistics for ANOVA.

The Case of Seasonal Depression

Psychologists are interested in determining whether there is a difference in the number of anxiety attacks during the different seasons of the year. In order to investigate it, they tracked 44 people for 12 months and asked them to record the number of anxiety attacks they had during each season.

1. What are the null and research hypotheses the psychologists are investigating?

2. What is the independent variable?

3. What are the levels of the independent variable?

4. What is the dependent variable?

5. Using Figures 9.43, 9.44, 9.45, and 9.46, what could the psychologists conclude?

Descriptives

Anxiety Attacks

	N	Mean	Std. Deviation	Std. Error	95% Confidence Interval for Mean		Minimum	Maximum
					Lower Bound	Upper Bound		
Winter	6	6.1667	1.47196	.60093	4.6219	7.7114	4.00	8.00
Spring	6	3.8333	.75277	.30732	3.0433	4.6233	3.00	5.00
Summer	6	3.8333	.75277	.30732	3.0433	4.6233	3.00	5.00
Fall	6	3.8333	.75277	.30732	3.0433	4.6233	3.00	5.00
Total	24	4.4167	1.38051	.28179	3.8337	4.9996	3.00	8.00

FIGURE 9.43. ANOVA descriptive statistics.

Test of Homogeneity of Variances

Anxiety Attacks

Levene Statistic	df1	df2	Sig.
1.997	3	20	.147

FIGURE 9.44. ANOVA tests of homogeneity of variance.

ANOVA

Anxiety Attacks

	Sum of Squares	df	Mean Square	F	Sig.
Between Groups	24.500	3	8.167	8.448	.001
Within Groups	19.333	20	.967		
Total	43.833	23			

FIGURE 9.45. Inferential statistics for ANOVA.

Multiple Comparisons

Anxiety Attacks

Bonferroni

(I) Season	(J) Season	Mean Difference (I-J)	Std. Error	Sig.	95% Confidence Interval	
					Lower Bound	Upper Bound
Winter	Spring	2.33333*	.56765	.003	.6718	3.9949
	Summer	2.33333*	.56765	.003	.6718	3.9949
	Fall	2.33333*	.56765	.003	.6718	3.9949
Spring	Winter	-2.33333*	.56765	.003	-3.9949	-.6718
	Summer	.00000	.56765	1.000	-1.6616	1.6616
	Fall	.00000	.56765	1.000	-1.6616	1.6616
Summer	Winter	-2.33333*	.56765	.003	-3.9949	-.6718
	Spring	.00000	.56765	1.000	-1.6616	1.6616
	Fall	.00000	.56765	1.000	-1.6616	1.6616
Fall	Winter	-2.33333*	.56765	.003	-3.9949	-.6718
	Spring	.00000	.56765	1.000	-1.6616	1.6616
	Summer	.00000	.56765	1.000	-1.6616	1.6616

*. The mean difference is significant at the 0.05 level.

FIGURE 9.46. Multiple-comparison tests for ANOVA.

The Case of Driving Away

Teachers, worried over the rising number of tardy students, noticed that students who had their own car were late more often than their peers who only had access to their parents' car. Both groups, they believed, were tardy more often than students who had no access to an automobile for their personal use. In order to investigate their theory, the teachers identified 10 students in each of these groups and tracked the number of times they were late for school during the month.

1. What are the null and research hypotheses the teachers are investigating?

2. What is the independent variable?

3. What are the levels of the independent variable?

4. What is the dependent variable?

5. By looking at Figures 9.47, 9.48, 9.49, and 9.50, what could the teachers conclude?

Descriptives

Times Late

	N	Mean	Std. Deviation	Std. Error	95% Confidence Interval for Mean		Minimum	Maximum
					Lower Bound	Upper Bound		
Owns Car	10	3.8000	1.22927	.38873	2.9206	4.6794	2.00	6.00
Has Access to Car	10	5.3000	.94868	.30000	4.6214	5.9786	4.00	7.00
No Access to Car	10	6.8000	1.13529	.35901	5.9879	7.6121	5.00	9.00
Total	30	5.3000	1.64317	.30000	4.6864	5.9136	2.00	9.00

FIGURE 9.47. ANOVA descriptive statistics.

Test of Homogeneity of Variances

Times Late

Levene Statistic	df1	df2	Sig.
.385	2	27	.684

FIGURE 9.48. ANOVA tests of homogeneity of variance.

ANOVA

Times Late

	Sum of Squares	df	Mean Square	F	Sig.
Between Groups	45.000	2	22.500	18.243	.000
Within Groups	33.300	27	1.233		
Total	78.300	29			

FIGURE 9.49. Inferential statistics for ANOVA.

Multiple Comparisons

Times Late

Bonferroni

(I) Access	(J) Access	Mean Difference (I-J)	Std. Error	Sig.	95% Confidence Interval	
					Lower Bound	Upper Bound
Owns Car	Has Access to Car	-1.50000*	.49666	.016	-2.7677	-.2323
	No Access to Car	-3.00000*	.49666	.000	-4.2677	-1.7323
Has Access to Car	Owns Car	1.50000*	.49666	.016	.2323	2.7677
	No Access to Car	-1.50000*	.49666	.016	-2.7677	-.2323
No Access to Car	Owns Car	3.00000*	.49666	.000	1.7323	4.2677
	Has Access to Car	1.50000*	.49666	.016	.2323	2.7677

*. The mean difference is significant at the 0.05 level.

FIGURE 9.50. Multiple-comparison tests for ANOVA.

The Case of Climbing

Mount Rainier is the largest glaciated mountain in the United States, rising some 14,400 feet above sea level. While thousands attempt to summit this massive mountain every year, not all of them make it and so they turn back short of their goal. A group of climbers was discussing this one day and argued that success depended on the route the climbers took, as well as on whether or not they were with a professional guide. To try to settle their argument, they decided to ask the next 64 people coming down from the mountain three things: the route they took, the altitude they reached (remember, 14,400 is the summit), and whether they used a guide service.

1. What the three main-effect null hypotheses the climbers are investigating?

2. What are the three research hypotheses the climbers are investigating?

3. What are the independent variables?

4. What are the levels of the independent variables?

5. What is the dependent variable?

6. Using the information in Figures 9.51 and 9.52, what conclusion would the climbers come to?

Descriptive Statistics

Dependent Variable:Altitude

Route	Guide	Mean	Std. Deviation	N
Ingraham Direct	Yes	10124.50	3253.08	8
	No	9232.87	1700.15	8
	Total	9678.68	2549.40	16
Gibraltar Rock	Yes	8387.50	2903.36	8
	No	9070.12	3167.77	8
	Total	8728.81	2956.51	16
Disappointment Cleaver	Yes	9039.50	3240.93	8
	No	9034.75	2467.10	8
	Total	9037.12	2782.46	16
Nisqually Ice Cliff	Yes	11019.62	2962.54	8
	No	8195.62	3120.26	8
	Total	9607.62	3281.15	16
Total	Yes	9642.78	3113.89	32
	No	8883.34	2580.69	32
	Total	9263.06	2862.66	64

FIGURE 9.51. Descriptive statistics for factorial ANOVA.

Tests of Between-Subjects Effects

Dependent Variable:Altitude

Source	Type III Sum of Squares	df	Mean Square	F	Sig.	Partial Eta Squared
Corrected Model	4.699E7	7	6712984.964	.801	.590	.091
Intercept	5.491E9	1	5.491E9	655.302	.000	.921
Route	1.005E7	3	3349004.125	.400	.754	.021
Guide	9227925.062	1	9227925.062	1.101	.299	.019
Route * Guide	2.772E7	3	9238652.438	1.102	.356	.056
Error	4.693E8	56	8380065.661			
Total	6.008E9	64				
Corrected Total	5.163E8	63				

a. R Squared = .091 (Adjusted R Squared = -.023)

FIGURE 9.52. Inferential statistics for factorial ANOVA.

CHAPTER 10

The Chi-Square Tests

Introduction

Up to this point we have dealt with quantitative data and, because of that, we have been using parametric statistical tools. As I said in the very beginning, this is normal for introductory statistics students. Other than the topic of this chapter, the chi-square test, it is unusual for beginning students to use other nonparametric statistical tests.

In order to begin understanding the purpose of the chi-square, as well as how it works, think back to our chapter on the analysis of variance. In it, we saw the one-way analysis of variance, where we investigated an independent variable with three or more levels, and the factorial analysis of variance (e.g., a two-way ANOVA) where we compared more than one independent variable, regardless of the number of levels within each of the variables. The chi-square tests work using basically the same principles but instead of each level representing a quantitative value (e.g., test scores), the levels of the chi-square test contain counts of values (i.e., nominal data). Let's look at an example of both a "one-way" and "factorial" chi-square to help us understand what we are getting into.

▓ The One-Way Chi-Square Test

The one-way chi-square test, or the chi-square *goodness of fit* as it is more commonly called, helps us determine if a distribution of data values we observe matches how we expect the data to be distributed. The following example will explain what I mean.

Suppose we are in a situation where we have posed this question to a random group of 300 people walking down the street: "What do you think of allowing 16-year-old children to vote?" Suppose, when you looked at your data, you found that 95 people supported the idea, 105 didn't like the idea, and 100 didn't care; these are called the *observed* values in each of the cells. This is shown in Table 10.1.

TABLE 10.1. Observed Values of Vote Data

	It is all right with me.	I do not like the idea.	I do not care.
Observed	95	105	100

Once we have collected our observed values, the question then becomes, "We just defined goodness of fit as the comparison of a set of observed values to a set of expected values; how do we determine the expected values?" That's easy; since we are testing goodness of fit, we are interested in comparing the observed values to values we expect. For example, if we are interested in comparing our results to national averages showing that 40% of people like the idea and the rest are evenly split (i.e., 30% each) between not liking the idea or not caring, this results in Table 10.2.

TABLE 10.2. Observed and Expected Values of Vote Data

	It is all right with me.	I do not like the idea.	I do not care.
Observed	95	105	100
Expected	120 (i.e., 40%)	90 (i.e., 30%)	90 (i.e., 30%)

In order to test your hypothesis, you would compare the proportion of data values in the table to determine if the differences between what you observed and what you expected are significant.

Before we move forward, it is important to know that the chi-square test is not designed to be used with a small number of expected values. Because of that, it is important to ensure you expect at least five occurrences of each of the cell values; trying to do any calculations with less than that will adversely affect your results.

The Factorial Chi-Square Test (the Chi-Square Test of Independence)

The factorial chi-square, or as it is more often called, the *chi-square test of independence*, is very akin to the factorial ANOVA in that you're collecting data where you have more than one independent variable. For example, imagine you're the ultimate authority charged with deciding whether sex education should be taught in your district's elementary schools. Before you make your decision, you decide that you need to take into consideration the will of the public. In order to do that, you randomly select 60 male and 60 female parents within the community (i.e., gender is the first independent variable). You ask them to tell you their opinion, yes or no, whether they agree with the proposed new curriculum (i.e., their opinion about the subject is the second independent variable).

We've just seen something that might look counterintuitive. In this case, it seems natural that "opinion" would be a dependent variable in that we would be interested in measuring how a person feels about a subject. Remember, however, in this example we are not interested in how any one person feels about the subject; what we really want to know is how many people with a given opinion fall into each category. Given that, "opinion" is the independent variable, and the count of each opinion is the dependent variable where we are collecting nominal data.

In testing your hypothesis, your feeling is that there will be an equal distribution of "yes" and "no" answers (i.e., the expected values) between the genders. For the sake of discussion, I have gone ahead and entered a set of equal expected values, along with a set of observed values, in Table 10.3.

TABLE 10.3. Observed and Expected Values of Vote Data by Gender

	Yes		No	
	Observed	Expected	Observed	Expected
Females	35	30	25	30
Males	25	30	35	30

Now, in order to test our hypothesis, we will need to compare our observed values to our expected values. Before we can do that, we first need to have a better understanding of the concepts underlying the chi-square.

Computing the Chi-Square Statistic

In the tests we have examined up to this point, we have compared a statistic generated from data we collected to a value from a table appropriate to a given statistical test (e.g., the z table, the t table, or F table). By doing so, we have been able to determine if a significant difference existed between the statistic we computed and a specific point on a given data distribution.

At first glance, the chi-square procedure works in much the same manner except, when using it, we are not interested in comparing a single statistic computed from a specific data value to a table value. Instead we are trying to determine if the distribution of the variable in question is similar to the distribution we are expecting. We do this by comparing a computed value of chi-square, based on the entire observed distribution, to a critical value of chi-square based on the expected distribution. When we do, we are able to tell if a distribution of data we have observed is similar to or significantly different from the distribution we are expecting.

Because it is very easy to understand, let's go back and use our example of the goodness-of-fit test where we asked for opinions about persons as young as 16 being allowed to vote. Table 10.4 is the same one we used as part of that discussion.

TABLE 10.4. Observed and Expected Values of Vote Data

	It is all right with me.	I do not like the idea.	I do not care.
Observed	95	105	100
Expected	120	90	90

Our null hypothesis would read:

■ *There will not be a significant difference in the number of responses expected in each cell and the number of responses observed in each cell.*

Obviously the research hypothesis would read:

▪ *There will be a significant difference in the number of responses expected in each cell and the number of responses observed in each cell.*

In short, I have hypothesized that the observed distribution of responses will not be different from the expected distribution of responses. Testing this hypothesis will involve comparing a chi-square value computed from the observed responses to a critical chi-square value based on a given degrees of freedom value. As we are used to doing, we will reject our null hypothesis if the computed value of chi-square is greater than the critical value of chi-square. If we do reject the null hypothesis, we are saying nothing more than the distribution we observed is significantly different from the distribution we expected. The formula for computing the chi-square statistic is

$$\chi^2 = \Sigma \frac{(O-E)^2}{E}$$

First, χ is the Greek letter "chi"; of course we will be computing its squared value. As you can see in Table 10.5, for each row we are going to:

1. Subtract the expected value (i.e., E) in each row from the observed value (i.e., O) in each row. This is shown in the second column.
2. Square the value from Step 1. This is shown in the third column.
3. Divide the value from Step 2 by the expected value (i.e., E) for that row; this gives us the chi-square value for that row. This is shown in the fourth column.
4. Add all of the chi-square values from each row (Step 3) to get the computed chi-square value. This is shown at the bottom of the fourth column of Table 10.5.

TABLE 10.5. Computing the Chi-Square Statistic

	Observed – expected	Observed – expected squared	Chi-square = Column 2 ÷ number expected
It is all right with me.	95 – 120 = –25	625	625/120 = 5.21
I do not like the idea.	105 – 90 = 15	225	225/90 = 2.5
I do not care.	100 – 90 = 10	100	100/90 = 1.11
			Computed chi-square = 8.82

Now all we have to do is compare our computed value to the table (i.e., critical) value. To do that, we need to know what the distribution looks like.

The Chi-Square Distribution

As has been the case throughout this book, we will use a formula to create the chi-square distribution. Like the F distribution, we will base this distribution on squared values (i.e., chi-square), so the distribution starts at zero and all values in the distribution are positive. This distribution is also based on the number of degrees of freedom from the data we have collected, but how we compute the degrees of freedom is going to depend on the particular chi-square test we are using. The fewer the degrees of freedom, the more peaked the distribution; a larger number of degrees of freedom will result in a flatter distribution that is more skewed to the right. This idea is shown in Figure 10.1.

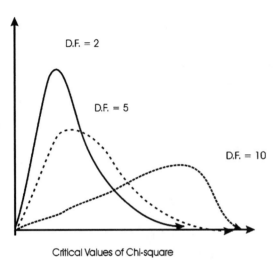

Critical Values of Chi-square

FIGURE 10.1. Shape of chi-square distribution for different degrees of freedom.

As I said earlier, we will compute degrees of freedom differently for each of the chi-square tests but let's focus on the goodness-of-fit test.

$$df = (\text{number of entries}) - 1$$

The number of entries in the formula represents the number of levels of the independent variable; in this case we have three (e.g., "It is all right with me") so we have 2 degrees of freedom (i.e., $3 - 1 = 2$). We will use this value, with our chi-square table, to determine the critical chi-square value. Just so you are aware: many chi-square tables show critical values for a range of different alpha values. Since we have primarily used the traditional alpha value of .05 throughout the book, I have only shown that part of Table 10.6; there is a complete table in Appendix F.

TABLE 10.6. Critical Values of Chi-Square

Degrees of freedom	Alpha = .05
1	3.84
2	5.99
3	7.82
4	9.49
5	11.07
6	12.59
7	14.07
8	15.51
9	16.92
10	18.31

Here I have only shown critical values of chi-square up to 10 degrees of freedom; I did that for two reasons. First, since computing the degrees of freedom is a function of the number of cells and rows, you will rarely wind up with more than 10 degrees of freedom. Second, as you can see, the critical value of chi-square with only 10 degrees of freedom is 18.31; that's a really large value. In order to compute a chi-square value larger than that, you would need observed and expected values that were totally out of proportion. Needless to say, that does not happen very often.

Knowing that, let's get back to testing our null hypothesis:

> ■ *There will not be a significant difference in the number of responses expected in each cell and the number of responses observed in each cell.*

Looking at the table, since we have 2 degrees of freedom, our critical chi-square value is 5.99. Since our computed value of 8.82 is greater than our critical value, we reject our null hypothesis; there appears to be a significant difference in the observed number of responses for each answer and the expected number of responses for each value. We can see this in Figure 10.2.

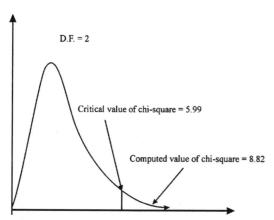

FIGURE 10.2. Testing a hypothesis using the computed and critical values of chi-square.

In order to use SPSS, let's first enter our data. As you can see, we have one variable, Vote. In order for the chi-square test to work, we would need to enter an appropriate value for each of our 300 voters. This means we would enter, for example, 95 values of 1 for people who support the idea, 105 values of 2 for those who do not like the idea, and 100 values of 3 for people who do not care (Figure 10.3).

In order to run the actual test, we would select Analyze, Nonparametric Tests, Legacy Dialogs, and Chi-Square; this is seen in Figure 10.4.

We would then identify Vote as our Test Variable and include Expected Values of 120, 90, and 90. We've also asked for descriptive statistics to be computed as shown in Figure 10.5.

SPSS would generate two tables for us. In Figure 10.6, we can see the number of votes we observed in each category, the number of votes we expected in each category, and the difference (i.e., the Residual) between the two. In Figure 10.7, we can see a chi-square value of 8.82, just as we calculated, and a *p* value (i.e., Asymp. Sig.) of .012, which also supports our decision to reject the null hypothesis.

What about the Post-Hoc Test?

At this point, you would expect us to run a post-hoc test to determine where the significant difference lies. Unfortunately, running post-hoc tests with a chi-square test is very difficult to both explain and calculate; SPSS does not do that for us. Because of

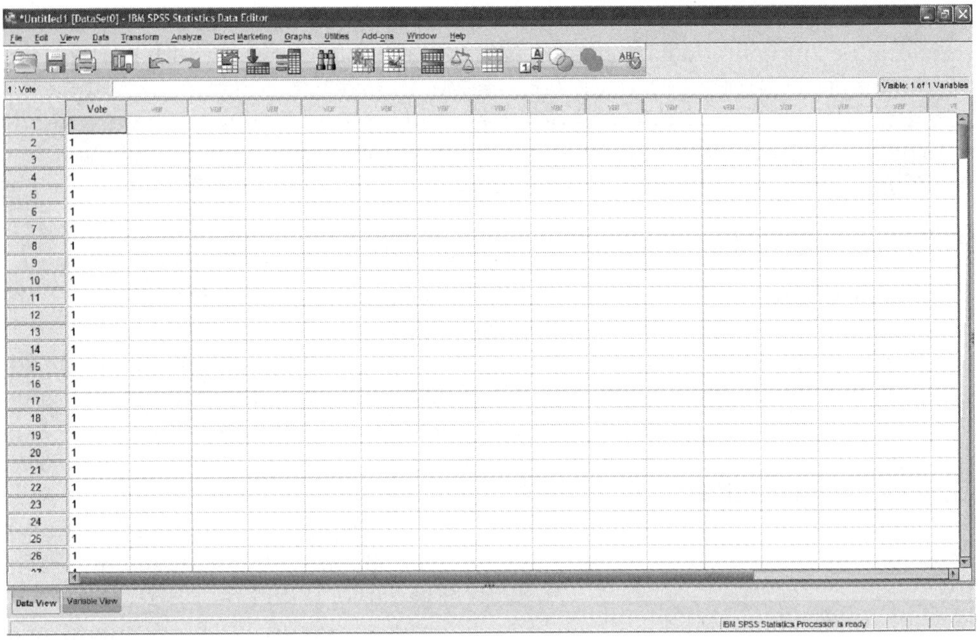

FIGURE 10.3. Vote data in the Data View spreadsheet.

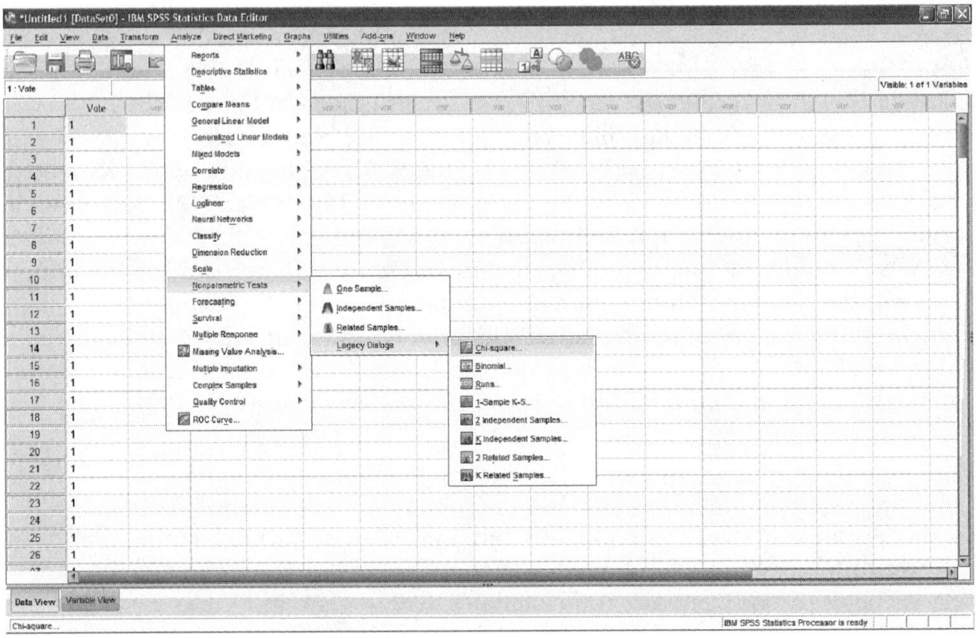

FIGURE 10.4. Selecting the chi-square test.

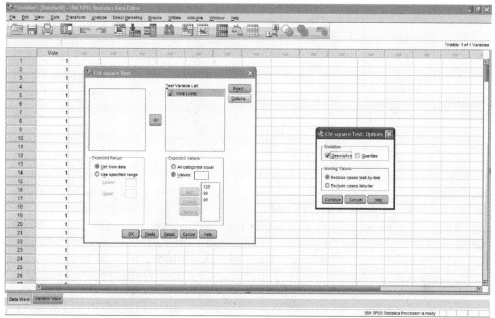

FIGURE 10.5. Identifying the Test Variable and the Expected Values.

that we are happy knowing that a significant difference does exist; where that difference lies is obvious.

Working with an Even Number of Expected Values

In this case, we knew the expected values for our formula. At other times, if you have an even distribution of expected values, you can have SPSS calculate them. For example, if we were expecting an equal number of responses for each question, SPSS would have computed the expected value for each cell by dividing the total number of observed values by the number of cells. In this case, we would wind up with an expected frequency of 100 for each response, 300 observed values divided by 3 possible answers. We can see how we would set that up in Figure 10.8; instead of inputting the values we wanted, we've simply told SPSS that all categories are equal. The resulting observed and expected values are shown in Table 10.7.

As you can see in Figures 10.9 and 10.10, since our observed values are not much

Vote

	Observed N	Expected N	Residual
It is all right with me	95	120.0	-25.0
I do not like the idea	105	90.0	15.0
I do not care	100	90.0	10.0
Total	300		

FIGURE 10.6. Observed and expected values for vote data.

Test Statistics

	Vote
Chi-square	8.819
Df	2
Asymp. Sig.	.012

FIGURE 10.7. Inferential statistics for the chi-square goodness-of-fit test.

different from the expected values, the computed value of p (i.e., .779) would be larger than our traditional alpha value of .05. This means, of course, that we wouldn't reject our null hypothesis if we expected an equal number of responses for each of the answers.

The Case of the Belligerent Bus Drivers

We will better learn how to interpret the results of the chi-square in just a few pages, but for now let's go ahead and put the six-step model to work for us. Let's suppose a statistician was recently hired to act as an expert witness in a court case. The case involved a local school system that was suing the bus driver's union, charging that the bus drivers were malingering. This, the school system stated, was evidenced by the fact that far too many bus drivers called in sick on Mondays and Fridays, effectively extending a normal weekend into a three-day holiday. The school system presented Table 10.8 showing total absences for given workdays throughout the school year.

TABLE 10.7. Observed and Expected Values of Vote Data with All Categories Equal

	It is all right with me.	I do not like the idea.	I do not care.
Observed	95	105	100
Expected	100	100	100

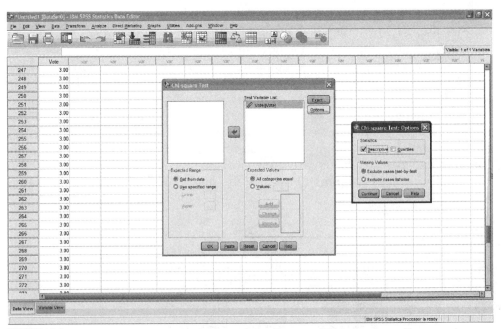

FIGURE 10.8. Identifying the Test Variable and the Expected Values when all categories are equal.

vote

	Observed N	Expected N	Residual
It is all right with me	95	100.0	-5.0
I do not like the idea	105	100.0	5.0
I do not care	100	100.0	.0
Total	300		

FIGURE 10.9. Observed and expected values for the vote data.

Test Statistics

	vote
Chi-square	.500
df	2
Asymp. Sig.	.779

FIGURE 10.10. Inferential statistics for the chi-square goodness-of-fit test.

TABLE 10.8. Observed Values of Absence by Day of Week

Monday	25
Tuesday	10
Wednesday	16
Thursday	19
Friday	30

The school system pointed out to the driver's union that they felt the number of absences should be about equally distributed over the course of the week (e.g., 20% of the absences should occur on Monday, 20% on Tuesday). The driver's union agreed that the numbers seemed higher on Mondays and Fridays but felt it was just a coincidence. They dismissed the concerns of the school system and told the administrators they would have to support their allegations before they would take any action. At this point, the school system had no resort other than to attempt to make its point.

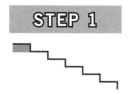

STEP 1

Identify the Problem

In this case, only the school board is interested in this problem from a practical perspective; the bus driver's union is happy with the status quo. The school system has already collected the numeric data it needs since it had the time, and so on, necessary to do so; there was nothing unethical about its collecting the data. Again, although the drivers may disagree, this is something the schools should investigate.

> ■ *Providing school bus transportation is integral to student attendance and ultimately, achievement. Administrators are attempting to determine if the drivers are manipulating the system to allow more time away from the job.*

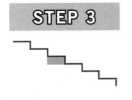

State a Hypothesis

By reading this scenario, the null hypothesis the school board wants to investigate is apparent:

- *There will be no significant difference in the number of absences expected on a given day and the actual number of absences observed on a given day.*

The research hypothesis would be

- *There will be a significant difference in the number of absences expected on a given day and the actual number of absences observed on a given day.*

Identify the Independent Variable

Having stated this, it is easy to see that our independent variable is "day of week." We have five levels: Monday, Tuesday, Wednesday, Thursday, and Friday.

STEP 4

Identify and Describe the Dependent Variable

The dependent variable is the number of absences for each day. Notice, in Figure 10.11, the total number of observed absences is 100. This means our expected value for each day of the week is 20 since the school district believes absences should be evenly distributed.

Since we are dealing with nominal data, it does not make sense to compute most of the descriptive statistics; the most important value, the mode, is easily seen in Table 10.8—the most absences occurred on Fridays. We could, however, use SPSS to create the bar chart in Figure 10.12 if we wanted; it verifies what we already know.

Day of Week

	Observed N	Expected N	Residual
Monday	25	20.0	5.0
Tuesday	10	20.0	-10.0
Wednesday	16	20.0	-4.0
Thursday	19	20.0	-1.0
Friday	30	20.0	10.0
Total	100		

FIGURE 10.11. Observed and expected values for day of the week data.

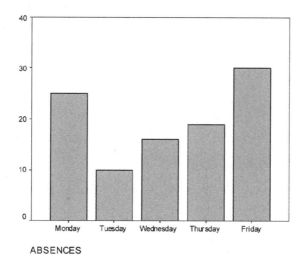

ABSENCES

FIGURE 10.12. Bar chart for day of the week data.

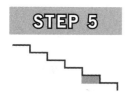

STEP 5

Choose the Right Statistical Test

In this case, we have one independent variable with five levels; we have one dependent variable representing nominal data. Because of this, we will use the chi-square goodness-of-fit test (i.e., the one-way chi-square test).

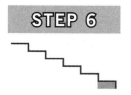

STEP 6

Use Data Analysis Software to Test the Hypothesis

Just by looking at the graphical output and the descriptive statistics, we had a fairly good idea that a problem existed. By looking at Figure 10.13, we can see there is a significant difference between the number of absences the school district expects each day of the week and the actual number of absences the drivers take; this decision is supported by a p value (i.e., .017) less than an alpha value of .05. Based on this, even though we do not have a post-hoc test, the results are obvious; calling in sick 25 times on Monday and 30 times on Friday, the drivers are ruining their case.

Test Statistics

	Day of Week
Chi-square	12.100[a]
Df	4
Asymp. Sig.	.017

FIGURE 10.13. Inferential statistics for the chi-square goodness-of-fit test.

The Case of the Irate Parents

In this case, we have been hired to investigate the contention of a group of parents that there is gender discrimination in a local elementary school's gifted program. Our job is to try to determine if the number of boys and girls in the program represents what we could expect in the general population. Let's use our six-step model:

STEP 1

Identify the Problem

This is certainly a significant, practical problem and one that is definitely within the ability and purview of the school to investigate; to not do so would be highly unethical. Investigating it would simply involve counting the number of each gender in the program and comparing it to the percentage of each gender in the population. Our problem statement could read:

■ *Because of perceived inequalities, this study will investigate whether the number of males and females in the gifted program is proportionate to the gender distribution throughout the rest of the school.*

STEP 2

State a Hypothesis

Our null hypothesis is easy enough:

■ *There will not be a significant difference in the percentage of boys and girls in the gifted program and the percentage of boys and girls in the general population.*

It would lead to the following research hypothesis:

■ *There will be a significant difference in the percentage of boys and girls in the gifted program and the percentage of boys and girls in the general population.*

STEP 3

Identify the Independent Variable

Our independent variable is gender, and the two levels are boys and girls.

STEP 4

Identify and Describe the Dependent Variable

Our dependent variable is the number of each gender in the gifted program. Let's assume there are 48 boys and 52 girls. Obviously, the mode is "girls"; it doesn't make a lot of sense to compute any other descriptive statistics.

STEP 5

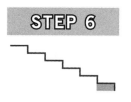

Choose the Right Statistical Test

In this case, we have an independent variable with two levels: boys and girls. We also have a dependent variable that represents nominal data. We will use the computer to compute the chi-square goodness-of-fit test but, first, let's decide on our expected values. Remember, we could either expect an equal number of boys and girls or we could base our expected values on some other criteria. Suppose we weren't sure so we did a little research and found that approximately 45% of the elementary school students in the United States are males (i.e., 45 out of 100 students would be expected to be males); that would give us the following observed and expected values shown in Figure 10.14.

Gender

	Observed N	Expected N	Residual
Male	48	45.0	3.0
Female	52	55.0	-3.0
Total	100		

FIGURE 10.14. Observed and expected values for gender.

There doesn't appear to be a lot of difference between our observed and expected values but, as always, let's use SPSS to check.

STEP 6

Use Data Analysis Software to Test the Hypothesis

As can be seen in Figure 10.15, our p value (i.e., .546) is much greater than an alpha value of .05; this means we fail to reject the null hypothesis. Because of this, parents of neither gender should complain; the proportion of boys and girls in the gifted classes is not significantly different from the national average.

Test Statistics

	Gender
Chi-square	.364
df	1
Asymp. Sig.	.546

FIGURE 10.15. Inferential statistics for the chi-square goodness-of-fit test.

The Chi-Square Test of Independence

At the start of the chapter I said that the chi-square test of independence is much like the factorial ANOVA in that we are interested in looking at data based on two or more independent variables. Again, the only difference is that with the chi-square test we are using nominal rather than quantitative data.

Suppose, for example, we are back to the group we just visited in the last case. This time they are complaining about an unequal distribution of gender within ability groups in the local school system. They point out that if one looks at the number of females in each of the ability groups, it seems there is a disproportionate number of males in the highest group. They use Table 10.9 to show their point.

TABLE 10.9. Observed Values of Gender within Ability Group

	Male	Female	Total
Low-ability group	40	24	64
Middle-ability group	35	24	59
High-ability group	56	21	77
Total	131	69	200

By looking at the chart, we can see where the parents might be concerned. There are more males than females in the low- and middle-ability groups but there are nearly three times as many males as females in the high-ability group. The parents may really have something to complain about this time!

Since we have used two independent variables to identify nominal-level data, we will use a two-way chi-square test, often called the *chi-square test of independence*. This test is used in situations where we want to decide if values represented by one independent variable are not influenced, or are independent of, the same values represented by another independent variable. In this case, we want to see if the total number of students, when we look at it from a gender perspective, is independent of the same total number of students when we look at it from an ability group perspective. We can investigate this question using the following research hypothesis:

■ *Gender has a significant effect on membership in an ability group (i.e., they are not independent of one another).*

Our null hypothesis is:

■ *Gender has no significant effect on membership in an ability group (i.e., they are independent of one another).*

By looking at these hypotheses you can see we have two independent variables: ability group and gender. Ability group has three levels: low, middle, and high; and gender has two: male and female. Just like with the factorial ANOVA, this means we are dealing with a 3 × 2 chi-square table.

Our dependent variable is the actual number of occurrences of each of these values. As we just saw, there is a total of 200 students—131 males and 69 females broken into the three separate ability groups. There are 40 males and 24 females in the low-ability group, 35 males and 24 females in the middle-ability group, and 56 males and 21 females in the high-ability group.

Computing Chi-Square for the Test of Independence

Since we have more than one variable, we have to compute chi-square a bit differently. First, we will look at how we determine our expected values; after that we will look at computing the cell and overall chi-square values, and then we will worry about the degrees of freedom.

Computing Expected Values for the Test of Independence

We will use our table of observed values to compute our expected values using the following formula:

$$E = \frac{(\text{Row Total})(\text{Column Total})}{\text{Sample Size}}$$

Using this equation, we can compute the expected value for the first cell (i.e., low-ability group males) by inserting our row total of 64, our column total of 131, and then our sample size of 200 into the formula. We then go through the following three steps.

1. $E = \dfrac{(64)(131)}{200}$

2. $E = \dfrac{8384}{200}$

3. $E = 41.9$

Once we computed the expected value for each observed value, the resulting table, Table 10.10, is sometimes called a *contingency table* since we want to decide if placement in one group is contingent on placement in a second group. If the expected proportions in each cell are not significantly different from the actual proportions, then the groups are independent; in this case, it would mean that gender does not have an effect on ability group. If the differences are significant, then we will say the two variables are dependent on one another. That would mean gender does have an effect on ability group (Table 10.10).

TABLE 10.10. Contingency Table for the Chi-Square Test of Independence

	Male	Female	Total
Low-ability group	Observed = 40 Expected = 41.9	Observed = 24 Expected = 22.1	Observed = 64 Expected = 64.0
Middle-ability group	Observed = 35 Expected = 38.6	Observed = 24 Expected = 20.4	Observed = 59 Expected = 59
High-ability group	Observed = 56 Expected = 50.4	Observed = 21 Expected = 26.6	Observed = 77 Expected = 77
Total	Observed = 131 Expected = 131	Observed = 69 Expected = 69	Observed = 200 Expected = 200

Computing the Chi-Square Value for the Test of Independence

Since all of the observed values are fairly close to the expected values, things might not be as bad as some parents think. In order to be sure, we will compute the chi-square values for each of the cells and then add those together to get the overall chi-square value. We will use the same formula we used with the goodness-of-fit test:

$$\chi^2 = \Sigma \frac{(O-E)^2}{E}$$

Unlike the goodness-of-fit test, now we will focus on each individual cell rather than an entire row. This is shown in Table 10.11.

1. Subtract the expected value in each cell from the observed value in each cell (column A in the table).

2. Square that value (column B).

3. Divide that value by the expected value for that cell; this gives us the chi-square value for that cell (column C).

4. Add the chi-square values for each cell to get the overall chi-square value (the bottom of column C); this is shown in Table 10.11.

TABLE 10.11. Computing the Chi-Square Statistic

	Observed – expected	Observed – expected squared	Chi-square = Column 2 ÷ number expected
Low-ability males	40 – 41.9 = –1.9	3.61	3.61/41.9 = .08
Low-ability females	24 – 22.1 = 1.9	3.61	3.61/22.1 = .16
Middle-ability males	35 – 38.6 = –3.6	12.96	12.96/38.6 = .36
Middle-ability females	24 – 20.4 = 3.6	12.96	12.96/20.4 = .64
High-ability males	56 – 50.4 = 5.6	31.36	31.36/50.4 = .62
High-ability females	21 – 26.6 = –5.6	31.36	31.36/26.6 =1.18
			Computed chi-square = 3.02

Determining the Degrees of Freedom for the Test of Independence

Here is the formula for the degrees of freedom for the test of independence:

$$df = (R - 1)(C - 1)$$

In this formula, C represents the number of columns in the contingency table; in this case we have two, one for female and one for male. R represents the number of rows; we have three, one for each ability group. When we subtract one from each of these values and then multiply them together, we wind up with 2 degrees of freedom.

We Are Finally Going to Test Our Hypothesis

As usual, to test our hypothesis, we have to compare our computed value of chi-square to the appropriate critical value. In the table, the critical value for 2 degrees of freedom is 5.99, greater than our computed value of 3.02. This means our p value would be greater than or equal .05. As a result, we fail to reject the null hypothesis. Apparently, ability group and gender are independent of one another.

You might be thinking, "This is great, but what does it mean? We have failed to reject the null, but what's the bottom line? What does this mean to the complaining parents?" The answer is very straightforward; all we are saying is that gender and their ability level are independent of one another. There is about the same proportion of males and females in each of the ability groupings. To get even more specific, let's look at the percentages of each group in Table 10.12.

TABLE 10.12. Percentage of Each Gender within Ability Group

	Male	Female	Total
Low-ability group	62.5%	37.5%	100%
Middle-ability group	59.3%	40.7%	100%
High-ability group	72.7%	27.3%	100%
Total	65.5%	34.5%	100%

Here, the actual number of boys and girls in each ability group has been replaced by the percentage of the particular group they represent. For example, out of all of the students in the low-ability group, 62.5% of them are males; obviously the rest are females. In the high-ability group, 72.7% are males while only 27.3% are females. When you first look at this, it does seem like there is some inequality. You cannot, however, go by just the cell values; you also have to compare your cell percentages to the overall percentages.

For example, our sample of 200 kids is 65.5% male; that means you would expect about 65.5% of the students in each group to be male. As you can see, the percentage of males in each group is close to 65.5%. At the same time, the overall percentage of females in the study is 34.5%; again, the cell percentages are close to this value. Because of this, we can say that the two variables, gender and ability level, are independent of one another and any differences we see are purely due to chance. In other words, if we know something about a person's gender, it doesn't tell us a thing about her ability level.

Suppose, however, the p value had been less than .05 and we were able to reject the null hypothesis, thereby saying there was some dependence between the two variables. Have we seen a situation like that before? Sure we have. In essence, what we would be stating is that the two variables interact with one another. This is the same type of phenomenon we saw with the factorial ANOVA, but this time we are using nominal data.

Checking What We Just Computed with SPSS

In order to use SPSS to compute the chi-square test of independence, we first set up our data as shown in Figure 10.16. In this case we used a value of 1 for male and 2 for female and used values from 1 to 3 to represent low-, middle-, and high-ability groups. At that point, we select Analyze, Descriptive Statistics, and Crosstabs.

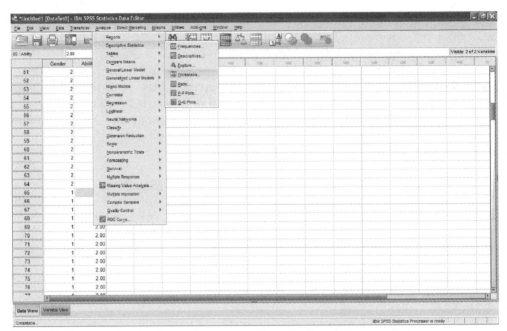

FIGURE 10.16. Selecting the Crosstabs command on the Data View spreadsheet.

Following that, as shown in Figure 10.17, we select Gender as our Row and Ability Group as our Column; we also ask for Descriptive Statistics. Figures 10.18 and 10.19 give us the output.

Because of rounding issues, the chi-square value here is slightly different from that we computed manually, but the p value of .220, when divided by 2 (i.e., .110), still shows that we fail to reject our null hypothesis. This means, as we said earlier, that gender and ability level are independent of one another; any differences we see are purely due to chance.

The Corporal Punishment Conundrum

The principal of a high school in a small town has recently started using corporal punishment in his school. Many of the residents are opposed to this plan because they feel that minority children are at the highest risk for being punished in this manner. Because many minority students are already at risk for failure or dropout, concerned community members have asked him to investigate their assertion. The principal assures them there is no problem but agrees to investigate.

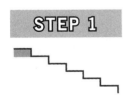

Identify the Problem

Obviously, this is a problem that should be investigated, and it's certainly ethical to collect numeric data representing the number of cases of corporal punishment within each ethnic group. It's within the principal's authority to investigate this issue, and

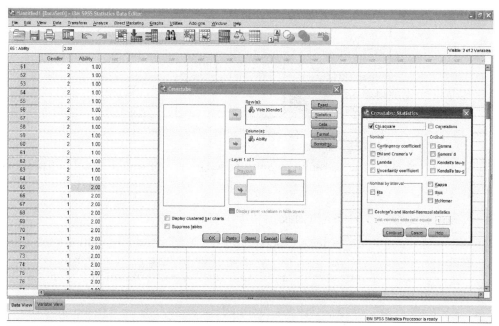

FIGURE 10.17. Identifying the rows and columns for the Crosstabs command.

it is in the best interest of the students. Given that, let's move forward by stating the problem:

> *Corporal punishment has been shown to negatively affect achievement,*
> *motivation, and persistence in school. To exacerbate the problem, studies*
> *have shown that it is disproportionately applied to minority students, many*
> *of whom are already at risk for failure. It is felt that the relationship between*
> *the frequency of corporal punishment and minority groups be investigated at*
> *the high school in question.*

Gender * Ability Group Crosstabulation

Count

		Ability Group			Total
		Low Ability	Middle Ability	High Ability	
Gender	Male	40	35	56	131
	Female	24	24	21	69
Total		64	59	77	200

FIGURE 10.18. Counts of gender within ability group.

Chi-Square Tests

	Value	df	Asymp. Sig. (2-sided)
Pearson Chi-Square	3.031[a]	2	.220
Likelihood Ratio	3.073	2	.215
N of Valid Cases	200		

a. 0 cells (.0%) have expected count less than 5. The minimum expected count is 20.36.

FIGURE 10.19. Output of the Crosstabs command used for the chi-square test of independence.

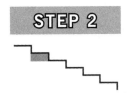

State a Hypothesis

In order to investigate our hypothesis, we will ask 20 students from three different ethnic groups whether they have been corporally punished. Our null hypothesis will be:

> ■ *Membership in an ethnic group has no significant effect on being corporally punished (i.e., they are independent of one another).*

Our research hypothesis would read:

> ■ *Membership in an ethnic group has a significant effect on being corporally punished (i.e., they are not independent of one another).*

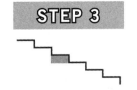

Identify the Independent Variable

Here we have two independent variables. In the first, our independent variable is ethnic group; for the sake of our discussion we'll call them A, B, and C. The second independent variable is whether or not the student has been punished.

Identify and Describe the Dependent Variable

We can call our dependent variable "Punishment" and, again, there are two possible responses—yes or no. We will use the data in Table 10.13 to represent our participants' responses.

If we wanted to, we could use a histogram to display our data; in this case, we can use the histogram in Figure 10.20 to help us see the relationship between the ethnic groups and their level of support.

TABLE 10.13. Punishment within Ethnic Group

		Yes	No	Total
Group A	Count	5	15	20
Group B	Count	14	6	20
Group C	Count	6	14	20
Total	Count	25	35	60

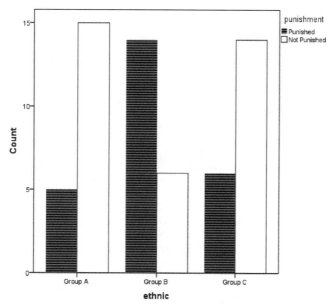

FIGURE 10.20. Bar chart showing counts of punishment within ethnic group.

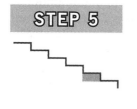

Choose the Right Statistical Test

Herein we have two independent variables, ethnic group and punishment. Because the dependent variable represents the number of occurrences of each level within the variables, we will use the chi-square test of independence (i.e., the factorial chi-square).

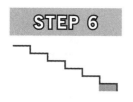

Use Data Analysis Software to Test the Hypothesis

Figure 10.21 provides us with the contingency table showing the observed counts, while Figure 10.22 shows us the results of the chi-square test.

Our p value of .007 is well below our alpha value of .05, so we must reject our null hypothesis; the use of corporal punish-

Count

		punishment		Total
		Punished	Not Punished	
Ethnic	Group A	5	15	20
	Group B	14	6	20
	Group C	6	14	20
Total		25	35	60

FIGURE 10.21. Counts of punishment within ethnic group.

Chi-Square Tests

	Value	df	Asymp. Sig. (2-sided)
Pearson Chi-Square	10.011	2	.007
Likelihood Ratio	10.141	2	.006
Linear-by-Linear Association	.101	1	.750
N of Valid Cases	60		

FIGURE 10.22. Inferential statistics for the chi-square test of independence.

ment is not independent of ethnic group. This just means that corporal punishment is used inordinately among different ethnic groups.

Post-Hoc Tests Following the Chi-Square

As we said earlier, explaining and running the post-hoc tests for the chi-square tests is somewhat difficult. Some statistical software packages will do it for us by running tests that determine the nonproportional cells by using techniques called simultaneous confidence-interval procedures, while others will use a modification of the Bonferroni technique (the same tool we talked about when we discussed the ANOVA). If you're lucky enough to have software that does this, you do not have a problem.

In most instances, however, the software will not do this, so you have to compare, using the chi-square test, each cell to each of the others to find the nonproportional scales. In our last example, this would have involved making six separate comparisons. Not only is this time consuming, but it also increases the probability of a Type I error (i.e., rejecting the null hypothesis when we shouldn't) since, as we said when we discussed the ANOVA, the alpha values of .05 are cumulative—the more tests we do, the larger alpha gets.

There is some good news, however. We can look at the table and begin to see where the significant differences lie. For example, we can see that groups A and C have far lower levels of punishment than does Group B. Obviously, that means that groups A and C have a much larger number of students who say they have not been corporally punished. While not as exact as a post-hoc test, these observations will go a long way toward helping the researcher decide how and why the p value was so small.

The Case of Type of Instruction and Learning Style

Throughout the world, more and more colleges and universities are introducing distance learning programs as a viable alternative for students who need educational programs that are time and place independent. Some people are critical of these programs because they feel they may not equally support all learning styles and therefore may lead to failure in such a learning environment. In order to investigate their concerns, let's use learning style and attrition information from a sample of 80 students. By analyzing the interaction of learning style and attrition, we will be able to determine if the concerns about distance education are warranted.

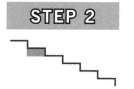

STEP 1

Identify the Problem

This is definitely a critical problem since I work primarily in a distance learning environment. I am interested in this, I have the ability, time, resources, and so on, to thoroughly investigate it, and it is certainly ethical. All criteria seem to be met, so let's state our problem and see if I get to keep my job!

■ *In an effort to meet the needs of students who cannot be time and place dependent, distance education programs are offered at most institutions of higher education throughout the United States. Many critics believe that a student's chance of failure from these programs may be higher because of a conflict between their learning style and the manner by which the course is delivered.*

STEP 2

State a Hypothesis

Let's use the following null hypothesis:

■ *Learning style has no significant effect on performance in different learning environments (i.e., they are independent of one another).*

The research hypothesis is

■ *Learning style has a significant effect on performance in different learning environments (i.e., they are not independent of one another).*

STEP 3

Identify the Independent Variable

Let's use learning style as our first independent variable. For the purpose of this example, we will assume we are using a popular learning style inventory where students answer a series of questions and are then classified into one of four learning styles: Convergers, Divergers, Assimilators, and Accommodators. We won't get into the meaning of these four styles; we will just consider them to be the four levels of our independent variable. Our second independent variable, status, has two levels: success or failure.

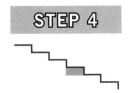

STEP 4

Identify and Describe the Dependent Variable

In this case, we will define success in the program as having graduated. Let's assume we have already created the dataset and we have created the following contingency table shown in Figure 10.23 and the bar chart shown in Figure 10.24.

		Learning Style				Total
		Accomodator	Assimilator	Diverger	Converger	
Graduate	Graduate	9	16	12	10	47
	Non-graduate	5	12	6	10	33
Total		14	28	18	20	80

FIGURE 10.23. Counts of graduation status within learning style.

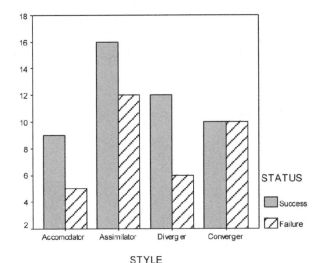

STYLE

FIGURE 10.24. Bar chart showing counts of graduation within learning style.

Although we can't be sure, it looks like there might be a difference in success rates. Since their success and failure rates are equal, it seems the Convergers have proportionately a larger number of students failing to finish the program than students with other learning styles. Let's move forward and see if our suspicions are right.

STEP 5

Choose the Right Statistical Test

In this case, our dependent variable represents nominal data: the count of each student for a given learning style and their graduation status. Since we have two independent variables, we should use the chi-square test of independence.

STEP 6

Use Data Analysis Software to Test the Hypothesis

The results of the chi-square test are shown in Figure 10.25.

Despite our observations, because of our relatively large p value, we cannot reject our null hypothesis; success in the distance education program appears to be independent of a student's learning style. Interestingly, this is data I have collected from distance education students over the past 15 years. Believe it or not, and contrary to what we might think, it seems that learning style doesn't make a difference in distance education.

Chi-Square Tests

	Value	df	Asymp. Sig. (2-sided)
Pearson Chi-Square	1.304	3	.728
Likelihood Ratio	1.308	3	.727
Linear-by-Linear Association	.377	1	.539
N of Valid Cases	80		

FIGURE 10.25. Inferential statistics for chi-square test of independence.

Summary

The chi-square test, more than likely, is the only nonparametric test that most beginning statisticians will ever use. The key points to remember are that we are analyzing nominal (i.e., categorical) data and, unlike the earlier tests we have looked at, we are not going to use the chi-square test to compare means or variances. Instead, we are going to compare the frequency of values in a data distribution; we will either compare this observed frequency to an expected frequency that we specify or an expected frequency that has been predetermined. As was the case with the analysis of variance, we can use the chi-square test for either one independent variable (i.e., the goodness-of-fit test) or multiple independent variables (i.e., the test of independence).

Do You Understand These Key Words and Phrases?

chi-square goodness-of-fit test

contingency table

goodness of fit

one-sample chi-square test

chi-square test of independence

expected value

observed value

Do You Understand These Formulas?

Computed value of chi-square:

$$\chi^2 = \Sigma \frac{(O-E)^2}{E}$$

Degrees of freedom for goodness of fit:

$$df = (\text{number of entries}) - 1$$

Degrees of freedom for chi-square test of independence:

$$df = (R - 1)(C - 1)$$

Expected values for chi-square test of independence:

$$E = \frac{(\text{Row Total})(\text{Column Total})}{\text{Sample Size}}$$

Quiz Time!

As you've just seen, working with the chi-square test is very easy. Before we move forward, however, take a look at each of the following scenarios. After you've read through these cases and answered the questions, you can check your work at the end of the book.

The Case of Prerequisites and Performance

A psychology professor in a graduate school was concerned that the grade distribution for his class was different for students whose undergraduate degree was in psychology and for students whose major was in other areas. He collected and analyzed data that he had collected from his classes and wound up with the results shown in Figures 10.26 and 10.27.

		Grade					
		A	B	C	D	F	Total
College Major	Psychology	24	10	7	6	7	54
	Other	10	15	13	11	12	61
Total		34	25	20	17	19	115

FIGURE 10.26. Counts of grades within major.

Chi-Square Tests

	Value	df	Asymp. Sig. (2-sided)
Pearson Chi-Square	10.966	4	.027
Likelihood Ratio	11.172	4	.025
Linear-by-Linear Association	6.697	1	.010
N of Valid Cases	115		

FIGURE 10.27. Inferential statistics for chi-square test of independence.

1. What are the null and research hypotheses the professor is investigating?
2. What are the independent variables and their levels?
3. What is the dependent variable?
4. Which chi-square test would he use to test the hypothesis?
5. What does this tell the professor about undergraduate major and performance in a graduate school psychology class?

The Case of Getting What You Asked For

I once met a professor who only used a three-letter grading system. Everyone with an average numeric grade greater than 1 standard deviation above the mean makes an A; averages falling between the mean ± 1 standard deviation are assigned a B, and anything lower than that receives a C. Using the empirical rules, this means that about 68% of students should receive a B, 16% should receive an A, and 16% should receive a C. One day I heard his class complaining about what they perceived as an inordinate number of B grades. The professor responded by showing the students the output presented in Figures 10.28 and 10.29.

Grade

	Observed N	Expected N	Residual
A	10	16.0	-6.0
B	67	68.0	-1.0
C	23	16.0	7.0
Total	100		

FIGURE 10.28. Counts of grades.

Test Statistics

	Grade
Chi-square	5.327
df	2
Asymp. Sig.	.070

FIGURE 10.29. Inferential statistics for chi-square goodness of fit.

1. What are the null and research hypotheses the professor is testing?

2. What is the independent variable and its levels?

3. What is the dependent variable?

4. Which chi-square test would he use to test the hypothesis?

5. What does this tell his class about the distribution of grades according to his grading scheme?

The Case of Money Meaning Nothing

In this case, imagine we are working with a group that wants to build a clinic to work with drug-addicted clients. After picking the property and going to the city for approval they find that many local citizens have complained that they do not want such a facility in their neighborhood. "After all," they said, "that means we will have a lot of indigent people wandering around; they are the ones with drug problems!" In order to investigate their concerns, let's imagine we have collected relevant data and have produced Figures 10.30 and 10.31.

		Income					Total
		< $20,000	$20,000 to $39,999	$40,000 to $59,999	$60,000 to $79,999	$80,000 and higher	
Drug Use	Yes	22	27	33	21	24	127
	No	5	6	7	9	7	34
Total		27	33	40	30	31	161

FIGURE 10.30. Counts of drug use within income group.

Chi-Square Tests

	Value	Df	Asymp. Sig. (2-sided)
Pearson Chi-Square	2.055	4	.726
Likelihood Ratio	1.960	4	.743
Linear-by-Linear Association	.715	1	.398
N of Valid Cases	161		

FIGURE 10.31. Inferential statistics for chi-square test of independence.

1. What are the null and research hypotheses we are investigating?

2. What are the independent variables and their levels?

3. What is the dependent variable?

4. Which chi-square test would we use to test this hypothesis?

5. What would we tell the persons concerned with building the clinic in their neighborhood?

The Case of Equal Opportunity

A clothing manufacturer, always aware that it is both federally mandated as well as morally the right thing to do, wants to ensure that gender distribution among employees is about the same as the national average. With a little research, he finds that females represent 45% of the population but only 41% of workers in the company. Rather than just push it aside as a fluke, the manager decides to look at the SPSS output shown in Figures 10.32 and 10.33.

Employee Gender

	Observed N	Expected N	Residual
Female	41	45.0	-4.0
Male	59	55.0	4.0
Total	100		

FIGURE 10.32. Observed and expected values of gender.

Test Statistics

	Employee Gender
Chi-square	.646
df	1
Asymp. Sig.	.421

FIGURE 10.33. Inferential statistics for chi-square goodness-of-fit test.

1. What are the null and research hypotheses the manager is investigating?

2. What is the independent variable and its levels?

3. What is the dependent variable?

4. Which chi-square test would the manager use to test this hypothesis?

5. What would the manager learn about gender representation in the company?

CHAPTER 11

The Correlational Procedures

Introduction

When we discussed descriptive statistics, we talked about the idea of using scatter-plots to look at the relationship between two sets of quantitative data. We saw cases where two values, votes and campaign spending for example, might have a positive relationship—higher levels of spending leads to a greater number of votes. In another case, we might collect data about the amount of time people exercise on a daily basis and their blood pressure. In this case, we might find a negative relationship: when one value goes up (i.e., the amount of exercise) the other value goes down (i.e., blood pressure). In this chapter we're going to expand on that idea by learning how to numerically describe the degree of relationship by computing a correlation coefficient. We'll continue by showing how to use scatterplots and correlation coefficients together to investigate hypotheses. The operative word here is "investigate"; that word will be important in just a few minutes.

■ Understanding the Idea of Correlations

As we saw when we discussed scatterplots, when we look at the relationship between two sets of data, one of three things can happen. First, as the values in one dataset get larger, so can the values in the other dataset. Second, as the values in one data-set get larger, the values in the other dataset can get smaller. Third, as the values in one dataset change, there might be no apparent pattern in the change of values in the other dataset. This idea, called the *correlation* between two variables, is shown in Table 11.1.

In order to better understand correlation, let's look at a case we used when we discussed scatterplots. In this case, we investigated whether the amount of tuition reimbursement spent by a company in a given year's time was correlated with the number of resignations during that same time period. As a reminder, the management of the company was concerned that more and more employees were taking advantage of that benefit, becoming better educated and moving to another company for a higher salary. Using the following data, the company's manager decided to investigate whether such a relationship really existed. In Table 11.2, each row represents a given year, the amount of tuition for that year, and the number of resignations for that same year.

TABLE 11.1. Understanding Correlation

Both values go up		One value goes up, the other goes down		There is no clear pattern	
Variable 1	Variable 2	Variable 1	Variable 2	Variable 1	Variable 2
1	5	1	9	1	3
2	6	2	8	2	5
3	7	3	7	3	2
4	8	4	6	4	4
5	9	5	5	5	1

TABLE 11.2. Tuition and Resignations by Year

Year	Tuition (thousands)	Resignations
1	$32	20
2	$38	25
3	$41	28
4	$42	25
5	$50	30
6	$60	32
7	$57	40
8	$70	45
9	$80	45
10	$100	90

You can see that there is no easily identified dependent variable. Since the correlation is a descriptive tool, we are interested in using our numeric statistics to look at the relationship between two variables, tuition and resignations. As we just said, management is worried that the more money they spend, the greater the number of resignations. As we said, though, that is only one of three possibilities. First, if they are right, as the amount of tuition reimbursement goes up, so will the number of resignations. A second possibility is that, when the amount of tuition money goes up, the number of resignations goes down. The final option is that there will be no obvious pattern between the number of resignations and the amount of tuition money.

In situations like this, we can investigate these relationships with any of several correlational statistics. The key in determining which correlation tool to use depends on the type of data we have collected. In this case, both resignations and tuition spent are quantitative (i.e., interval or ratio data). Because of that, we will use the *Pearson product–moment correlation coefficient* (most people call it *Pearson's r*). Karl Pearson was interested in relationships of this type and developed a formula that allows us to calculate a correlation coefficient that numerically describes the relationship between two variables. Here is his formula; we will go through it step by step and soon be using it with no problem.

$$r = \frac{n\left(\sum xy\right) - \left(\sum x * \sum y\right)}{\sqrt{\left[n\sum x^2 - \left(\sum x\right)^2\right]\left[n\sum y^2 - \left(\sum y\right)^2\right]}}$$

First, let's look at the data we'll use in Table 11.3, specifically columns 2 and 3.

TABLE 11.3. Tuition and Resignations

Tuition (thousands)	Resignations
$32	20
$38	25
$41	28
$42	25
$50	30
$60	32
$57	40
$70	45
$80	45
$100	90

If you closely examine these data, you can see that, generally speaking, as tuition goes up, so do resignations. In order to continue with our calculations, we will modify Table 11.3 and create Table 11.4 and refer to tuition as *x* and resignations as *y*; we will label these as columns A and C.

TABLE 11.4. Values Needed for Computing the Pearson Correlation Coefficient

x	*y*
32	20
38	25
41	28
42	25
50	30
60	32
57	40
70	45
80	45
100	90
(A)	(C)

Let's make another modification and create Table 11.5 by adding two columns where we can enter the values when we square each of the x and y values and one column where we can put the product when we multiply each value of x by each value of y. We are also going to put a row in at the bottom where we can sum everything in each of the columns; these new columns will be labeled B, D, and E.

TABLE 11.5. Values Needed for Computing the Pearson Correlation Coefficient

x	x^2	y	y^2	xy
32	1024	20	400	640
38	1444	25	625	950
41	1681	28	784	1148
42	1764	25	625	1050
50	2500	30	900	1500
60	3600	32	1024	1920
57	3249	40	1600	2280
70	4900	45	2025	3150
80	6400	45	2025	3600
100	10000	90	8100	9000
$\Sigma x = 570$ (A)	Σx^2 36562 (B)	3 $\Sigma y = 380$ (C)	$\Sigma y^2 = 18108$ (D)	$\Sigma xy = 25238$ (E)

In this case, remember that n is the number of pairs of data in the equation, in this case 10. Knowing that, we now have everything we need to include in the formula and we can compute Pearson's r using the following steps. Notice we are just taking the values from the bottom row of Table 11.5 and inserting them into the formula; again, we have these columns labeled A–E.

1. $r = \dfrac{10(E) - (A*C)}{\sqrt{[10*B - (A)^2][10*D - (C)^2]}}$

2. $r = \dfrac{10(25238) - (570*380)}{\sqrt{[10*36562 - (570)^2][10*18108 - (380)^2]}}$

3. $r = \dfrac{252380 - 216600}{\sqrt{[365620 - 324900][181080 - 144400]}}$

4. $r = \dfrac{(252380 - 216600)}{\sqrt{40720*36680}}$

5. $r = \dfrac{35780}{\sqrt{1493609600}}$

6. $r = \dfrac{35780}{38647.25}$

7. $r = .926$

We could, of course, use SPSS to analyze the same data; we would start by inputting our data and selecting Analyze, Correlate, and Bivariate (i.e., two variables) as shown in Figure 11.1.

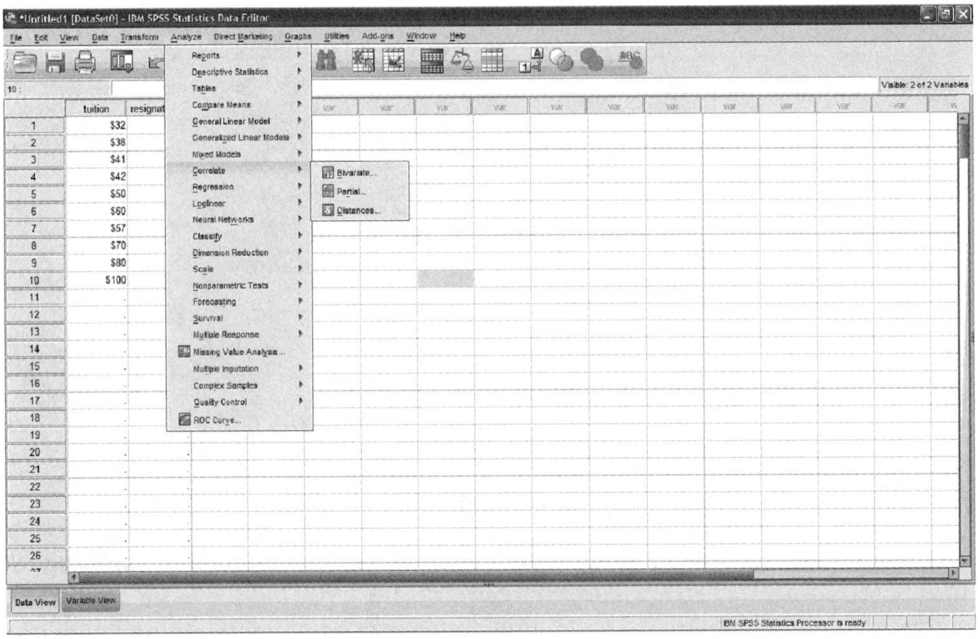

FIGURE 11.1. Using the Correlate and Bivariate options on the Data View spreadsheet.

In Figure 11.2, we identify the two fields we want to correlate—tuition and resignations. Obviously, since we're dealing with two quantitative variables, we select Pearson's correlation.

These commands would produce the table in Figure 11.3. As you can see in the figure, Pearson's r is exactly what we calculated by hand. You can also see the software has given us a p value (i.e., Sig. 2-tailed) of zero; this indicates a significant correlation. We'll discuss that shortly, but for now let's focus on the r value itself.

▨ Interpreting Pearson's *r*

In this case our r value is .926 but it could have ranged from –1 to +1. When you're computing a correlation, three things can happen:

1. If both values go up, or if both values go down, it is a positive correlation. The greater the positive correlation, the closer r is to +1.

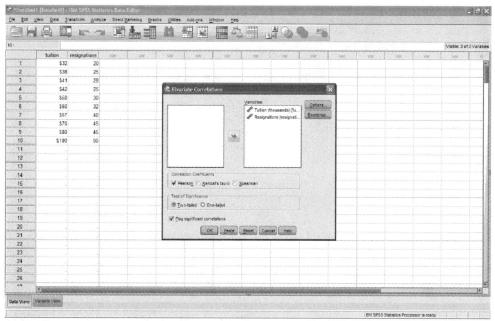

FIGURE 11.2. Selecting the variables to be correlated on the Data View spreadsheet.

Correlations

		Tuition (in thousands of dollars)	Resignations
Tuition (in thousands of dollars)	Pearson Correlation	1	.926**
	Sig. (2-tailed)		.000
	N	10	10
Resignations	Pearson Correlation	.926**	1
	Sig. (2-tailed)	.000	
	N	10	20

**. Correlation is significant at the 0.01 level (2-tailed).

FIGURE 11.3. Positive Pearson correlation coefficients.

2. If one value goes up and the other goes down, they are negatively correlated. The greater the negative correlation, the closer r moves to –1.

3. If r is close to zero, it is indicative of little or no correlation; there is not a clear relationship between the values.

A Word of Caution

A positive correlation exists when the values of two variables both go up or if both values go down. Only when one variable goes up and the other goes down do we say there is a negative correlation. Do not make the mistake of labeling a case in which both variables go down as a negative correlation. The three possibilities are shown below in Table 11.6.

TABLE 11.6. Relationship between Positive and Negative Correlations

↓↑	↓↓	↑↑
Negative correlation	Positive correlation	Positive correlation
One value goes up, the other value goes down.	Both values go down.	Both values go up.
Pearson's r is between zero and –1.	Pearson's r is between zero and +1.	Pearson's r is between zero and +1.

In the preceding case, we have a very high r value of .926, but what does that really mean? How high does the r value have to be for us to consider the relationship between the variables to be meaningful? The answer to this question is that the interpretation of the r value depends on what the researcher is looking for.

Any time we analyze data, an r value of .90 or greater would be excellent for a positive correlation and a value less than –.90 would be great for a negative correlation. In both instances we could clearly see that a relationship exists between the variables being considered. At other times, we might consider an r value in the .60s and .70s (or –.60 to –.70) to be sufficient; it would just depend on the type of relationship we were looking for. Obviously, if our correlation coefficient is smaller than this, especially anything between zero and .50 for a positive correlation or between –.50 and zero for a negative correlation, it is apparent that a strong relationship apparently does not exist between the two variables.

An Even More Important Word of Caution!

We just saw that a strong relationship existed between the amount of tuition money spent and the number of resignations. Many beginning statisticians make the mistake of thinking that a positive correlation means one thing caused the other to happen. While you might be able to make an inference, you cannot be absolutely sure of a cause-and-effect relationship; it might be due to chance.

For example, did you know there is a large correlation between the number of churches in a city and the number of homicides in the same city? While you may be thinking "blasphemy," it's true—a large number of churches is highly correlated with a large number of homicides. Think carefully about what I just said, though. Should people insist that churches be torn down in order to lower the homicide rate? Of course not! While the values are correlated, they certainly aren't causal. The real reason underlying the correlation is the size of the town or city. Larger cities have a greater number of both churches and homicides; smaller towns have a smaller number of each. They are related to each other but not in any causal way.

This is a perfect example of being a good consumer of statistics. You can see quite a few of these noncausal correlations being presented in various newspapers, magazines, and other sources; in a lot of cases, people are trying to use them to bolster their point or get their way about something. What's the lesson? Simple, always look at the big picture.

Now, let's use the same dataset but create Table 11.7 by changing the resignation numbers just a bit. Again, just by looking, we can see that an apparent relationship exists. In this case, it appears that the number of resignations goes down as the amount of tuition money spent goes up. Let's check that using the information given in Figure 1.4.

TABLE 11.7. Tuition and Resignation Data

Tuition (thousands)	Resignations
$32	40
$38	35
$41	28
$42	25
$50	20
$60	22
$57	15
$70	25
$80	10
$100	30

Correlations

		Tuition	Resignations
Tuition	Pearson Correlation	1	-.419
	Sig. (2-tailed)		.228
	N	10	10
Resignations	Pearson Correlation	-.419	1
	Sig. (2-tailed)	.228	
	N	10	10

FIGURE 11.4. Negative Pearson correlation coefficients.

Here SPSS computed an r value of $-.419$ and, since it is between $-.5$ and zero, it is a rather small negative correlation. While there is a correlation between the money spent and the resignations, it is not very strong. We can also see that the r value is negative. This indicates an inverse relationship: when one of the values goes up, the other value goes down. In this case we can see that, as the amount of tuition money spent gets larger, there are fewer resignations. Remember, although a correlation exists here, there may or may not be a causal relationship; be careful how you interpret these values.

Now look at Table 11.8; do you see a pattern emerging? Does there seem to be a logical correlation between the two sets of values? If you look closely, you'll see there does not appear to be. Sometimes a value in the left column will have a much greater value in the right column (e.g., 41 and 58), and sometimes the value in the right column will be much lower (e.g., 32 and 20). As you can see in the SPSS output in Figure 11.5, because of the lack of a pattern, Pearson's r is only .013; this supports our observation that there does not seem to be a meaningful correlation between the two sets of values. Remember, though, this might be purely coincidental; you cannot use these results to infer cause and effect.

TABLE 11.8. Tuition and Resignation Data

Tuition (thousands)	Resignations
$32	20
$38	35
$41	58
$42	25
$50	22
$60	72
$57	4
$70	15
$80	90
$100	5

Correlations

		Tuition	Resignations
Tuition	Pearson Correlation	1	.013
	Sig. (2-tailed)		.971
	N	10	10
Resignations	Pearson Correlation	.013	1
	Sig. (2-tailed)	.971	
	N	10	10

FIGURE 11.5. Pearson correlation coefficients showing a weak positive correlation.

■ A Nonparametric Correlational Procedure

It stands to reason that, if we can compute a correlation coefficient for quantitative data, we should be able to do the same for nonparametric data. For beginning statisticians, the only nonparametric correlation we generally use is the *Spearman rank-difference correlation* (most often called *Spearman's* rho). Given its name, it only works with rank-level (i.e., ordinal) data. Let's use the data in Table 11.9 as an example.

TABLE 11.9. Employee and Management Rank of Issues

Issue	Employee rank	Management rank
A	3	3
B	4	7
C	7	9
D	8	1
E	1	5
F	5	2
G	2	6
H	6	4
I	10	10
J	9	8

In this case, let's imagine we're members of the board of directors of a company where several issues have arisen that seem to be affecting employee morale. We've decided that our first step is to ask both the management team and employees to rank these issues in terms of importance. We then want to determine if there is agreement between the two groups in terms of the importance of the issues labeled A–J.

Here we have three columns: the first column represents the 10 issues facing the company (i.e., A–J) that employees and management are asked to rank. The second column shows how the issues were ranked by the employees. The third column shows how the issues were ranked by the management. We can use the Spearman rank-difference correlation formula to help us determine if the employees and management ranked them in a similar order:

$$r_s = 1 - \frac{6\left(\sum d^2\right)}{n(n^2 - 1)}$$

This formula is more straightforward than the Pearson formula, but there are components we have not used up to this point. First, r_s is the symbol for Spearman's rho; that's the correlation coefficient we are interested in computing. Notice, we're using the same symbol as Pearson's r except we're adding a lower-case s to differentiate it. Second, the number 6 is exactly that: a constant value of 6. Next, the lower-case n is the number of objects we are going to rank; in this case we have 10.

Computing the square value of all of the rankings (i.e., d^2) involves first determining the difference between the employee and management rankings; this is shown in the column labeled d in Table 11.10. For example, the employees' ranking for issue D is an 8; management's ranking for that same issue is 1; the difference between the two rankings is 7. The d^2 value for that issue is 49 (i.e., 7 * 7); this is nothing more than difference value squared.

We can compute our Spearman's rho coefficient using the following steps:

1. $r_s = 1 - \dfrac{6(108)}{10(100 - 1)}$

2. $r_s = 1 - \dfrac{648}{990}$

3. $r_s = 1 - .655$

4. $r_s = .345$

TABLE 11.10. Necessary Values for Spearman's Rho

Issue	Employee rank	Management rank	d	d^2
A	3	3	0	0
B	4	7	−3	9
C	7	9	−2	4
D	8	1	7	49
E	1	5	−4	16
F	5	2	3	9
G	2	6	−4	16
H	6	4	2	4
I	10	10	0	0
J	9	8	1	1
				$\Sigma d^2 = 108$

Interpreting Spearman's rho is exactly the same as interpreting Pearson's *r*. The value of rho can range from −1 to +1, with values of rho approaching +1 indicating a strong agreement in rankings; values approaching −1 indicate an inverse agreement in rankings. In Figure 11.6 we have a rho value of .345 indicating there is very little agreement in the rankings of the issues between employees and management. As usual, we can use SPSS to verify our manual computations:

Correlations

			Employees	Management
Spearman's rho	Employees	Correlation Coefficient	1.000	.345
		Sig. (2-tailed)	.	.328
		N	10	10
	Management	Correlation Coefficient	.345	1.000
		Sig. (2-tailed)	.328	.
		N	10	10

FIGURE 11.6. Spearman's rho correlation coefficient showing a moderate positive correlation.

The p Value of a Correlation

Pay attention also to the p value shown in the table. Sometimes this confuses the beginning statistician because he wants to use it to reject or fail to reject the hypothesis stated in the second step of the six-step process; that's not what it is for. Remember, the research hypothesis stated in Step 2 for a correlation is simply looking for a meaningful relationship between the two variables under consideration; we use our subjective interpretation of the r value and scatterplot to help us make that decision. The p value shown as part of the output for a correlation tests the hypothesis, "the correlation coefficient in the population is significantly different from zero." By using this hypothesis, statisticians are able to make decisions about a population r value by using sample data. This is very, very rarely done by beginning statisticians, so we won't go into any further detail. Knowing that, let's move forward to our first case study.

The Case of the Absent Students

It seems we have talked a lot about students missing class, haven't we? One would guess I think going to class is important! Perhaps, if I think that way, then I would enjoy looking at the correlation between the number of times students are absent during a semester and their score on the final exam. In order to do that, I could use our six-step procedure.

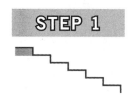

Identify the Problem

In this case, it's apparent that I'm concerned about the number of absences and their relationship to the final grade. Since that meets all of the criteria for a problem I might investigate (e.g., within my control, ethical, etc.), we can state the following problem.

■ *Research has shown a relationship between a student's number of absences and class achievement. This study will investigate whether there is such a relationship between grades and the number of absences in my classes.*

Before we move forward, look at what I just wrote. Is this really a problem? No it may not be; since the correlation measures relationships that aren't necessarily causal, we're simply trying to determine if a problem may exist.

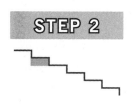

State a Hypothesis

My hypothesis will be that a negative relationship exists between the number of absences and the grade on the final exam. In other words, the fewer days a student misses, the higher their final exam grade will be. Given that, our research hypothesis would read:

> ▪ *There will be a negative correlation between the number of times a student is absent and their grade on the final exam.*

Right off the bat, you can see something different about this hypothesis in that it does not include the word "significant." This is because of three things. First, the correlational procedures are really descriptive statistics. When we learned to calculate the correlations earlier, we saw that they are used to tell us whether or not a linear relationship exists between two variables. Second, despite the fact they are descriptive in nature, many people will state a hypothesis and then use the correlation to investigate rather than test it. Third, because we are not involved in testing a null hypothesis, we are going to subjectively evaluate our research hypothesis. Because of these things, the wording and processes of our six-step model are going to change slightly.

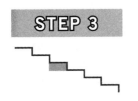

STEP 3 — Identify the Independent Variable

Here's one of those changes I just mentioned. By reading the case, did you notice anything else different about the hypothesis we are investigating? If not, consider the fact that, for the first time, we are not looking at a scenario where we are trying to determine if differences exist between groups of data. Rather, we are trying to decide if a relationship exists between two sets of data. Here, we are trying to determine if it is true that absences from class are directly related to scores on the final exam.

As we said earlier, when we look at relationships of this type, we are looking at the correlation between the two sets of values. In this case, we are trying to determine if a negative correlation exists; is it true that a higher number of absences is related to a lower grade on the final exam, or vice versa? In other cases, we are trying to decide if a positive correlation exists. We saw a good example in the earlier chapters where we looked at the relationship between the amount of tuition money spent and the number of resignations in a given year. In both cases we are trying to see if we can predict the values in one dataset by knowing the values in another.

Since, in these cases, we are not looking at cause and effect, we have to get away from the idea of having an independent variable and a dependent variable. Since we are trying to determine if a relationship exists between two variables, instead of independent and dependent variables, we use the terms *predictor* (i.e., what we are using to help us make a prediction) variable and *criterion* (i.e., what we are trying to predict) variable. In this case, our predictor variable is the number of absences for each student.

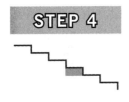

STEP 4 — Identify and Describe the Dependent Variable

Again, since we technically do not have a dependent variable, our criterion variable will be each student's grade on the final examination. Usually at this point we would show the descriptive statistics for that variable, but there's something else that's going to be a bit different about this procedure. Since both our predictor variable and our criterion variable are quantitative, we can show the descriptive statistics for both. Let's use Table 11.11 showing data from 20 students.

TABLE 11.11. Absences and Score Data

Absences	Score	Absences	Score	Absences	Score	Absences	Score
1	100	4	85	7	70	12	70
2	95	6	82	9	90	6	80
4	90	7	80	11	55	7	80
5	90	2	97	2	94	2	93
6	80	7	77	1	90	3	88

Just by looking closely at this table, it appears that our hypothesis may be right on target; it does seem as if the kids with the lower number of absences have higher test scores. Since we have already seen, step by step, how to calculate the descriptive statistics, let's just look at the results from SPSS shown in Figure 11.7.

Descriptive Statistics

	Mean	Std. Deviation	N
Number of Absences	5.2000	3.18880	20
Final Exam Score	84.3000	10.84872	20

FIGURE 11.7. Descriptive statistics for absence and final exam data.

Overall, these do not look too bad; we have an average of 5.2 absences and an average grade of 84.30. Remember, though, these descriptive statistics are not as meaningful to us as when we were looking for significant differences. Let's move forward and try to resolve our hypothesis.

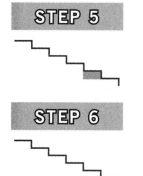

STEP 5

Choose the Right Statistical Test

Since we have two sets of quantitative data and we want to determine if a relationship exists between the two, we will use Pearson's r to test our hypothesis.

STEP 6

Use Data Analysis Software to Test the Hypothesis

In Figure 11.8, we can see we have a Pearson r value of –.836, which indicates a strong negative correlation; it appears that the number of absences is inversely related to a student's final exam score.

In Figure 11.9, the data values have been plotted on the scatterplot with grades plotted across the x (bottom) axis and the corresponding number of absences plotted along the y (left) axis. We instructed SPSS to plot a line of

Correlations

		Number of Absences	Final Exam Score
Number of Absences	Pearson Correlation	1	-.836
	Sig. (2-tailed)		.000
	N	20	20
Final Exam Score	Pearson Correlation	-.836	1
	Sig. (2-tailed)	.000	
	N	20	20

FIGURE 11.8. Negative Pearson correlation coefficients for absence and exam scores.

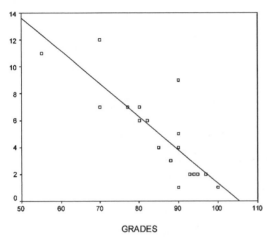

FIGURE 11.9. Scatterplot showing the negative correlation between absences and exam scores.

best fit and it shows a nearly 45-degree angle upward to the left. When we see a line such as this, we immediately know the relationship we have plotted is negative; when one value goes up, the other goes down. In this case, since the line is nearly at a 45-degree angle, we know the relationship is fairly strong. This verifies what we saw with Pearson's r.

Always keep in mind, however, that while both the computed value of r and the scatterplot indicate a strong relationship, don't be fooled. Correlations are just descriptive measures of the relationship; that doesn't necessarily mean that one caused the other to happen.

Another Example: The Case against Sleep

We are not, in every instance, looking for a strong negative or positive relationship; in some cases we want to make sure a relationship doesn't exist. Let's suppose, for example, a group of overzealous parents has determined the number of hours children sleep is directly related to their performance in class. After all, they surmise, how can a sleepy child do well in class? Knowing this, the parents have set up a rigid standard that requires their high school students to get at least 9 hours of sleep per night. Naturally, the students do three things. First, they get aggravated about this "stupid" requirement; second, many of them disobey the edicts of their parents; and third, they will do anything they can, even resort to the use of statistics, to help show there is no relationship between hours slept and class performance!

STEP 1

Identify the Problem

In this case, the students are trying to do something I've already warned you about—they want to try to "prove" there is no relationship between the number of hours they sleep and their performance in class. Remembering that correlations are both descriptive and in many instances do not represent a causal relationship, they should be very careful with what they discover! At the same time, the students do have the ability to investigate this problem, and to them it is important, ethical, and so on. Here is their problem statement:

> A group of parents believes there is a strong relationship between the number of hours per night a student sleeps and the student's academic performance; because of that, they have mandated a minimum number of hours their students must sleep per night. Students do not think this is realistic or fair and want to investigate the validity of their parents' claim.

STEP 2

State a Hypothesis

To begin their argument, the students start with the following hypothesis:

> There will be no correlation between the number of hours slept per night and grade point average.

STEP 3

Identify the Independent Variable

Again, we do not have an independent variable. Instead we are looking at a predictor variable called "number of hours slept."

STEP 4

Identify and Describe the Dependent Variable

Our criterion variable is "grade point average." In Table 11.12, we have data for 32 students. In Figure 11.10, we can use our software to describe both the predictor (i.e., hours) and the criterion (i.e., GPA) variables.

Here we have a B grade point average (i.e., 3.0725). The average hours slept is somewhat low, but remember, those two things aren't important. We are interested in the relationship between the two variables. By looking at the dataset, we can see that there is no apparent relationship between the two. For example, some students who slept 4 hours had a grade point average of 4.00, while others who slept the same amount of time had a grade point average of 2.00. The lowest grade point average (1.15) was shared by two students; one of them slept 4.5 hours, while the other slept over 7 hours. Given this, we will be expecting a small correlation coefficient and a flat line on the scatterplot. Let's see what happens.

TABLE 11.12. Hours Slept and GPA Data

Hours	GPA	Hours	GPA	Hours	GPA	Hours	GPA
4	4.00	5.5	3.00	6.5	3.88	7	3.60
4	2.00	6	4.00	7	1.99	7	3.21
4.5	3.98	6	3.01	7	3.98	7	2.88
4.5	1.15	6	3.12	7	4.00	8	4.00
5	3.33	6	2.88	7	2.98	8	1.84
5	3.46	6	2.75	7	1.15	10	2.83
5	3.80	6.5	2.00	7	3.10	12	3.87
5.5	2.15	6.5	3.45	7	3.84	13	3.09

Descriptive Statistics

	Mean	Std. Deviation	N
GPA	3.0725	.83556	32
Hours Slept	6.6719	1.97814	32

FIGURE 11.10. Descriptive statistics for GPA and hours slept data.

Choose the Right Statistical Test

We have two datasets, each of which contains quantitative data, and we're looking to see if a relationship exists between the two datasets. Given this, we will use Pearson's r.

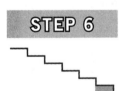

Use Data Analysis Software to Test the Hypothesis

When we enter the data, we will get the output shown in Figures 11.11 and 11.12.

This doesn't look very promising for the parents, does it? We can see the line of best fit is almost parallel with the bottom

Correlations

		GPA	Hours Slept
GPA	Pearson Correlation	1	.084
	Sig. (2-tailed)		.646
	N	32	32
Hours Slept	Pearson Correlation	.084	1
	Sig. (2-tailed)	.646	
	N	32	32

FIGURE 11.11. Pearson correlation coefficient showing a weak correlation between GPA and hours slept.

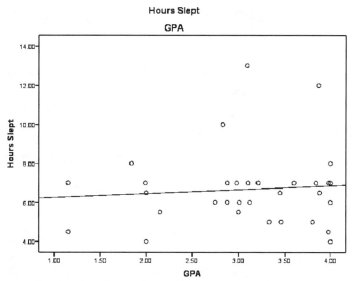

FIGURE 11.12. Scatterplot showing the weak correlation between GPA and hours slept.

of the graph and is nowhere near the 45-degree angle we have seen in the other examples. We can also see, just as we predicted, a very low value for Pearson's r of .084. This means there is almost no correlation between the number of hours students sleep and their grade point average. By using this output and looking back at the data table, this conclusion becomes obvious. Perhaps some of those who slept 4 hours used their time to study, while some others used the time to watch television; we will never know by looking at this data. The only thing we do know is that, despite the fact that correlations never prove cause and effect, the students will be using these results as "scientific proof" that they should be able to stay up later.

The Case of Height versus Weight

Let's look at one last example that's dear to many of us, the relationship between height and weight. We all know that, generally speaking, taller people weigh more than shorter people, but that doesn't always hold true.

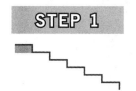

Identify the Problem

In this case, we're interested in looking at something that is usually pretty obvious: tall people generally weigh more than short people. While this is not really a problem per se, it is the type of thing that correlations are used for a lot of times.

▪ *The purpose of this study is to examine the relationship between height and weight.*

State a Hypothesis

This one is easy.

■ *There will be a positive correlation between a person's height and weight.*

Identify the Independent Variable

In this case our predictor variable (i.e., the independent variable) is height. It is quantitative, and, for our purposes, it ranges from 50 inches to 80 inches (i.e., 4′ 2″ to 6′ 8″).

STEP 4

Identify and Describe the Dependent Variable

In this case, our criterion variable is weight. Again, let's arbitrarily use 80 to 320 pounds. We can use the data in Table 11.13 to show the heights and weights of 20 people. The resultant descriptive statistics are shown in Figure 11.13.

TABLE 11.13. Height and Weight Data

Weight	Height	Weight	Height	Weight	Height	Weight	Height
100	60	190	70	120	58	145	68
120	62	150	65	110	63	200	72
130	58	135	69	170	60	180	74
150	70	130	62	160	68	210	76
140	78	170	66	150	69	170	68

Descriptive Statistics

	Mean	Std. Deviation	N
Weight	151.5000	29.82846	20
Height in Inches	66.8000	5.78200	20

FIGURE 11.13. Descriptive statistics for weight and height.

We can see that the 20 participants have an average height of 66.8 inches (i.e., 5′ 6.8″) and an average weight of 151.5 pounds.

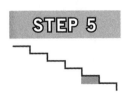

Choose the Right Statistical Test

We have two variables, both representing quantitative data, and we are looking for the relationship between them. Obviously, we are going to use Pearson's r.

STEP 6

Use Data Analysis Software to Test the Hypothesis

SPSS would produce the results shown in Figure 11.14.

It appears we have a fairly strong positive relationship (i.e., $r = .629$); this is verified by the scatterplot where we have a line sloping up to the right at an approximately 60-degree angle (Figure 11.15).

Correlations

		Weight	Height in Inches
Weight	Pearson Correlation	1	.629
	Sig. (2-tailed)		.003
	N	20	20
Height in Inches	Pearson Correlation	.629	1
	Sig. (2-tailed)	.003	
	N	20	20

FIGURE 11.14. Pearson correlation coefficient showing a moderately high correlation between weight and height.

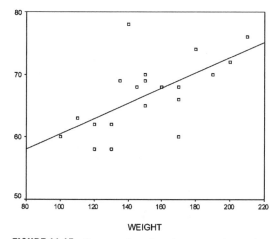

WEIGHT

FIGURE 11.15. Scatterplot showing a moderately high correlation between weight and height.

Based on what we just saw, it seems we can somewhat support our hypothesis. We do have a moderately strong r value, and our line of best fit is sloping up to the right. Again, it is not a perfect 45-degree angle, but it is a lot closer to that than it is to being a flat line. The bottom line is that height seems to be a fairly good predictor of weight but there are obviously other factors that contribute as well.

The Case of Different Tastes

Recently I visited the Coca-Cola museum in Atlanta. During the tour, I was amazed to learn that the formula for Coke differs throughout the world; the museum had dispensers that allowed us to taste different varieties of their product from different countries. I tried some and I was astonished at how bad some of them tasted. "How," I asked, "could anyone enjoy drinking this? Wouldn't it be better if everyone drank the same brand of Coke that we enjoy in the United States?"

Of course, my statistical mind went immediately to work. I thought, "If I could only get two groups of people, some from the United States and some from foreign countries, I could get them to rank samples from different parts of the world. I will bet, if everyone were exposed to OUR formula, they would agree that it is best!"

STEP 1 — Identify the Problem

Here, again, there's really not a problem from my perspective. I'm just interested in knowing if people from throughout the world rank the different varieties of Coke in the same way. It might, however, be in the best interests of Coca-Cola to investigate my idea.

■ *Developing, manufacturing, and distributing soft drinks throughout the world is a timely and expensive proposition. If there is no difference in the rankings of different products, it is possible that manufacturing and advertising costs could be lessened by focusing on a smaller number of products.*

STEP 2 — State a Hypothesis

If I really wanted to conduct such a study, I would use the following hypothesis:

■ *There will be no difference in the preference rankings for persons from the United States and persons from outside the United States.*

STEP 3 — Identify the Independent Variable

In this case, our predictor variable is going to be a person's country of residence. Either he is from the United States or he is from a foreign country.

STEP 4 — Identify and Describe the Dependent Variable

Let's suppose I asked each group to rank their preferences for 10 different formulas of Coke. Our dependent variable is the rankings for each of the groups. Let's use the data in Table 11.14 to get started.

The chart has three columns. The first column represents the 10 different formulas we are interested in ranking. Just for the sake of our discussion, we will assume that formula A is that from the United States. The second column shows the average ranking for each brand by persons from the United States; the third column shows the same thing for persons from outside the United States.

TABLE 11.14. Ranking of Coke Products from Different Countries

Formula	United States	Foreign
A	5	10
B	3	5
C	1	6
D	1	3
E	6	7
F	7	9
G	9	1
H	10	2
I	4	8
J	8	4

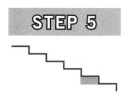

STEP 5 · Choose the Right Statistical Test

Because we are using ordinal data (i.e., the rankings from the two groups) and we are interested in computing a correlation between the two groups, we will be using the Spearman rho formula.

STEP 6 · Use Data Analysis Software to Test the Hypothesis

SPSS would generate Figures 11.16 and 11.17.

Our value of rho (i.e., –.316) indicates that a small negative correlation exists; people from foreign countries have a slight tendency to rank the sodas in exactly the opposite order from those in the United States.

The scatterplot verifies this; the line is going slightly down from left to right. If the same line was flat, it would indicate a great deal of disagreement in the rankings (e.g., one group ranked an item tenth and the other first, one ranked an item second,

Correlations

			US Ranking	Foreign Ranking
Spearman's rho	US Ranking	Correlation Coefficient	1.000	-.316
		Sig. (2-tailed)	.	.374
		N	10	10
	Foreign Ranking	Correlation Coefficient	-.316	1.000
		Sig. (2-tailed)	.374	.
		N	10	10

FIGURE 11.16. Spearman's rho showing a moderate negative correlation in rankings between the United States and foreign countries.

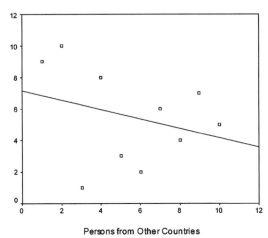

FIGURE 11.17. Scatterplot showing a moderate negative correlation in rankings between the United States and foreign countries.

and the other ninth, etc.). A line with a perfect 45-degree angle would indicate a nearly perfect agreement on the rankings. As we have said all along, that will probably never happen.

Once We Have a Linear Relationship, What Can We Do with It?

We have learned how to determine whether or not a linear relationship exists, now it is time to put it to work. To do so, think back to our use of Pearson's r in the last example. It was fairly easy to use it to determine if a relationship existed between two datasets. Again, the relationship wasn't perfect, but there was a relationship between height and weight.

Given that, what would happen if we wanted to use the relationship based on our sample data to try to predict values for other people in the population? For example, suppose we wanted to estimate a person's weight, as closely as possible, based on their height? Fortunately for us, we have another correlational tool, *linear regression*; that allows us to do exactly that.

■ Linear Regression

If we had a dataset of height and weight data and wound up with an r value of 1.00, we would be able to state, with 100% accuracy, either a person's height based on knowing his weight or his weight based on knowing his height. Unfortunately, we rarely, if ever, have that degree of accuracy.

At the same time, unless the r value is zero, we know there is some degree of relationship between height and weight. Knowing that, and depending on the value of r, we should be able to predict, with some accuracy, one of the variables based on knowing the other variable. The key to doing so is a thorough understanding of the line of best fit. Up to this point we have referred to it as a line that shows the trend of a data distribution, but now we need to know exactly what it is, how to compute it, and how to use it.

The Regression Equation

Those of you who paid more attention than I did in algebra might remember the formula for plotting a line on a graph; this is known as a *regression equation*.

$$y = a + bx$$

First, y is the value we are trying to predict. For example, if we wanted to predict weight based on height, we would change our equation to read:

$$\text{Weight} = a + bx$$

Lower-case x represents the value of the predictor variable. In this case, we know a person's height, so let's enter that into our equation.

$$\text{Weight} = a + b(\text{height})$$

The next symbol, b, represents the slope; this tells us how steep or flat the best-fit line is. We can enter it into our equation.

$$\text{Weight} = a + \text{slope}(\text{height})$$

Finally, a is the point on the y-axis where the line of best fit crosses for a given value of x; we call this the *intercept*.

$$\text{Weight} = \text{intercept} + \text{slope (height)}$$

Since we do not know the value for weight but we do know the value for height, all we need to get this equation to work is to determine how to compute the intercept and the slope.

Computing the Slope

Before we actually compute the slope, it is probably best to understand exactly what we are trying to compute. In order to do this, let's define the slope as the change in the y value on a graph based on every incremental change in the x value on the same graph. In order to get a better conceptual understanding of this, look at Figure 11.18.

You can see the line labeled A starts at the zero and runs up, to the right, at a 45-degree angle (i.e., a positive slope). In this case, as x goes up 1 point, so does y. For example, you can see the line passes through the points ($x = 1$, $y = 1$), ($x = 2$, $y = 2$), and so on.

In the following picture, we have the opposite. The slope is negative; for each value that x goes up, the value for y goes down. This means, as shown in Figure 11.19, that there is a negative correlation.

Finally, in Figure 11.20, you can see the slope is equal to zero; as the value of x goes up, the value of y stays the same.

It stands to reason then that if we want to use regression to estimate one value based on another, we need to

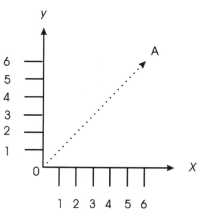

FIGURE 11.18. Positive slope.

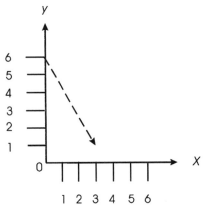

FIGURE 11.19. Negative slope.

FIGURE 11.20. Zero slope.

know the slope of the line. Since we have moved beyond the point where we can refer to it in general terms, we need a formula to compute the slope of a line based on data values we have collected. Luckily for us, such a formula exists:

$$slope = \frac{n\left(\sum xy\right) - \left(\sum x\right)\left(\sum y\right)}{n\left(\sum x^2\right) - \left(\sum x\right)^2}$$

As you can see, this formula looks quite a bit like the formula for Pearson's r. Since we have seen how to do all of those computations, this will prove to be no trouble at all. First, let's set up Table 11.15 for us to use. You can see that I have already done some of the basic math (e.g., squared each of the height and weight values) that we will need to use in our equation. We can put these into our regression equation and work through the following four steps:

TABLE 11.15. Values Needed to Compute the Slope

Height	x^2	Weight	y^2	xy
60	3600	100	10000	6000
62	3844	120	14400	7440
58	3364	130	16900	7540
70	4900	150	22500	10500
78	6084	140	19600	10920
70	4900	190	36100	13300
65	4225	150	22500	9750
69	4761	135	18225	9315
62	3844	130	16900	8060
66	4356	170	28900	11220
58	3364	120	14400	6960
63	3969	110	12100	6930
60	3600	170	28900	10200
68	4624	160	25600	10880

Height	x^2	Weight	y^2	xy
69	4761	150	22500	10350
68	4624	145	21025	9860
72	5184	200	40000	14400
74	5476	180	32400	13320
76	5776	210	44100	15960
68	4624	170	29000	11560
$\Sigma x = 1336$	$\Sigma x^2 = 89880$	$\Sigma y = 3030$	$\Sigma y^2 = 475950$	$\Sigma xy = 204465$

1. $slope = \dfrac{20(204465) - (1336)(3030)}{20(89880) - (1336)^2}$

2. $slope = \dfrac{4089300 - 4048080}{1797600 - 1784896}$

3. $slope = \dfrac{41220}{12704}$

4. $slope = 3.24$

Computing the Intercept

As was the case with the slope, it is important to understand what the intercept is prior to using it in our formula. By definition, it is the value of y when the value of x is zero. Let's use our pictures from above to help us better understand this. In Figure 11.21, when x is zero, so is y; this means our intercept is zero. In Figure 11.22, when x is zero, y (i.e., our intercept) is 6. Finally, in Figure 11.23, the slope in the diagram is 3; when x is zero, the line of best fit crosses the y-axis at 3.

As was the case with the slope, it is not always easy to look at a scatterplot and determine the intercept. We can calculate it, however, so that does not present a problem.

$$Intercept = \overline{y} - slope(\overline{x})$$

We have the slope value and using the data from our height and weight table, we can easily compute the mean of x and the mean of y, so let's put these values into our equation.

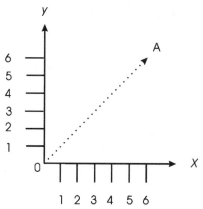

FIGURE 11.21. Intercept of zero.

$$Intercept = 151.5 - 3.24(66.8)$$

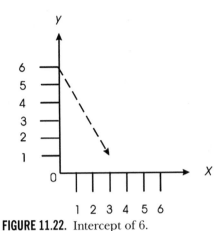

FIGURE 11.22. Intercept of 6.

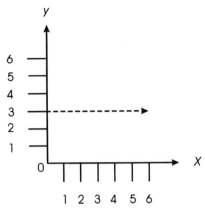

FIGURE 11.23. Intercept of 3 with zero slope.

This gives us an intercept value of –65.242. Notice that this number is negative; that is perfectly normal and happens from time to time. Using it, we have everything we need for our regression formula so let's pick a value of x (i.e., the height), 65 for example, and predict what our y value (i.e., the weight) should be.

$$Weight = -65.242 + 3.24(65)$$

When we do the math, we find that, for a person who is 65 inches tall, we are predicting a weight of 145.68. In order to verify this figure, look at the point on the scatterplot in Figure 11.24 where 65 inches and 145 inches intersect. When you find that point, you'll see that it is very near the line of best fit. This means our regression formula is a pretty good estimator of weight based on height (see Figure 11.24).

Let's use this same data as input into the SPSS regression function, shown in Figure 11.25, to make sure we get the same results. First, we'll create two variables, Height and Weight. Following that, we'll enter the same data we used to compute the correlation; this is shown in Figure 11.26. As seen in Figure 11.27, in order to run the regression, we'll select Analyze, Regression, and Linear. We follow this by identifying our independent and dependent variables as well as the statistics we want to have generated (Figure 11.28).

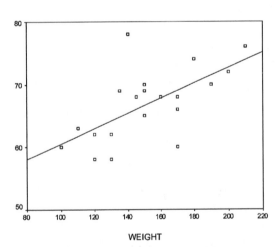

FIGURE 11.24. Regression plot for height and weight.

Figure 11.29 shows that SPSS would provide the same descriptive statistics as for the correlation procedure and would create the following output from the regression calculation. On the left, you can see a box containing the words "Constant" and "Height in inches"; their actual values fall into the column labeled "Unstandardized Coefficients," meaning they are

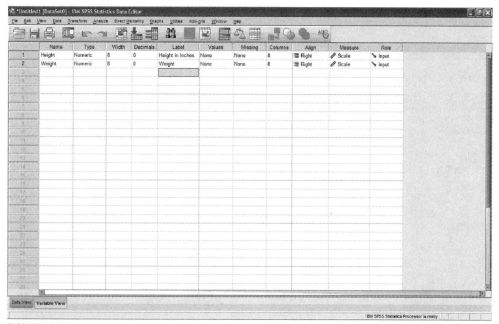

FIGURE 11.25. Creating the height and weight variables in the Variable View spreadsheet.

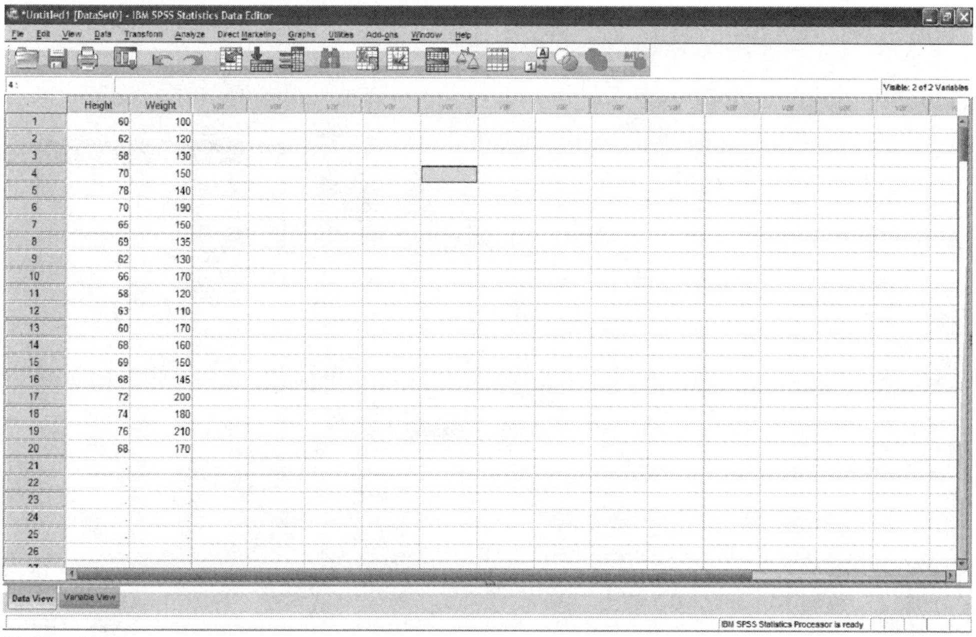

FIGURE 11.26. Height and weight data in the Data View spreadsheet.

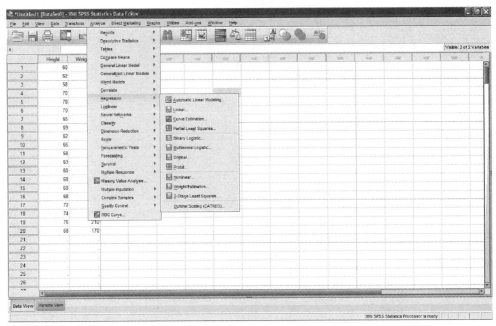

FIGURE 11.27. Selecting the Linear Regression command.

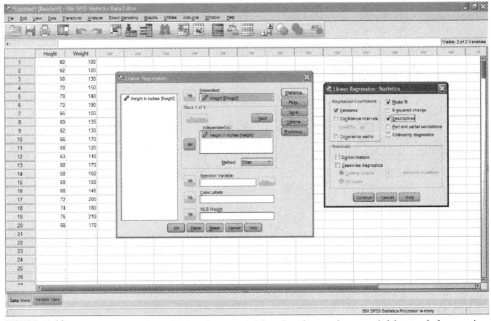

FIGURE 11.28. Identifying the independent variable, the dependent variable, and the statistics in the Linear Regression command.

Coefficients^a

Model		Unstandardized Coefficients		Standardized Coefficients	T	Sig.
		B	Std. Error	Beta		
1	(Constant)	-65.242	63.373		-1.030	.317
	Height in Inches	3.245	.945	.629	3.432	.003

a. Dependent Variable: Weight

FIGURE 11.29. Descriptive statistics from the linear regression.

in their original scale. Within that column, there are two smaller columns. The "B" column contains the actual coefficients that go into the regression equation; the intercept is –65.24 and the slope is 3.245. In both instances, the computed values are exactly the same as those we computed by hand. To the right of these is the standard error for both values.

The next column is labeled "Standardized Coefficients"; these are more commonly known as "Beta coefficients." In this column, we will see the standardized values for all independent (i.e., predictor) values in our particular regression equation. Since, in this case, we are dealing with only one independent variable, there is only the one value. To the right of that, the t values and p values shown are used to test the hypothesis that the slope and intercept values are zero in the population from which the data points are drawn. They help the user estimate how likely it is that the computed coefficients occurred by chance.

Why Wasn't It Exactly Right?

A lot of folks are surprised when we get to this point. After all of the computations, they want to know why our regression formula did not exactly predict the person's weight based on their height. The answer is simple: the correlation coefficient between height and weight is less than 1.00. Because of that, error is introduced into the process, just as it was in the other statistical tests we have used. This is verified by three things.

First, we can look at the scatterplot and see that all of the values do not fall directly onto the line of best fit. Second, we can see that the standard errors for both the independent variable and the slope are greater than zero. Third, we can compute a coefficient of determination by squaring the Pearson's r value from the correlation (i.e., .629 * .629 = .395). This coefficient tells us that 39.5% of the change in the criterion variable (i.e., weight) is caused by the predictor variable (i.e., height). The remaining 71.50% is either due to error or other variables not considered in the equation.

To better understand this concept, let's imagine we had a dataset where there was a perfect correlation between the independent and dependent variables. In Table 11.16, you can see that, for every 1-inch difference in height, there is a 5-pound increase in weight.

If we computed Pearson's r for the two values, as shown in Figure 11.30, this would mean it would be exactly 1.00, indicating a perfect positive linear relationship.

If we look at Figure 11.31, we first see that our standard error is zero; given that, we do not expect error to interfere with our ability to use the regression procedure to accurately use the predictor variable to predict the criterion variable. We can also see our intercept (i.e., constant) and our slope (i.e., height). In this case if we know a person's height, we can accurately state their weight.

TABLE 11.16. Height and Weight Data

Height	Weight	Height	Weight	Height	Weight	Height	Weight
60	100	65	125	70	150	75	175
61	105	66	130	71	155	76	180
62	110	67	135	72	160	77	185
63	115	68	140	73	165	78	190
64	120	69	145	74	170	79	195

Correlations

		Height in inches	Weight
Height in inches	Pearson Correlation	1	1.000
	Sig. (2-tailed)		.000
	N	20	20
Weight	Pearson Correlation	1.000	1
	Sig. (2-tailed)	.000	
	N	20	20

FIGURE 11.30. Pearson correlation coefficients between height and weight.

Coefficients[a]

Model		Unstandardized Coefficients		Standardized Coefficients	t	Sig.
		B	Std. Error	Beta		
1	(Constant)	-200.000	.000		.	.
	Height in inches	5.000	.000	1.000	.	.

a. Dependent Variable: Weight

FIGURE 11.31. Inferential statistics from the linear regression.

If everything we have said up to this point is true, then we should be able to use the slope and height, along with a given value of the predictor variable, to accurately predict a criterion variable. Let's use a height of 64 and enter all of these data into our equation:

1. Weight = –200 + 5(64)
2. Weight = –200 + 320
3. Weight = 120

As we can see, for a height of 64 inches, we are expecting the weight to be 120; we can verify that by looking back at the table. We could further verify it by computing the coefficient of determination (i.e., 1 * 1 = 1). This tells us 100% of the change in the criterion variable is caused by the predictor variable. We can also see this quite clearly by looking at the scatterplot in Figure 11.32.

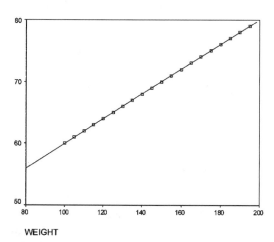

WEIGHT

FIGURE 11.32. Scatterplot of the height and weight data.

As I said, however, this perfect prediction can only come true if there is a perfect linear relationship between the predictor and criterion variables. This perfect relationship, of course, happens very rarely. In the vast majority of cases, we can use our regression equation to help predict values, but we have to be aware of the error inherent in the process.

Using the Six-Step Model: The Case of Age and Driving

Like all of the statistical tools we have talked about, we can use regression as part of our six-step process. In order to do that, let's imagine an argument between an elderly rights group and a group interested in highway safety. The safety group wants drivers to be tested every year after they reached the age of 65; they contend this will cut down on the number of highway accidents. The elderly rights group claims there is no relationship between drivers' age and their safety record. They feel that drivers who are 85 years old are just as safe as those who are 65. You decide to collect data and to help settle the argument.

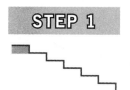

STEP 1

Identify the Problem

Here again we have an implied problem. The safety group believes one thing, the elderly rights group another. Since both sides are basing their contentions on feelings rather than fact, we need a fairly generic problem statement.

▪ *Safety groups have stated they believe that drivers reaching the age of 65 should be tested each year; their reasoning is that, once a driver reaches that age, their ability to safely drive an automobile diminishes. Elderly rights groups feel there is no manner by which the given number of accidents in a year's time can be predicted by age. This study will collect and analyze data to see if such a relationship exists.*

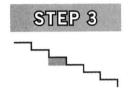

State a Hypothesis

Remember, we are going to leave the word "significant" out of this.

■ *There is no relationship between the age of elderly drivers and the number of automobile accidents they are involved in during a given year.*

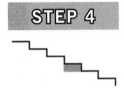

Identify the Independent Variable

Our predictor variable is age. Since we are dealing with "elderly" people, let's use a range from 65 to 85.

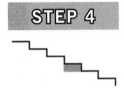

Identify and Describe the Dependent Variable

Our criterion variable is the number of accidents an elderly driver is involved in for a given year. Let's use Table 11.17 for our example. The descriptive statistics and the correlation matrix for our dataset are shown in Figures 11.33 and 11.34.

TABLE 11.17. Age and Accident Data

Age	Accidents	Age	Accidents	Age	Accidents	Age	Accidents
67	4	66	4	82	3	85	2
72	3	82	0	83	2	76	4
65	6	79	2	70	5	65	3
78	2	77	5	73	4	74	3
80	1	73	3	67	5	79	3

Descriptive Statistics

	Mean	Std. Deviation	N
Number of Accidents	3.2000	1.47256	20
Driver's Age	74.6500	6.41770	20

FIGURE 11.33. Descriptive statistics for accident and age data.

An average of 3.2 accidents per year is somewhat disturbing. What is even more worrisome is that our correlation coefficient is negative; it seems there is an inverse relationship between age and the number of accidents. Remember, although we are getting a good picture of what seems to be happening, these are descriptive statistics. Let's move forward and see what happens from here.

Correlations

		Number of Accidents	Driver's Age
Pearson Correlation	Number of Accidents	1.000	-.694
	Driver's Age	-.694	1.000
Sig. (1-tailed)	Number of Accidents	.	.000
	Driver's Age	.000	.
N	Number of Accidents	20	20
	Driver's Age	20	20

FIGURE 11.34. Pearson correlation coefficients between accidents and age.

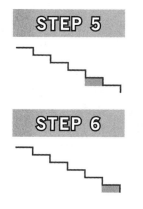

STEP 5   Choose the Right Statistical Test

Given the type of data we've collected and our desire to predict one value given the other, we will use the linear regression function to analyze our data.

STEP 6   Use Data Analysis Software to Test the Hypothesis

Using the linear regression function, SPSS will produce Figures 11.35 and 11.36.

The bad feeling we had after Step 3 is getting worse. We have a negative correlation coefficient; this means there is an inverse relationship. This idea is supported by the best-fit line sloping downward from right to left. Let's move to the next step and see what happens.

Based on the correlation coefficient, it is clear that we have a linear, albeit negative, relationship between age and the number of accidents in which a driver is involved. Let's investigate that further by picking a couple of ages from our dataset, 82 and 65, insert them into our regression formula, and see what happens. First, for the 82-year-old driver:

1. Accidents = 15.086 + −.159(82)
2. Accidents = 15.086 − 13.038
3. Accidents = 2.084

Coefficients

Model		Unstandardized Coefficients		Standardized Coefficients		
		B	Std. Error	Beta	t	Sig.
1	(Constant)	15.086	2.917		5.171	.000
	Driver's Age	-.159	.039	-.694	-4.089	.001

FIGURE 11.35. Inferential statistics from the linear regression.

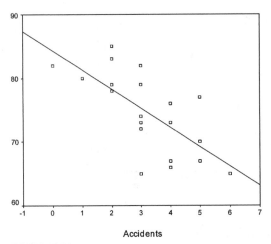

FIGURE 11.36. Scatterplot of the age and accident data.

Then for the 65-year-old driver:

1. Accidents = 15.086 + –.159(65)
2. Accidents = 15.086 – 10.335
3. Accidents = 4.751

Unfortunately, for the highway safety folks, the results were exactly the opposite of what they believed: the 85-year-old driver is estimated to be in about two accidents per year, while the 65-year-old drivers will be in over four accidents during that same time period!

Remember, although it is implied, we are not proving cause and effect here. Our coefficient of determination is only .481. This tells us that only 48.1% of the change in the criterion variable is attributable to age. This means there is error involved; obviously, other things predict the number of accidents. In this case, we might have to consider other factors such as number of miles driven per year, number of other drivers on the road, and prevalent weather conditions in a given driver's location.

Summary

Most of my students agree that the correlational procedures are extremely easy to understand. We used basic math skills to help us decide if a linear relationship exists between two sets of data. If we do find a linear relationship, we then determine if we can use one set of data to predict values in the other set. Although we did not talk about them, we could consider many other tools of this type: correlations where there are more than two variables; regression formulas with more than one predictor variable (i.e., multiple regression); and regression equations where you can use your data to predict a nominal value (i.e., logistic regression). Use of these tools is beyond the scope of this book, but the good news is that once you understand the purpose and use of the tools in this chapter, understanding the others is made a lot easier.

Do You Understand These Key Words and Phrases?

coefficient of determination

correlation coefficient

criterion variable

intercept

line of best fit

Pearson's *r*

predictor variable · · · · · · · · · · · · · · · slope

Spearman's rho · · · · · · · · · · · · · · · *x*-axis

y-axis

Do You Understand These Formulas?

Intercept:

$$Intercept = \bar{y} - slope(\bar{x})$$

Regression equation (equation of a line):

$$y = a + bx$$

Slope of a line:

$$slope = \frac{n\left(\sum xy\right) - \left(\sum x\right)\left(\sum y\right)}{n\left(\sum x^2\right) - \left(\sum x\right)^2}$$

Quiz Time!

We covered everything we need to cover in this chapter but, before we go, let's make sure you completely understand everything we have done. Answer each of these questions and then turn to the end of the book to check your work.

The Case of "Like Father, Like Son"

In Figures 11.37, 11.38, and 11.39 there is output from a regression analysis trying to predict a son's age of death based on the father's age of death:

1. What does Pearson's *r* tell us?

2. What is the coefficient of determination? What does it tell us?

3. Given a father's age at death of 65, how long would we expect the son to live?

4. Generally speaking, in which direction would the line of best fit flow?

Descriptive Statistics

	Mean	Std. Deviation	N
Son's Age	79.1000	3.75500	10
Father's Age	74.3000	4.52278	10

FIGURE 11.37. Descriptive statistics for father and son age data.

Correlations

		Son's Age	Father's Age
Pearson Correlation	Son's Age	1.000	.790
	Father's Age	.790	1.000
Sig. (1-tailed)	Son's Age	.	.003
	Father's Age	.003	.
N	Son's Age	10	10
	Father's Age	10	10

FIGURE 11.38. Positive Pearson correlation coefficients between father and son age data.

Coefficients[a]

Model		Unstandardized Coefficients		Standardized Coefficients	t	Sig.
		B	Std. Error	Beta		
1	(Constant)	30.387	13.403		2.267	.053
	Father's Age	.656	.180	.790	3.641	.007

a. Dependent Variable: Son's Age

FIGURE 11.39. Inferential statistics from the linear regression.

The Case of "Can't We All Just Get Along?"

Figure 11.40 shows the correlation between how Democrats and Republicans ranked the items in a list of the most pressing issues in America:

1. Why would you use Spearman's rho to analyze this data?

2. What does Spearman's rho tell us?

3. What would you say to the president if he asked you the results?

Correlations

			Democrat	Republican
Spearman's rho	Democrat	Correlation Coefficient	1.000	-.515
		Sig. (2-tailed)	.	.128
		N	10	10
	Republican	Correlation Coefficient	-.515	1.000
		Sig. (2-tailed)	.128	.
		N	10	10

FIGURE 11.40. Spearman rho correlation coefficients for rankings between political parties.

The Case of More Is Better

Figures 11.41, 11.42, and 11.43 show the output of a regression procedure using highest level of education attained and average salary in the workplace.

1. If you were asked to speak to a group of potential high school dropouts, how would you interpret Pearson's r?

2. In comparison to the overall average, what would you say to someone who was thinking about dropping out in the ninth grade?

3. What is the coefficient of determination? What does it tell us?

4. Generally speaking, in which direction would the line of best fit flow?

Descriptive Statistics

	Mean	Std. Deviation	N
Income	14,692.31	6.5774	1430
Highest Year	13.07	3.090	1430

FIGURE 11.41. Descriptive statistics for income and highest year of education data.

Correlations

		Income	Highest Year
Pearson Correlation	Income	1.000	.437
	Highest year	.437	1.000
Sig. (1-tailed)	Income	.	.000
	Highest year	.000	.
N	Income	1430	1430
	Highest year	1430	1430

FIGURE 11.42. Moderate Pearson correlation coefficient between income and highest year of education.

Coefficients[a]

Model		Unstandardized Coefficients		Standardized Coefficients		
		B	Std. Error	Beta	T	Sig.
1	(Constant)	4721.569	563.170		8.206	.000
	Highest Year	770.527	41.934	.437	18.375	.000

a. Dependent Variable: Income

FIGURE 11.43. Inferential statistics from the linear regression.

The Case of More Is Better Still

Suppose you use Figures 11.44, 11.45, and 11.46 to follow up the prior case. At this point, you tell potential dropouts that students who drop out of school earlier tend to get married earlier and have more children.

1. What would you tell the students about Pearson's r?

2. What is the coefficient of determination? What does it tell us?

3. In comparison to the overall average, what would you say to someone who was thinking about dropping out in the eighth grade?

4. Generally speaking, in which direction would the line of best fit flow?

5. Combining what you know from Scenario 3, what would you say to the students, in terms of living in poverty?

Descriptive Statistics

	Mean	Std. Deviation	N
Children	1.85	1.68	1430
Highest Year	13.07	3.090	1430

FIGURE 11.44. Descriptive statistics for the number of children and highest year of education.

Correlations

		Children	Highest Year
Pearson Correlation	Children	1.000	-.259
	Highest year	-.259	1.000
Sig. (1-tailed)	Children	.	.000
	Highest year	.000	.
N	Children	1430	1430
	Highest year	1430	1430

FIGURE 11.45. Small negative Pearson correlation coefficient between number of children and highest year of education.

Coefficients^a

Model		Unstandardized Coefficients		Standardized Coefficients	T	Sig.
		B	Std. Error	Beta		
1	(Constant)	3.357	.184		19.195	.000
	Highest Year	-.129	.014	-.237	-9.405	.000

a. Dependent Variable: Children

FIGURE 11.46. Inferential statistics from the linear regression.

Conclusion: Have We Accomplished What We Set Out to Do?

Looking back, while writing this book, I had several goals in mind. First, I wanted the reader to put any past statistical prejudices behind them and look at this topic from a new perspective. Second, I wanted the reader to understand that, in the huge majority of cases, you are going to be using a relatively easy-to-understand set of statistical tools. Third, even though we worked through the manual calculations, we also saw we could use SPSS, one of the many easily used software packages, for our calculations. Last, I wanted the reader to understand there is a straightforward manner in which they can approach these cases. Did I accomplish these goals? Let's take a moment and see.

Statistics in a New Light

Many of you may have approached this book with some trepidation. Simply put, you really do not like statistics, and, if it were up to you, you would have avoided even thinking about taking a statistics course. Do you still feel that way? I hope not because, hopefully, you've seen that statistics are no more than two things. First, they are a set of tools that help us "describe" data we collect. Second, they are tools that help us make decisions about that same data. As you saw, every statistical tool you've learned about is based on basic math: addition, subtraction, multiplication, and division. You can do those basic things, and now you can do this.

A Limited Set of Statistical Techniques

There are literally hundreds of statistical tests and techniques we could have talked about in this book. You may have heard of tools such as multivariate regression, canonical correlations, structural equation modeling, and the Komolgorov–Smirnoff test (no, that's not a joke) tossed around but noticed we didn't talk about them here. Why not? Easy; as I said earlier, as a beginning statistician and a consumer of statistics, you do not need many of these tools except in limited situations. What you do need, in the vast majority of cases, is the group of statistical tests we talked about in this book. I firmly believe these will suit your needs at least 90% of the time. Again, if you need something else, there are plenty of people willing and able to help you. With their expert guidance, you'll find that most statistical techniques are fairly straightforward—just like those we have talked about.

The Use of Statistical Software Packages

A lot of us statistical "old-timers" remember the days when we had no choice; we had to calculate these statistics manually. Although the underlying math is basic, it is easy to make a mistake, especially when you're dealing with large groups of numbers. Because of that, there are many statistical software packages, usually based on a common spreadsheet, that can be used to run these different tests. Here we used SPSS but, as I said, the differences between the packages are not so great that you cannot quickly learn them.

A Straightforward Approach

As I said at the beginning of the book, I can remember being faced with situations when I first started learning about statistics where I was not sure which statistical technique I should be using. This anxiety, combined with the other issues I have already discussed, led to a great deal of reluctance on my part, as well as that of my classmates, to get really excited about what we were supposed to be doing. As you saw in this text, however, we were able to identify a set of six steps that helped us navigate through the statistical fog! As I have said before, these six steps do not cover every statistical situation you might find yourself in. They should, however, cover most cases.

At Long Last

So, there we have it, it seems we have met our goals. I hope you are now able to not only identify and interpret the statistics you'll need most of the time, but that you're also a lot more comfortable as a consumer of statistics. We have taken what many of you considered to be a huge, anxiety-provoking problem, and we have shown how easy it is solving it if we only approach it using small, logic-driven steps. Why can't the rest of life be so simple?

APPENDIX A

Area under the Normal Curve Table (Critical Values of z)

z score	0.00	0.01	0.02	0.03	0.04	0.05	0.06	0.07	0.08	0.09
0.0	.0000	.004	.0080	.0120	.0160	.0199	.0239	.0279	.0319	.0359
0.1	.0398	.0438	.0478	.0517	.0557	.0596	.0636	.0675	.0714	.0753
0.2	.0793	.0832	.0871	.0910	.0948	.0987	.1026	.1075	.1103	.1141
0.3	.1179	.1217	.1255	.1293	.1331	.1368	.1406	.1443	.1480	.1517
0.4	.1554	.1591	.1628	.1664	.1700	.1736	.1772	.1808	.1844	.1879
0.5	.1915	.1950	.1985	.2019	.2054	.2088	.2123	.2157	.2190	.2224
0.6	.2257	.2291	.2324	.2357	.2389	.2422	.2454	.2486	.2517	.2549
0.7	.2580	.2611	.2642	.2673	.2704	.2734	.2764	.2794	.2823	.2852
0.8	.2881	.2910	.2939	.2967	.2995	.3023	.3051	.3078	.3106	.3133
0.9	.3159	.3186	.3212	.3238	.3264	.3289	.3315	.3340	.3365	.3389
1.0	.3413	.3438	.3461	.3485	.3508	.3531	.3554	.3577	.3599	.3621
1.1	.3643	.3665	.3686	.3708	.3729	.3745	.3770	.3790	.3810	.3830
1.2	.3849	.3869	.3888	.3907	.3925	.3944	.3962	.3980	.3997	.4015
1.3	.4032	.4049	.4066	.4082	.4099	.4115	.4131	.4147	.4162	.4177
1.4	.4192	.4207	.4222	.4236	.4261	.4265	.4279	.4291	.4306	.4319
1.5	.4332	.4345	.4357	.4370	.4382	.4394	.4406	.4418	.4429	.4441
1.6	.4452	.4463	.4474	.4484	.4495	.4505	.4515	.4525	.4535	.4545
1.7	.4554	.4564	.4573	.4582	.4591	.4590	.4608	.4616	.4625	.4633
1.8	.4641	.4649	.4645	.4664	.4671	.4678	.4686	.4693	.4699	.4706
1.9	.4713	.4719	.4726	.4732	.4738	.4744	.4750	.4756	.4761	.4767
2.0	.4772	.4778	.4783	.4788	.4793	.4798	.4803	.4808	.4812	.4817
2.1	.4821	.2826	.4830	.4834	.4838	.4842	.4846	.4850	.4854	.4857
2.2	.4861	.4874	.4868	.4871	.4875	.4887	.4881	.4884	.4887	.4890
2.3	.4893	.4896	.4998	.4901	.4904	.4906	.4909	.4911	.4913	.4916
2.4	.4918	.4920	.4922	.4925	.4927	.4929	.4931	.4932	.4934	.4936
2.5	.4938	.4940	.4941	.4943	.4945	.4946	.4948	.4949	.4951	.4952
2.6	.4953	.4955	.4956	.4957	.4959	.4960	.4961	.4962	.4963	.4964
2.7	.4965	.4966	.4967	.4468	.4969	.4970	.4971	.4972	.4973	.4974
2.8	.4974	.4975	.4976	.4977	.4977	.4978	.4979	.4979	.4980	.4981
2.9	.4981	.4982	.4982	.4983	.4985	.4984	.4985	.4985	.4986	.4986
3.0	.4987	.4987	.4987	.4988	.4988	.4989	.4989	.4989	.4990	.4990

APPENDIX B

Critical Values of *t*

df	Alpha = .10	Alpha = .05	Alpha = .025	df	Alpha = .10	Alpha = .05	Alpha = .025
1	3.078	6.314	12.706	18	1.330	1.734	2.101
2	1.886	2.920	4.303	19	1.328	1.729	2.093
3	1.638	2.353	3.182	20	1.325	1.725	2.086
4	1.533	2.132	2.776	21	1.323	1.721	2.080
5	1.476	2.015	2.571	22	1.321	1.717	2.074
6	1.440	1.943	2.447	23	1.319	1.714	2.069
7	1.415	1.895	2.365	24	1.318	1.711	2.064
8	1.397	1.860	2.306	25	1.316	1.708	2.060
9	1.383	1.833	2.262	26	1.315	1.706	2.056
10	1.372	1.812	2.228	27	1.314	1.703	2.052
11	1.363	1.796	2.201	28	1.313	1.701	2.048
12	1.356	1.782	2.179	29	1.311	1.699	2.045
13	1.350	1.771	2.160	30	1.310	1.645	2.042
14	1.345	1.761	2.145	40	1.303	1.684	2.021
15	1.341	1.753	2.131	60	1.296	1.671	2.000
16	1.337	1.746	2.120	120	1.289	1.658	1.980
17	1.333	1.740	2.110	∞	1.282	1.645	1.960

Critical Values of F When Alpha $= .01$

Within-group degrees of freedom	Between-groups degrees of freedom									
	1	2	3	4	5	6	7	8	9	10
1	4052	5000	5403	5625	5764	5859	5928	5982	6022	6056
2	98.50	99	99.17	99.25	99.30	99.33	99.36	99.37	99.39	99.40
3	34.12	30.82	29.46	28.71	28.24	27.91	27.67	27.49	27.35	27.23
4	21.20	18.00	16.69	15.98	15.52	15.21	14.98	14.80	14.66	14.55
5	16.26	13.27	12.06	11.39	10.97	10.67	10.46	10.29	10.16	10.05
6	13.75	10.92	9.78	9.15	8.75	8.47	8.26	8.10	7.98	7.87
7	12.25	9.55	8.45	7.85	7.46	7.19	6.99	6.84	6.72	6.62
8	11.26	8.65	7.59	7.01	6.63	6.37	6.18	6.03	5.91	5.81
9	10.56	8.02	6.99	6.42	6.06	5.80	5.61	5.47	5.35	5.26
10	10.04	7.56	6.55	5.99	5.64	5.39	5.20	5.06	4.94	4.85
11	9.65	7.21	6.22	5.67	5.32	5.07	4.89	4.74	4.63	4.54
12	9.33	6.93	5.95	5.41	5.06	4.82	4.64	4.50	4.39	4.30
13	9.07	9.7	5.74	5.21	4.86	4.62	4.44	4.30	4.19	4.10
14	8.86	6.51	5.56	5.04	4.69	4.46	4.28	4.14	4.03	3.94
15	8.68	6.36	5.42	4.89	4.56	4.32	4.14	4.00	3.89	3.80
16	8.53	6.23	5.29	4.77	4.44	4.20	4.03	3.89	3.78	3.69
17	8.40	6.11	5.18	4.67	4.34	4.10	3.93	3.79	3.68	3.59
18	8.29	6.01	5.09	4.58	4.25	4.01	3.84	3.71	3.60	3.51
19	8.18	5.93	5.01	4.50	4.17	3.94	3.77	3.63	3.52	3.43
20	8.10	5.85	4.94	4.43	4.10	3.87	3.70	3.56	3.46	3.37
21	8.02	5.78	4.87	4.37	4.04	3.81	3.64	3.51	3.40	3.31
22	7.95	5.72	4.82	4.31	3.99	3.76	3.59	3.45	3.35	3.26
23	7.88	5.66	4.76	4.26	3.94	3.71	3.54	3.41	3.30	3.21
24	7.82	5.61	4.72	4.22	3.90	3.67	3.50	3.36	3.26	3.17
25	7.77	5.57	4.68	4.18	3.85	3.63	3.46	3.32	3.22	3.13
26	7.72	5.53	4.64	4.14	3.82	3.59	3.42	3.29	3.18	3.09
27	7.68	5.49	4.60	4.11	3.78	3.56	3.39	3.26	3.15	3.06
28	7.64	5.45	4.57	4.07	3.75	3.53	3.36	3.23	3.12	3.03
29	7.60	5.42	4.54	4.04	3.73	3.50	3.33	3.20	3.09	3.00
30	7.56	5.39	4.51	4.02	3.70	3.47	3.30	3.17	3.07	2.98
40	7.31	5.18	4.31	3.83	3.51	3.29	3.12	2.99	2.89	2.80
60	7.08	4.98	4.13	3.65	3.34	3.12	2.95	2.82	2.72	2.63
120	6.85	4.79	3.95	3.48	3.17	2.96	2.79	2.66	2.56	2.47
∞	6.63	4.61	3.78	3.32	3.02	2.80	2.64	2.51	2.41	2.32

Within-group degrees of freedom	Between-groups degrees of freedom								
	12	15	20	24	30	40	60	120	∞
1	6106	6157	6209	6235	6261	6287	6313	6339	6366
2	99.42	99.43	99.45	99.46	99.47	99.47	99.48	99.49	99.50
3	27.05	26.87	26.69	26.60	26.50	26.41	26.32	26.62	26.13
4	14.37	14.20	14.02	13.93	13.84	13.75	13.65	13.56	13.46
5	9.89	9.72	9.55	9.47	9.38	9.29	9.20	9.11	9.02
6	7.72	7.56	7.40	7.31	7.23	7.14	7.06	6.97	6.88
7	6.47	6.31	6.16	6.07	5.99	5.91	5.82	5.74	5.65
8	5.67	5.52	5.36	5.28	5.20	5.12	5.03	4.95	4.86
9	5.11	4.96	4.81	4.73	4.65	4.57	4.48	4.40	4.31
10	4.71	4.56	4.41	4.33	4.25	4.17	4.08	4.00	3.91
11	4.40	4.25	4.10	4.02	3.94	3.86	3.78	3.69	3.60
12	4.16	4.01	3.86	3.78	3.70	3.62	3.54	3.45	3.36
13	3.96	3.82	3.66	3.59	3.51	3.43	3.34	3.25	3.17
14	3.80	3.66	3.51	3.43	3.35	3.27	3.18	3.09	3.00
15	3.67	3.52	3.37	3.29	3.21	3.13	3.05	2.96	2.87
16	3.55	3.41	3.26	3.18	3.10	3.02	2.93	3.84	2.75
17	3.46	3.31	3.16	3.08	3.00	2.92	2.83	2.75	2.65
18	3.37	3.23	3.08	3.00	2.92	2.84	2.75	2.66	2.57
19	3.30	3.15	3.00	2.92	2.84	2.76	2.67	2.58	2.49
20	3.23	3.09	2.94	2.86	2.78	2.69	2.61	2.52	2.42
21	3.17	3.03	2.88	2.80	2.72	2.64	2.55	2.46	2.36
22	3.123	2.98	2.83	2.75	2.67	2.58	2.50	2.40	2.31
23	3.07	2.93	2.78	2.70	2.62	2.54	2.45	2.35	2.26
24	3.03	2.89	2.74	2.66	2.58	2.49	2.40	2.31	2.21
25	2.99	2.85	2.70	2.62	2.54	2.45	2.36	2.27	2.17
26	2.96	2.81	2.66	2.58	2.50	2.42	2.33	2.23	2.13
27	2.93	2.78	2.63	2.55	2.47	2.38	2.29	2.20	2.10
28	2.90	2.75	2.60	2.52	2.44	2.35	2.26	2.17	2.06
29	2.87	2.73	2.57	2.49	2.41	2.33	2.23	2.14	2.03
30	2.84	2.70	2.55	2.47	2.39	3.30	2.21	2.11	2.01
40	2.66	2.52	2.37	2.29	2.20	2.11	2.02	1.92	1.80
60	2.50	2.35	2.20	2.12	2.03	1.94	1.84	1.73	1.60
120	2.34	2.19	2.03	1.95	1.86	1.76	1.66	1.53	1.38
∞	2.18	2.04	1.88	1.79	1.70	1.59	1.47	1.32	1.00

APPENDIX D

Critical Values of *F* When Alpha = .05

Within-group degrees of freedom	Between-groups degrees of freedom									
	1	2	3	4	5	6	7	8	9	10
1	161.4	199.5	215.7	224.6	230.2	234.0	236.8	238.9	240.5	241.9
2	18.51	19.0	19.16	19.25	19.30	19.33	19.35	19.37	19.38	19.40
3	10.13	9.55	9.28	9.12	9.01	8.94	8.89	8.85	8.81	8.79
4	7.71	6.94	6.59	6.39	6.26	6.16	6.09	6.04	6.00	5.96
5	6.61	5.79	5.41	5.19	5.05	4.95	4.88	4.82	4.77	4.74
6	5.99	5.14	4.76	4.53	4.39	4.28	4.21	4.15	4.10	4.06
7	5.59	4.74	4.35	4.12	3.97	3.87	3.79	3.73	3.68	3.64
8	5.32	4.46	4.07	3.84	3.69	3.58	3.5	3.44	3.39	3.35
9	5.12	4.26	3.86	3.63	3.48	3.37	3.29	3.23	3.18	3.14
10	4.96	4.10	3.71	3.48	3.33	3.22	3.14	3.07	3.02	2.98
11	4.84	3.98	3.59	3.36	3.20	3.09	3.01	2.95	2.90	2.85
12	4.75	3.89	3.49	3.26	3.11	3.00	2.91	2.85	2.80	2.75
13	4.67	3.81	3.41	3.18	3.03	2.92	2.83	2.77	2.71	2.67
14	4.60	3.74	3.34	3.11	2.96	2.85	2.76	2.70	2.65	2.60
15	4.54	3.68	3.29	3.06	2.90	2.79	2.71	2.64	2.59	2.54
16	4.49	3.63	3.24	3.01	2.85	2.74	2.66	2.59	2.54	2.49
17	4.45	3.59	3.20	2.96	2.81	2.70	2.61	2.55	2.49	2.45
18	4.41	3.55	3.16	2.93	2.77	2.66	2.58	2.51	2.46	2.41
19	4.38	3.52	3.13	2.90	2.74	2.63	2.54	2.48	2.42	2.38
20	4.35	3.49	3.10	2.87	2.71	2.60	2.51	2.45	2.39	2.35
21	4.32	3.47	3.07	2.84	2.68	2.57	2.49	2.42	2.37	2.32
22	4.30	3.44	3.05	2.82	2.66	2.55	2.46	2.40	2.34	2.30
23	4.28	3.42	3.03	2.80	2.64	2.53	2.44	2.37	2.32	2.27
24	4.26	3.40	3.01	2.78	2.62	2.51	2.42	2.36	2.30	2.25
25	4.24	3.39	2.99	2.76	2.60	2.49	2.40	2.34	2.28	2.24
26	4.23	3.37	2.98	2.74	2.59	2.47	2.39	2.32	2.27	2.22
27	4.21	3.35	2.96	2.73	2.57	2.46	2.37	2.31	2.25	2.20
28	4.20	3.34	2.95	2.71	2.56	2.45	2.36	2.29	2.24	2.19
29	4.18	3.33	2.93	2.70	2.55	2.43	2.35	2.28	2.22	2.18
30	4.17	3.32	2.92	2.69	2.53	2.42	2.33	2.27	2.21	2.16
40	4.08	3.23	2.84	2.61	2.45	2.34	2.25	2.18	2.12	2.08
60	4.00	3.15	2.76	2.53	2.37	2.25	2.17	2.10	2.04	1.99
120	3.92	3.07	2.68	2.45	2.29	2.17	2.09	2.02	1.96	1.91
∞	3.84	3.00	2.60	2.37	2.21	2.10	2.01	1.94	1.88	1.83

Within-group degrees of freedom	Between-groups degrees of freedom								
	12	15	20	24	30	40	60	120	∞
1	243.9	245.9	248.0	249.1	250.1	251.1	252.2	253.3	254.3
2	19.41	19.43	19.45	19.45	19.46	19.47	19.48	19.49	19.5
3	8.74	8.70	8.66	8.64	8.62	8.59	8.57	8.55	8.53
4	5.91	5.86	5.80	5.77	5.75	5.72	5.69	5.66	5.63
5	4.68	4.62	4.56	4.53	4.50	4.46	4.43	4.40	4.36
6	4.00	3.94	3.87	3.84	3.81	3.77	3.74	3.70	3.67
7	3.57	3.51	3.44	3.41	3.38	3.34	3.30	3.27	3.23
8	3.28	3.22	3.15	3.12	3.08	3.04	3.01	2.97	2.93
9	3.07	3.01	2.94	2.90	2.86	2.83	2.79	2.75	2.71
10	2.91	2.85	2.77	2.74	2.70	2.66	2.62	2.58	2.54
11	2.79	2.72	2.65	2.61	2.57	2.53	2.49	2.45	2.40
12	2.69	2.62	2.54	2.51	2.47	2.43	2.38	2.34	2.30
13	2.60	2.53	2.46	2.42	2.38	2.34	2.30	2.25	2.21
14	2.53	2.46	2.39	2.35	2.31	2.27	2.22	2.18	2.13
15	2.48	2.40	2.33	2.29	2.25	2.20	2.16	2.11	2.07
16	2.42	2.35	2.28	2.24	2.19	2.15	2.11	2.06	2.01
17	2.38	2.31	2.23	2.19	2.15	2.10	2.06	2.01	1.96
18	2.34	2.27	2.19	2.15	2.11	2.06	2.02	1.97	1.92
19	2.31	2.23	2.16	2.11	2.07	2.03	1.98	1.93	1.88
20	2.28	2.20	2.12	2.08	2.04	1.99	1.95	1.90	1.84
21	2.25	2.18	2.10	2.05	2.01	1.96	1.92	1.87	1.81
22	2.23	2.15	2.07	2.03	1.98	1.94	1.89	1.84	1.78
23	2.20	2.13	2.05	2.01	1.96	1.91	1.86	1.81	1.76
24	2.18	2.11	2.03	1.98	1.94	1.89	1.84	1.79	1.73
25	2.16	2.09	2.01	1.96	1.92	1.87	1.82	1.77	1.71
26	2.15	2.07	1.99	1.95	1.90	1.85	1.80	1.75	1.69
27	2.13	2.06	1.97	1.93	1.88	1.84	1.79	1.73	1.67
28	2.12	2.04	1.96	1.91	1.87	1.82	1.77	1.71	1.65
29	2.10	2.03	1.94	1.90	1.85	1.81	1.75	1.70	1.64
30	2.09	2.01	1.93	1.89	1.84	1.79	1.74	1.68	1.62
40	2.00	1.92	1.84	1.79	1..74	1.69	1.64	1.58	1.51
60	1.92	1.84	1.75	1.70	1.65	1.59	1.53	1.47	1.39
120	1.83	1.75	1.66	1.61	1.55	1.50	1.43	1.35	1.25
∞	1.75	1.67	1.57	1.52	1.46	1.39	1.32	1.22	1.00

Critical Values of F When Alpha $= .10$

Within-group degrees of freedom	Between-groups degrees of freedom									
	1	2	3	4	5	6	7	8	9	10
1	39.86	49.50	53.59	55.83	57.24	58.20	58.91	59.44	59.86	60.19
2	8.53	9.00	9.16	9.24	9.29	9.33	9.35	9.37	9.38	9.39
3	5.54	5.46	5.39	5.34	5.31	5.28	5.27	5.25	5.24	5.23
4	4.54	4.32	4.19	4.11	4.05	4.01	3.98	3.95	3.94	3.92
5	4.06	3.78	3.62	3.52	3.45	3.40	3.37	3.34	3.32	3.30
6	3.78	3.46	3.29	3.18	3.11	3.05	3.01	2.98	2.96	2.94
7	3.59	3.26	3.07	2.96	2.88	2.83	2.78	2.75	2.72	2.70
8	3.46	3.11	2.92	2.81	2.73	2.67	2.62	2.59	2.56	2.54
9	3.36	3.01	2.81	2.69	2.61	2.55	2.51	2.47	2.44	2.42
10	3.29	2.92	2.73	2.61	2.52	2.46	2.41	2.38	2.35	2.32
11	3.23	2.86	2.66	2.54	2.45	2.39	2.34	2.30	2.27	2.25
12	3.18	2.81	2.61	2.48	2.39	2.33	2.28	2.24	2.21	2.19
13	3.14	2.76	2.56	2.43	2.35	2.28	2.23	2.20	2.16	2.14
14	3.10	2.73	2.52	2.39	2.31	2.24	2.19	2.15	2.12	2.10
15	3.07	2.70	2.49	2.36	2.27	2.21	2.16	2.12	2.09	2.06
16	3.05	2.67	2.46	2.33	2.24	2.18	2.13	2.09	2.06	2.03
17	3.03	2.64	2.44	2.31	2.22	2.15	2.10	2.06	2.03	2.00
18	3.01	2.62	2.42	2.29	2.20	2.13	2.08	2.04	2.00	1.98
19	2.99	2.61	2.40	2.27	2.18	2.11	2.06	2.02	1.98	1.96
20	2.97	2.59	2.38	2.25	2.16	2.09	2.04	2.00	1.96	1.94
21	2.96	2.57	2.36	2.23	2.14	2.08	2.02	1.98	1.95	1.92
22	2.95	2.56	2.35	2.22	2.13	2.06	2.01	1.97	1.93	1.90
23	2.94	2.55	2.34	2.21	2.11	2.05	1.99	1.95	1.92	1.89
24	2.93	2.54	2.33	2.19	2.10	2.04	1.98	1.94	1.91	1.88
25	2.92	2.53	2.32	2.18	2.09	2.02	1.97	1.93	1.89	1.87
26	2.91	2.52	2.31	2.17	2.08	2.01	1.96	1.92	1.88	1.86
27	2.90	2.51	2.30	2.17	2.07	2.00	1.95	1.91	1.87	1.85
28	2.89	2.50	2.29	2.16	2.06	2.00	1.94	1.90	1.87	1.84
29	2.89	2.50	2.28	2.15	2.06	1.99	1.93	1.89	1.86	1.83
30	2.88	2.49	2.28	2.14	2.05	1.98	1.93	1.88	1.85	1.82
40	2.84	2.44	2.23	2.09	2.00	1.93	1.87	1.83	1.79	1.76
60	2.79	2.39	2.18	2.04	1.95	1.87	1.82	1.77	1.74	1.71
120	2.75	2.35	2.13	1.99	1.90	1.82	1.77	1.72	1.68	1.65
∞	2.71	2.30	2.08	1.94	1.85	1.77	1.72	1.67	1.63	1.60

Within-group degrees of freedom	Between-groups degrees of freedom								
	12	15	20	24	30	40	60	120	∞
1	60.71	61.22	61.74	62.00	62.26	62.53	62.79	63.06	63.33
2	9.41	9.42	9.44	9.45	9.46	9.47	9.47	9.48	9.49
3	5.22	5.20	5.18	5.18	5.17	5.16	5.15	5.14	5.13
4	3.90	3.87	3.84	3.83	3.82	3.80	3.79	3.78	3.76
5	3.27	3.24	3.21	3.19	3.17	3.16	3.14	3.12	3.10
6	2.90	2.87	2.84	2.82	2.80	2.78	2.76	2.74	2.72
7	2.67	2.63	2.59	2.58	2.56	2.54	2.51	2.49	2.47
8	2.50	2.46	2.42	2.40	2.38	2.36	2.34	2.32	2.29
9	2.38	2.34	2.30	2.28	2.25	2.23	2.21	2.18	2.16
10	2.28	2.24	2.20	2.18	2.16	2.13	2.11	2.08	2.06
11	2.21	2.17	2.12	2.10	2.08	2.05	2.03	2.00	1.97
12	2.15	2.10	2.06	2.04	2.01	1.99	1.96	1.93	1.90
13	2.10	2.05	2.01	1.98	1.96	1.93	1.90	1.88	1.85
14	2.05	2.01	1.96	1.94	1.91	1.89	1.86	1.83	1.80
15	2.02	1.97	1.92	1.90	1.87	1.85	1.82	1.79	1.76
16	1.99	1.94	1.89	1.87	1.84	1.81	1.78	1.75	1.72
17	1.96	1.91	1.86	1.84	1.81	1.78	1.75	1.72	1.69
18	1.93	1.89	1.84	1.81	1.78	1.75	1.72	1.69	1.66
19	1.91	1.86	1.81	1.79	1.76	1.73	1.70	1.67	1.63
20	1.89	1.84	1.79	1.77	1.74	1.71	1.68	1.64	1.61
21	1.87	1.83	1.78	1.75	1.72	1.69	1.66	1.62	1.59
22	1.86	1.81	1.76	1.73	1.70	1.67	1.64	1.60	1.57
23	1.84	1.80	1.74	1.72	1.69	1.66	1.62	1.59	1.55
24	1.83	1.78	1.73	1.70	1.67	1.64	1.61	1.57	1.53
25	1.82	1.77	1.72	1.69	1.66	1.63	1.59	2.56	1.52
26	1.81	1.76	1.71	1.68	1.65	1.61	1.58	1.54	1.50
27	1.80	1.75	1.70	1.67	1.64	1.60	1.57	1.53	1.49
28	1.79	1.74	1.69	1.66	1.63	1.59	1.56	1.52	1.48
29	1.78	1.73	1.68	1.65	1.62	1.58	1.55	1.51	1.47
30	1.77	1.72	1.67	1.64	1.61	1.57	1.54	1.50	1.46
40	1.71	1.66	1.61	1.57	1.54	1.51	1.47	1.42	1.38
60	1.66	1.60	1.54	1.51	1.48	1.44	1.40	1.35	1.29
120	1.60	1.55	1.48	1.45	1.41	1.37	1.32	1.26	1.19
∞	1.55	1.49	1.42	1.38	1.34	1.30	1.24	1.17	1.00

Critical Values of Chi-Square

Degrees of freedom	Alpha = .01	Alpha = .05	Alpha = .10
1	6.63	3.84	2.71
2	9.21	5.99	4.61
3	11.34	7.82	6.25
4	13.28	9.49	7.78
5	15.09	11.07	9.24
6	16.81	12.59	10.65
7	18.48	14.07	12.02
8	20.09	15.51	13.36
9	21.66	16.92	14.68
10	23.21	18.31	15.99
11	24.73	19.68	17.28
12	26.22	21.03	18.55
13	27.69	22.36	19.81
14	29.25	23.68	21.06
15	30.58	25.00	22.31

Selecting the Right Statistical Test

Number of independent variables	Number of levels in the independent variable	Number of dependent variables	Type of data the dependent variable represents	Statistical test to use	Alternate statistical test**
N/A*	N/A*	1	Quantitative	One-sample z-test or one-sample t-test	N/A
1	2	1	Quantitative	Dependent-sample t-test or independent-sample t-test	Wilcoxon t-test or Mann–Whitney U test
1	3 or more	1	Quantitative	Analysis of variance (ANOVA)	Kruskal–Wallis H test
2	2 or more	1	Quantitative	Factorial ANOVA	N/A
0***	0	1***	Quantitative	Pearson correlation	N/A
0***	0	1***	Ordinal	Spearman correlation	N/A
1 or more	2 or more	1 or more	Nominal	Chi-square	N/A

*Not all parametric tests have a nonparametric equivalent, or the test mentioned is already a nonparametric test.

**When you're working with one-sample z tests or one-sample t-tests, it is really not logical to point out the independent variable or its levels.

***Independent variables are called predictor variables in correlations. Dependent variables are known as criterion variables in correlations.

Glossary

a priori: Before an event occurs. For example, we decide our alpha value a priori. This means we decide our alpha value prior to the start of a study.

alpha value: The degree of risk we are willing to take when computing inferential statistics. Sometimes referred to as the Type I error rate.

alternate hypothesis: *See* "research hypothesis."

analysis of covariance (ANCOVA):   A version of the analysis of variance where initial differences in the dependent variable are taken into account prior to final calculations.

analysis of variance (ANOVA):   A statistical tool based on an f distribution. Varieties of the ANOVA include the one-way ANOVA, the factorial ANOVA, and the multivariate ANOVA (MANOVA).

area under the curve: A value ranging from 0 to 100% representing the percentage of values in a given range under one of the data distributions (e.g., the z distribution and the t distribution).

area under the normal curve table: *See* "z table."

assumptions: Characteristics of a dataset we assume to be true prior to using a given statistical procedure.

bar chart: A graph showing bars rising from the bottom that are proportionate to the number of occurrences of a given value in a dataset.

beta: Annotation used to designate the probability of making a Type II error.

between-groups variance: The total amount of variance between datasets representing the levels of an independent variable.

bimodal: A data distribution with two modes.

bivariate: Involving two datasets. For example, a correlation procedure using two datasets is a bivariate correlation.

Bonferroni test: A post-hoc test used with the analysis of variance. When significant overall differences are found, this test runs a series of modified t-tests to ascertain which levels of the independent variable are significantly different from one another.

categorical data: *See* "nominal data."

cell: The space formed at the intersection of a row and a column in a table. The cell is used to record frequencies of values meeting the criteria of the cell.

central limit theorem: Statement telling us that "If a sample size is sufficiently large, the sampling distribution of the mean will be approximately normal." A theorem is nothing more than a statement of an idea that has proven to be true.

chi-square goodness-of-fit test: A nonparametric test that compares an observed distribution of nominal data to an expected distribution of nominal data.

chi-square test of independence: A nonparametric test of association that determines if similar numbers of occurrences appear in groups of nominal-level data; sometimes called a "factorial chi-square test."

coefficient of determination: The result of squaring Pearson's r in a correlational procedure. It represents the strength of a correlation directly attributable to the independent variable.

computed range: The range of a quantitative dataset computed by subtracting the lowest observed value from the highest observed value.

computed value of F: The result of an analysis of variance. This value can be compared to a critical value of F to determine if a null hypothesis should be rejected.

computed value of p: The probability that datasets being compared are significantly different. This is done by comparing p to a predetermined alpha value.

computed value of t: An output value from both independent- and dependent-sample t-tests. In the days prior to the use of computers, this value was compared to a table (critical) value of t to determine if a null hypothesis should be rejected.

computed value of z: The value of a data point on a normal curve representative of the number of standard deviations above or below the mean.

confidence interval: A range of sample statistics in which there is a given probability that a population parameter exists.

consumer of statistics: A person who uses statistics to interpret data as well as understands data that others are presenting.

contingency table: A multidimensional table used in the chi-square tests when the rows and columns of the table are defined by the levels of the independent variables.

continuous data: A generic name for interval- or ratio-level data; also called "quantitative data."

correlation coefficient: An output value from any of the correlational procedures. It represents the strength of the relationship between two or more sets of data.

criterion variable: The value that is being predicted in a correlation procedure.

critical value: A table value to which one of the computed statistics (e.g., t, z, or F) is compared in order to test a null hypothesis. There are specific tables for each of the respective distributions.

degrees of freedom: The number of scores that are free to vary when describing a particular sample. The method by which they are calculated changes depending on the statistical test you are using, but for most calculations, the value is defined as one less than the number of data values.

dependent-sample *t*-test: Statistical tool used when a study involves one independent variable with two levels and one dependent variable that is measured with quantitative data. In this case, the levels of the independent variable must be related to or correlated with one another.

dependent variable: The "effect" that is being measured in a study.

descriptive statistics: Numeric and graphical statistical tools that help us "describe" the data so it can be better used for our decision making. Examples include the mean of a dataset or a pie chart.

deviation from the mean: The amount any given value in a dataset is from the mean of the dataset.

directional hypothesis: A hypothesis that implies a "greater than" or a "less than" relationship between the variables being studied; also called a "one-tailed hypothesis."

disordinal interaction: An interaction of two independent variables wherein the values of one independent variable have a dramatically opposite effect on a dependent variable than do values of the second independent variable.

effect size: In parametric statistics, a measure of the degree to which an independent variable affects a dependent variable.

empirical rule: The rule stating that, in a normal distribution, approximately 68% of values are within ±1 standard deviation from the mean, 95% of values are within ±2 standard deviations of the mean, and nearly all (99.7%) of all values are within ±3 standard deviations of the mean.

equal variances assumed/not assumed: A test used in parametric statistics to ensure that the variability within sets of data being compared is equitable. Significant differences in variance call for modification of how computed values in parametric statistics are calculated.

ethical research: A research study in which the researcher ensures that all participants participate voluntarily and that the participants not be harmed in any way—socially, physically, or mentally.

expected value: Value used in chi-square tests as a measure of the number of occurrences of each cell value that the researcher believes should appear. These expected values are compared to the actual number of occurrences in each cell.

experimental independent variable: *See* "manipulated (experimental) independent variable."

F distribution: The plot of *F* values computed from repeated samples of data.

F value: The value computed in an analysis of variance. It is compared to a critical value of *F* to interpret hypotheses.

factorial analysis of variance: Statistical tool used when a study involves more than one independent variable as well as one dependent variable that represents quantitative data.

fail to reject the null hypothesis: Your inability to reject the null hypothesis based on the results of your statistical test. This means you are unable to support your research hypothesis.

frequency distribution table: A table showing the number of occurrences of the various values in a dataset.

goodness of fit: The degree to which an observed distribution of data values fits the distribution that was expected.

graphical descriptive statistics: The use of tools such as pie charts, bar charts, and relative frequency diagrams to illustrate the characteristics of a dataset.

histogram: A graph showing the number of occurrences of the various values in a dataset.

hypothesis: A statement that reflects the researcher's beliefs about an event that has occurred or will occur.

independent-sample *t*-test: A statistical tool used when a study involves one independent variable with two levels and one dependent variable that is measured with quantitative data. In this case, the levels of the independent variable cannot be related to or correlated with one another.

independent variable: The "cause" being investigated in a study.

inferential statistics: Statistical tools used to make decisions or draw inferences about the data we have collected.

interaction effect: The simultaneous effect on a dependent variable of two or more independent variables.

intercept: The point at which the line of best fit crosses the *x*- or *y*-axis on a scatterplot.

interquartile range: The range of values between the first and third quartiles in a data distribution.

interval data: One of two types of data that are called quantitative or continuous. Interval data can theoretically fall anywhere within the range of a given dataset. The range can be divided into equal intervals, but this does not imply that the intervals can be directly compared. For example, a student scoring in the 80s on an examination is not twice as smart as a student scoring in the 40s. There is no absolute zero point in an interval-level dataset.

kurtosis: The degree to which a data distribution deviates from normal by being too "peaked" or too "flat."

latent independent variable: An independent variable that is examined "as is" and is not manipulated by the researcher. Examples include gender and ethnic group.

leptokurtosis: A bell-shaped distribution that is too peaked (i.e., too many values around the mean of a distribution) to be perfectly normally distributed.

levels of the independent variable: Different values of the independent variable that are investigated to determine if there is a differing effect on the dependent variable.

Levene's test: A statistical tool for testing the equality of variance in datasets being compared.

line of best fit: A graphical technique used in correlational procedures to show the trend of correlations being plotted. A line of best fit that emulates a 45-degree angle demonstrates a strong positive correlation. A line of best fit that appears flatter indicates a lesser degree of correlation.

logistic regression: A linear regression procedure used to predict the values of a nominal dependent variable.

lower confidence limit: The smallest value in a confidence interval.

main effect: The effect of a single independent variable on a dependent variable.

manipulated (experimental) independent variable: An independent variable where the researcher defines membership into the factors or levels. An example would show a researcher placing students in different groups to measure the effect of technology on learning. Sometimes called "experimental independent variable."

Mann–Whitney *U* test: Nonparametric alternative to the independent-sample *t*-test.

mean: A measure of central tendency reflecting the average value in a dataset. Only used with quantitative data (interval and ratio).

mean of means: The average score of a dataset created by computing and plotting the mean of repeated samples of a population.

mean square: A constant used in the calculation of an *F* value computed by dividing the Sum of Squares value for a specific group by the degrees of freedom for the same group.

measures of central tendency: Descriptive statistics that help us determine the middle of a dataset. Examples include the mean, the median, and the mode.

measures of dispersion: Descriptive statistics that help determine how spread out a data distribution is. Examples include the range, the standard deviation, and the variance.

measures of relative standing: Measures used to compare data points in terms of their relationship within a given dataset. Examples include z scores and percentiles.

median: A measure of central tendency that describes the midpoint of data that are ordinal, interval, or ratio level and have been sorted into ascending or descending sequence.

mode: A measure of central tendency reflecting the value that occurs most often in a dataset. Can be used with all types of data.

multimodal: A dataset having more than one mode.

multiple-comparison test: Tests used after the analysis of variance to determine exactly which means are significantly different from one another. An example is the Bonferroni test.

multiple regression: A regression procedure that uses more than one predictor variable.

multivariate: Referring to a statistical test that has more than one independent or dependent variable.

multivariate analysis of variance (MANOVA): Statistical tool used when a study involves any number of independent variables and more than one dependent variable that measures quantitative data.

negatively skewed: Skewed to the left (i.e., there are more values below the mean of a quantitative dataset than there are above the mean).

nominal data: Data that are categorical in nature. Examples include gender, ethnic group, and grade in school.

nondirectional hypothesis: A hypothesis that implies a difference will exist between the variables being studied but no direction is implied. Also called a "two-tailed hypothesis."

nonmanipulated (quasi-) independent variable: An independent variable where the levels are preexisting. An example would include a situation where the independent variable was gender; the levels are male and female.

nonparametric statistics: Inferential statistical tools used with nominal or ordinal data or with quantitative data where the distribution is very abnormally distributed (e.g., skewed or kurtotic).

normal distribution: A quantitative distribution wherein the distribution is bell-shaped.

null hypothesis: A hypothesis that states there will be no relationship between the variables being studied. The null hypothesis is the antithesis of the research hypothesis.

numeric descriptive statistics: The use of tools such as measures of central tendency, measures of dispersion, and measures of relative standing to illustrate the characteristics of a dataset.

observed value: The actual count of occurrences in a cell of nominal value. It is used in chi-square tests to compare to expected values.

one-sample chi-square test: The comparison of cell frequencies in a nominal distribution to those that would be expected according to a previously defined criterion. *See also* "chi-square goodness-of-fit test."

one-tailed hypothesis: *See* "directional hypothesis."

one-way analysis of variance (ANOVA):   Statistical tool used when a study involves one independent variable with three or more levels and one dependent variable that is measured with quantitative data.

ordinal data: Data that are rank-ordered. Examples include order of finish in a race and class standing.

ordinal interaction: An interaction effect where the influence of one independent variable remains in the same direction but varies in magnitude across levels of another independent variable

***p* value:**   The probability that groups being compared came from the same population.

paired-samples *t*-test: *See* "dependent-sample *t*-test."

parameter: Any value known about a dataset representing an entire population.

parametric: Pertaining to the use of quantitative data that form a mound-shaped distribution.

parametric statistics: Inferential statistical tools used with quantitative data.

Pearson's *r*:   Output from a correlation procedure using quantitative datasets. Pearson's *r* can range from –1.00 to +1.00 with extremely high values indicating a strong positive correlation; extremely low values indicate a strong negative correlation. Values around zero indicate a weak correlation.

percentile: A measure of relative standing that describes the percentage of other values in the dataset falling below it. For example, a test score of 50 that falls into the 80th percentile means that 80% of the other scores are less than 50.

pie chart: A circular chart divided into segments, with each representing the percentage of a given value in a dataset.

platykurtosis: A bell-shaped distribution that is too flat (i.e., fewer values around the mean of the distribution than is expected) to be perfectly normally distributed.

pooled standard deviation: An estimate of the population standard deviation based on a weighted average from a set of sample standard deviations.

population: All members of a group being investigated.

population parameter: Any value known about a population.

positively skewed: Skewed to the right (i.e., there are more values above the mean of a quantitative dataset than there are below the mean).

possible range: The set of values a dataset theoretically covers. For example, on most examinations, the possible range is 0 to 100.

post-hoc test: *See* "multiple-comparison test."

power: The ability of a statistical test to identify when there is a true significant difference in the values being compared. It is computed by subtracting beta from 1.00; this means that as power increases, the probability of making a Type II error decreases.

practical significance: The determination that the results of a research study are practically important or that they can be important or useful in real life.

predictor variable: The independent variable in a correlation procedure. It is used to predict values of the criterion variable.

problem statement: The specific issue, concern, or controversy that we want to investigate using inferential statistics.

qualitative data: Non-numeric data such as interviews and recordings.

quantitative data: A generic label for any data that are on an interval or ratio scale.

quartile: The range of data representing 25% of the distribution.

range: The set of values that a dataset actually covers. For example, if the lowest value in our dataset is 30 and the highest value is 70, our range is 40 (highest minus lowest).

rank data: *See* "ordinal data."

ratio data: One of two types of data (see "interval data") that can be classified as quantitative. Ratio-level data can be divided into equal intervals and allow for direct comparison. For example, a person who weighs 150 pounds is twice as heavy as a person who weighs 75 pounds. There is an absolute zero point in ratio-level data.

rejecting the null hypothesis: A decision made, based on the computed and critical value of a statistic, that indicates that differences between values being compared are true and not due to chance.

relative frequency: The percentage of occurrences of various values in a dataset.

research hypothesis: A synonym for the directional or nondirectional hypothesis being investigated. The research hypothesis is the antithesis of the null hypothesis.

sample: A subset of a population being studied.

sample statistic: Any value known about a sample of data.

sampling distribution of the means: The resulting distribution when repeated samples are drawn from a population and their mean is calculated and plotted. Given enough samples, this will result in a perfect normal distribution.

sampling error: The error inherent in measuring only a selected group (sample) from a population.

scatterplot: A graph that shows the relationship between sets of quantitative data.

significantly different: A difference between two variables being measured that cannot be attributed to sampling error. The values are different due to reasons other than chance and would cause a null hypothesis to be rejected.

simple frequency: The actual count of the number of occurrences of a given variable in a dataset.

skewed left: *See* "negatively skewed."

skewed right: *See* "positively skewed."

skewness: Degree to which a data distribution deviates from being a normal distribution (i.e., mound shaped) by having too many values in one end of the distribution or the other.

slope: The degree of inclination of a line of best fit.

Spearman's rho: A correlational procedure used with ordinal-level data. Rho can range from zero to +1.00. The closer rho is to +1.00, the stronger the correlation.

standard deviation: Measurement used to describe how spread out a dataset is. It can be thought of as the average distance of any data point in a distribution from the mean of the dataset. Obviously, the larger the standard deviation, the more spread out the dataset is. The standard deviation is the square root of the variance.

standard error of the mean: The standard deviation of a sampling distribution of the means.

standard error of mean difference: The standard deviation of the sampling distribution of the mean differences used in a t-test.

stanine: A measure of relative standing involving the division of a distribution of quantitative scores into nine parts.

statistic: Any value we know about a sample of data.

sum of squares: The sum of the within-group and between-groups sum of squares; used in the analysis of variance.

t distribution: The resulting distribution when repeated samples are drawn from a population and their t value is calculated and plotted. Given enough samples, this will result in a perfect normal distribution.

T-score: A measure of relative standing based on a scale of 20 to 80. Caution: This is not the t value computed as part of a t-test.

t table: A table showing the critical values of t for various alpha values and degrees of freedom.

theoretical significance: Research results that may be useful in testing a current theory or the development of a new theory.

two-tailed hypothesis: *See* "nondirectional hypothesis."

Type I error: Rejecting the null hypothesis when we should not reject it.

Type II error: Failing to reject the null hypothesis when we should reject it.

Type I error rate: The probability of rejecting the null hypothesis when we should not reject it (i.e., rejecting the null hypothesis when it is true). The Type I error rate is synonymous with our alpha value.

Type II error rate: The probability of failing to reject the null hypothesis when we should reject it; sometimes referred to as our "beta value."

univariate: Dealing with one independent and one dependent variable.

upper confidence limit: The highest value in a confidence interval.

variance: A measure of dispersion that is determined by squaring the standard deviation. Used in computations such as the analysis of variance where negative numbers are not allowed.

Wilcoxon t-test: Nonparametric alternative to the dependent-sample t-test.

within-group degrees of freedom: An integral part of the calculation of F values; computed by subtracting one from the number of groups (i.e., levels of the independent variable) in the analysis of variance.

within-group variance: The amount of variance that occurs within one of the groups of data represented by a level of an independent variable.

x-axis: The horizontal axis of a plot.

y-axis: The vertical axis of a plot.

z score: A measure of relative standing computed by subtracting the mean of a bell-shaped dataset from the observed score and dividing by the standard deviation.

z table: A table showing the critical value of z for all values in a normal distribution.

Answers to Quiz Time!

Chapter 1 (pp. 29–32)

1. This statement is not clear and concise. By reading it, it appears that the researcher wants to investigate some relationship between absenteeism and the particular school but one cannot say for certain.

2. In this case, there are two issues. First, this study is not within the scope of the teachers' expertise; mold inspection and the like is best left to experts. Second, all of the variables are not considered. The problem statement indicates that potential health hazards are to be investigated but it does not mention mold and asbestos. While they may be part of what could be considered health issues, there are quite a few other problems that might be identified as well.

3. This appears to be a valid problem. It meets all of the criteria and is stated in an appropriate manner.

4. This appears to be a good problem statement given that the attitudes of ranchers could be measured by an instrument representing numeric data. All other characteristics of the problem statement are met or assumed.

5. This has the potential to be a good problem statement, but there appears to be one major flaw; what is the scope of the problem? It would be better if the author included characteristics such as age, gender, and ethnic group that might be applicable to the study.

6. The major issue with this problem statement is one of scope. While the problem is clear, the goals of the lesson plan are clearly beyond their reach. Very few teachers at that level would have access to the types of tools necessary to conduct the research that is being suggested.

7. The problem statement seems to be very well written, but I worry about the scope. It would be better if the scope was limited to different makes of foreign cars. For example, the study might not be valid if the researcher were comparing large European luxury sedans to economical imports from Japan or Korea.

8. Again, there is the potential for a very good problem statement here. The author might better state the relationship between the variables, however. Given, weight gain is a part of prenatal lifestyle, but it is not the only component; other issues such as exercise, alcohol consumption, and tobacco use could also contribute to the health of newborn infants.

9. This problem statement begs to ask, "What is the problem?" The scope is not mentioned, it is not clear and concise, and not all of the variables seem to be included. In addition, it's not clear whether the researcher is comfortable with the subject area. In short, this is not a good problem statement.

The Case of Distance Counseling

Notice, in this case, that no direction is implied. The case study says "no one really knows if either one works better or worse than the other." Because of that, we need to state a nondirectional research hypothesis:

■ *There will be a significant difference in the number of sessions for treating depression between clients in a traditional setting and clients in a distance setting.*

The null hypothesis would be:

■ *There will not be a significant difference in the number of sessions for treating depression between clients in a traditional setting and clients in a distance setting.*

The Case of the New Teacher

In this case, a direction is implied. The teacher wants to determine if there are fewer parent–teacher conferences in schools in a less affluent area than in schools from a more affluent section of town. Her research hypothesis would be:

■ *There will be significantly fewer parent–teacher conferences in nonaffluent schools than in affluent schools.*

The null hypothesis would read:

■ *There will be no significant difference in the number of parent–teacher conferences between nonaffluent schools and affluent schools.*

The Case of Being Exactly Right

In this case, the engineers want to develop software to monitor the process to ensure the temperature is maintained at 155°. Because of that, the null and research hypotheses will reflect a comparison to an exact value. The engineer's research hypothesis would be:

■ *There will be a significant difference between the temperature of our process and 155°.*

The null hypothesis would read:

- *There will be no significant difference between the temperature of our process and 155°.*

The Case of "Does It Really Work?"

In this case, no direction is implied; the dean only wants to know if the students taking anatomy and physiology online do as well as their on-campus counterparts. Knowing that, the faculty would test the following research hypothesis:

- *There will be a significant difference in achievement between students taking anatomy and physiology in an online environment and students taking the same course on-campus.*

The null hypothesis would be:

- *There will be no significant difference in achievement between students taking anatomy and physiology in an online environment and students taking the same course on-campus.*

The Case of Advertising

In this case a direction is implied; the superintendent is interested in finding out if online advertising leads to a larger number of job applicants. Her research hypothesis would be:

- *There will be a significantly larger number of job applicants from Internet-based advertisements than from newspaper advertisements.*

The null hypothesis would be:

- *There will be no significant difference in the number of job applicants from Internet-based advertisements than from newspaper advertisements.*

The Case of Learning to Speak

In this case, teachers want to investigate the difference between two types of instruction: language immersion and formal instruction in the classroom. There is no direction implied, so our research hypothesis would read:

- *There will be a significant difference in language skills after one year of instruction between students taught in an immersion environment and students taught in the classroom.*

The null hypothesis would read:

- *There will be no significant difference in language skills after one year of instruction between students taught in an immersion environment and students taught in the classroom.*

Chapter 2 (pp. 58–59)

Hypothesis	Independent variable and levels	Dependent variable	Data type
1. There will be a significant difference in motivation scores between students in online programs and students in traditional programs.	Students: Those in traditional programs and those in online programs.	Motivation.	Quantitative (interval). There has to be some level of motivation, but ratios cannot be made.
2. There will be a significant difference in the number of males and females working as computer programmers in corporate America.	Programmers: Males and females.	Count of each gender.	Nominal.
3. There will be a significant difference in the number of females in computer science classes between students in the United States, France, and Russia.	Country: United States, France, and Russia.	Count of females in each country.	Nominal.
4. There will be a significant difference in weight gained during their freshman year between students who live at home and students who live away from home.	Residence: At home or away from home.	Weight gain.	Quantitative (ratio). There is a real zero point, and ratios can be made.
5. There will be a significant difference in first-year salaries between graduates of Ivy League schools and graduates of state universities.	University: Ivy League and state.	First-year salary.	Quantitative (ratio). There is a real zero (unemployment), and ratios can be made.
6. Administrative assistants who work in cubicles are significantly less productive than administrative assistants who work in enclosed offices.	Work location: Cubicles or enclosed office.	Productivity.	Quantitative (ratio) if measured on a numeric scale.
7. Primary-care patients who are treated by an osteopathic physician will have significantly fewer health problems than primary-care patients who are treated by an allopathic physician.	Patients: Those with osteopathic doctors and those with allopathic doctors.	Health problems.	Quantitative (ratio). There is a real zero, and ratios can be made.
8. Truck drivers who check their tire pressure frequently will have significantly higher miles per gallon than truck drivers who do not check their tire pressure frequently.	Truck drivers: Those who check their tire pressure frequently and those who do not.	Miles per gallon.	Quantitative (ratio). There is a real zero, and ratios can be made.
9. Insurance companies that use computer-dialing services to call prospective clients will have a significantly lower number of sales than insurance companies that use live agents to call prospective clients.	Method of solicitation: Computer-dialing and live agent.	Number of sales.	Quantitative (ratio). There is a real zero, and ratios can be made.
10. The rankings of favorite sporting activities will be significantly different between Mexico, the United States, and Canada.	Country: Mexico, the United States, and Canada.	Ranking of sporting activity.	Ordinal.

1. The average height is 68.07 inches.

2. The median class rank is 8; remember, this tells us nothing really. We need to see what grade or other measurement that relates to.

3. The mode of shoe size is 8.

4. There is no average class; it is a nominal value.

5. The median height is 69 inches.

6. The mode of height is 69.

7. The median shoe size is 9.

8. The mode class rank is 6; it occurs twice. Remember, we need to see what grade or other measurement that relates to.

9. Class is multimodal; there are four each of freshman, sophomore, and senior.

10. Class is a nominal value; the median should not be computed.

11. Class rank is ordinal; the average should not be computed.

12. The average shoe size is 9.6.

Chapter 3 (pp. 86–87)

1. The possible range for company A is 47; the possible range for company B is 47; the possible range for company C is 47.

2. The possible range for all companies taken together is 47.

3. The computed range of company A is 43; the computed range of company B is 25; and the computed range of company C is 40.

4. The range of all companies combined is 45.

5. For company A, the data are spread out since the computed range of 43 is 91.4% of the possible range. For company B, the data are less spread out since the computed range of 25 is only 53.2% of the possible range. For company C, the data are again very spread out since the computed range of 40 is 85.1% of the possible range. The computed overall range of 45 represents 95.7% of the possible range.

6. The standard deviation of company A is 14.89, the standard deviation for company B is 7.56, and the standard deviation for company C is 11.34.

7. For company A, on average, each value in the dataset is approximately 14.89 standard deviations away from the mean. For company B, on average, each value in the dataset is approximately 7.56 standard deviations away from the mean. For company C, on average, each value in the dataset is approximately 11.34 standard deviations away from the mean.

8. The standard deviation for age across all companies is 12.48.

9. For company A, the variance is 221.82, the variance for company B is 57.12, and the variance for company C is 128.67.

10. The variance of all companies combined is 155.82.

11. Since the variance is not in the scale of the original data, the variance is not very meaningful. A large value would indicate a broader range of data; smaller variance values would indicate a smaller range.

1.

TABLE 3.27. Data for Computation of z Scores, T-Scores, Stanines, and Ranked Mean Scores

Mean	Observed value	Standard deviation	z score	T-score	Stanine	Ranked mean score
30	33	2.00	1.50	65	8	100
48	52	5.00	.80	58	7	81
55	54	3.00	−.33	47	5	71
71	77	7.00	−.86	59	7	61
14	8	2.70	−2.22	28	1	55
23	35	5.00	2.40	74	9	48
61	48.6	2.90	−4.28	7	1	47
100	114	6.33	2.21	72	9	30
81	78.5	1.55	−1.61	34	2	23
47	60.00	12.0	1.08	61	8	14

25th percentile = 30
50th percentile = 51.5
75th percentile = 71

2. The z score for a score of 130 with a mean of 100 and a standard deviation of 15 is 2.00.

3. In this case, we would first have to determine the standard deviation. Since we know that it is the square root of the variance, it is easy to determine. The square root of $9,000 is $94.87. Since we are interested in a point 2 standard deviations below the mean, we then multiply that by 2: $94.87 * 2 = $189.74. In order to determine our cutoff point, we subtract that from $22,000 and get $21,810.26. Since the family we are interested in has an income of $19,500, they would qualify for governmental assistance.

Chapter 4 (pp. 115–116)

Since we said that we always use a computer to help us when we're graphically describing our dependent variable, we are in somewhat of a jam. Knowing we can't do that, let's use the following questions to understand conceptually where we are.

1. The easiest way to show the frequency of the values would be with a bar chart. Each bar would represent the actual number of each value in the dataset.

2. In this case there would seem to be no consistent relationship between the data values (i.e., the line of best fit would tend to be flat). This is based on the observation that, as some of the reading scores go up, their corresponding math scores do the same. In other cases, the opposite is true; as some of the reading scores go up, their corresponding math score goes down.

3. In this case, it appears to be a positive relationship. As each value of education goes up, so does the salary. The line of best fit would be sloped up from the left to the right.

4. The histogram is used with quantitative data in order to allow for fractional values; given this, nominal data are not plotted using a histogram. A histogram is mound-shaped if the mean, median, and mode are approximately equal and has a symmetrical distribution of values on either side of the center. If there are more values in either end of the distribution than would generally be found in a bell-shaped distribution, we say the distribution is skewed. If the mean is greater than the median, the distribution is positively skewed; if the mean is less than the median, negative skewness occurs. If there are more data values than expected in the middle of the dataset, we say the distribution is leptokurtic and the plotted distribution would be more peaked than normal. Fewer values in the center cause the distribution to "flatten out"; we call this distribution platykurtotic.

5. A perfect normal distribution has an equal mean, median, and mode with data values distributed symmetrically around the center of the dataset.

6. Since the mean is greater than the median, the distribution would be positively skewed.

7. The distribution will be leptokurtic; it will appear more peaked than a normal distribution.

8. No, there are many reasons a dataset with an equal mean and median does not have to be normally distributed. One primary reason would be the possibility of too few data values to give the symmetrical, mound-shaped appearance of a normal distribution.

9. No, most parametric inferential statistical tests are powerful enough to work with minor problems with skewness or kurtosis.

10. If the mean is less than the median, negative skewness occurs. Statistical software can be used to determine if this is problematic by returning a skewness coefficient. This value is only problematic if it is less than –2.00. If the mean is greater than the median, positive skewness occurs. Again, this is detrimental to decision making only if the software returns a value greater than +2.00.

11.

TABLE 4.10. Values for Quiz Time Question 11

z value 1	Area under the curve for z 1	z value 2	Area under the curve for z 2	Difference
3.01	49.87%	2.00	47.72%	2.15%
–2.50	49.38%	3.00	49.87%	99.25%
–1.99	47.67%	–0.09	3.59%	44.08%
1.50	43.32%	–1.50	43.32%	86.64%

Chapter 5

In-Chapter Exercises

How Are We Doing So Far? (p. 126)

1. z score:
    a. 1.347219
    b. –2.57196
    c. 1.837116
    d. 0.367423
    e. –0.36742
    f. –1.22474
    g. –0.24495
    h. –0.9798

2.

TABLE 5.6. Computing the Distance between *z* Scores

Value 1	z 1	Area under the curve for z 1	Value 2	z 2	Area under the curve for z 2	Area between z 1 and z 2
90	–1	.3413	95	1.5	.4332	.7745
89	–1.5	.4332	90	–1	.3413	–0.0919
91.5	–.25	.0987	93.5	.75	.2734	.3721
92	0	0	96	2	.4772	.4772

Do We Understand Everything We Need to Know about Confidence Intervals? (Table 5.13, p. 138)

TABLE 5.13. Learning to Compute the Width and Limits of a Confidence Interval

$\bar{x}$	Alpha	n	σ	Lower limit of CI	Upper limit of CI	Width of CI
100	.10	25	5	98.36	101.65	3.29
500	.05	50	25	493.07	506.93	13.86
20	.01	15	3	18.01	21.99	3.99
55	.05	20	7	51.93	58.07	6.14
70	.01	22	6	66.71	73.29	6.59
220	.10	40	10	217.40	222.60	5.20

Great News—We Will Always Use Software to Compute Our *p* Value
(Tables 5.19, 5.20, 5.21, pp. 153–154)

TABLE 5.19. Intelligence—Is There a Significant Difference?

Null hypothesis:	There will be no significant difference in the intelligence level between our class and the overall university level.
Research hypothesis:	The average intelligence level of our class will be significantly higher than the average intelligence level of all other classes in the university.
Population average:	100
Class Average:	105
SEM:	3
Alpha:	.05
Computed *z* value:	1.67
Critical *z* value:	1.645
Reject null hypothesis?	Yes
How do you know?	The computed value of *z* is greater than the critical value of *z*.
Is *p* less than .05?	Yes
How do you know?	Because the null is rejected, *p* must be less than alpha.

TABLE 5.20. Rainfall—Is There a Significant Difference?

Null hypothesis:	There will be no significant difference in the yearly amount of rainfall between Texas and Florida.
Research hypothesis:	There will be a significant difference in the yearly amount of rainfall between Texas and Florida.
Texas average:	70 inches
Florida average:	62 inches
SEM:	10 inches
Alpha:	.05
Computed *z* value:	−.8
Critical *z* value:	−1.645
Reject null hypothesis?	No
How do you know?	The computed value of *z* is not less than the critical value of *z*.
Is *p* less than .05?	Yes
How do you know?	Because the null is not rejected, *p* must be equal to or greater than alpha.

TABLE 5.21. Truancy—Is There a Significant Difference?

Null hypothesis:	There will be no significant difference in the number of truancy cases at our school and the district average number of truancy cases.
Research hypothesis:	There will be a significant difference in the number of truancy cases in our school and the district average number of truancy cases.
District average:	12 cases
Our school's average:	17 cases
SEM:	2 cases
Alpha:	.05
Computed *z* value:	2.5
Critical *z* value:	1.645
Reject null hypothesis?	Yes
How do you know?	The computed value of *z* is greater than the critical *z* value.
Is *p* less than .05?	Yes
How do you know?	Because the null is rejected, *p* must be less than alpha.

Chapter 5 Quiz Time! Answers (p. 157)

1. There is one independent variable, fish, with three levels—lakes, streams, and rivers. The number of fish caught is the dependent variable. Because you are collecting quantitative data, you would use a one-way ANOVA.

2. There is one independent variable, office, with two levels—well-lighted and dimly lighted. The level of productivity is the dependent variable. Because you are collecting quantitative data, you would use an independent-sample *t*-test.

3. Here there is one independent variable—country, with three levels—Canada, Mexico, and the United States. The number of health problems is the dependent variable; this would call for the use of a one-way ANOVA.

4. In this case, you have an independent variable with one level—foreign-manufactured cars. Since you want to compare miles per gallon, the dependent variable, to a known value of 19, you would use a one-sample *t*-test.

5. The number of children is the dependent variable; family is the independent variable and has three levels—lower, middle, and high socioeconomic status. In this case, you would use a one-way ANOVA.

6. The location of the show is the independent variable; the two levels are "on Broadway" and "touring." The degree of audience appreciation is a quantitative dependent variable; in this case, you would use an independent-sample *t*-test.

7. In this case, you have a predictor variable, number of siblings, and a criterion variable, annual income. Because of this, you would use the Pearson correlation.

8. Here there is one independent variable, political party, with three levels—Repub-

licans, Democrats, and Liberals. Since we have one dependent variable that is numeric, we would use a one-way ANOVA.

9. Here we have one independent variable, gender, with two levels—males and females. We are comparing the number of each in our office (i.e., the observed value) to the percentage of each nationwide (i.e., the expected value). Because of this, we would use a one-way chi-square test.

10. Herein, we're comparing the degree of procrastination of one group of authors, before and after a call from their editor. Because it's the same group being compared at two different times, we would use a dependent-sample t-test.

Chapter 6 (pp. 179–180)

1. In this case, there is a one-tailed "greater than" hypothesis stating that children without fluoride will have a significantly higher number of cavities than the national average of 2.0. The mean number of cavities for our sample group is 2.33 resulting in a mean difference of .333. While this is greater, given the p value of .215 (remember, you must divide the two-tailed p value of .430 by 2 when you have a directional hypothesis), the difference is not significant. We will fail to reject the null hypothesis and will not support the research hypothesis.

2. Here we have a one-tailed "less than" hypothesis stating that Ivy League-trained physicians will have significantly fewer malpractice lawsuits than the national mean of 11. In this case, the Ivy League physicians have an average of 5.20, giving a mean difference of –5.8. Our p value of .000 indicates this difference is significantly lower. We will reject the null hypothesis and support the research hypothesis.

3. In this case, we have a one-tailed "greater than" hypothesis in that we believe the mean graduation rate of students who went to community college for their first two years will be greater than the national average of 82%. The graduation rate of our sample is 79.6% and results in a p value of .0215; because of this, we will reject the null hypothesis. However, since the mean difference is opposite of what we hypothesized, we will not support the research hypothesis.

4. In this case, we've hypothesized that adult turkeys in the wild will weigh significantly less than 12.0 pounds. Given a mean difference of –1.467 and a p value of .008, we reject the null hypothesis and support the research hypothesis.

Chapter 7 (pp. 210–214)

The Case of the Homesick Blues

1. The researchers do not state that they expect a difference between the groups, so the research hypothesis must be nondirectional:

   ▪ *There will be a significant difference in the number of phone calls made during basic training between males and females.*

   The null hypothesis states:

> *There will be no significant difference in the number of phone calls made between males and females in basic training.*

2. The independent variable is gender; the two levels are male and female.

3. The dependent variable is "number of calls."

4. The mean value of male soldiers is much larger than that of the female soldiers. The p value for the Levene statistic is less than .05, so we have to use the "Equal variances not assumed" column. Since we have a nondirectional hypothesis, we have to divide the p value of .002 by two. In this case, the very low p value of .001 indicates there is a significant difference; male trainees call home a significantly greater number of times than female trainees.

The Case of the Cold Call

1. The management wants to see if working in cubicles leads to a higher number of phone calls, so they must state a "greater than" directional research hypothesis:

> *Employees in private offices will make significantly more phone calls than employees working in cubicles.*

The null hypothesis would read:

> *There will be no significant difference in the number of phone calls made between employees working in cubicles and employees working in private offices.*

2. The independent variable is "location," and the two levels are "cubicle" and "office."

3. The dependent variable is "number of calls."

4. In this case, the mean number of calls for employees in cubicles appears to be much greater than for those working in offices. Because of the significant Levene value, the "Equal variances not assumed" column must be used. The p value of .000 indicates that the null hypothesis must be rejected. Remember, check the means. In this case, contrary to the hypothesis, employees in cubicles, despite the noise level, make significantly more phone calls than their peers in private offices.

The Case of the Prima Donnas

1. The owners are investigating the female actresses' statement, so the research hypothesis must be a "greater than" directional research hypothesis:

> *Females, when placed on the marquee first, will lead to significantly higher ticket sales than when males are placed first.*

The null hypothesis would be:

> *There will be no significant difference in ticket sales when females are first on the marquee or when males are first on the marquee.*

2. The independent variable is "Name," and the two levels are "Female" and "Male."

3. The dependent variable is "ticket sales."

4. Here the females do have a higher number of ticket sales, but we need to determine if the difference is significant. Using the "Equal variances assumed" column, the low p value (i.e., 439) shows that the differences between male and female actors are different but not significantly so.

The Case of the Wrong Side of the Road

1. In this case, I am hypothesizing that the gas stations on the way out of town have prices that are significantly higher than those coming into town. Because of that, I state a one-tailed directional research hypothesis:

   ▓ *Gas stations going out of town have significantly higher prices than gas stations going into town.*

   The null hypothesis would read,

   ▓ *There will be no significant difference in gas prices between stations going into town and those coming out of town.*

2. The independent variable is "Gas Station," and the levels are "Into town" and "Out of town."

3. The dependent variable is "Price."

4. Here we have mean values that are very close; despite that, we must test them for significance. Here we can use the "Equal variances assumed" column and see that we have a very large p value, .721. Given that, we cannot reject the null hypothesis; even though the prices are different, they are not significantly different.

Chapter 8 (pp. 236–240)

The Case of Technology and Achievement

1. In this case, the proponents of technology believe that computers are beneficial to students. Because of that, they want to test a directional "greater than" research hypothesis:

   ▓ *Students will have significantly higher grades after the use of technology in the classroom than before they used technology in the classroom.*

   The null hypothesis will be:

   ▓ *There will be no significant difference in grades before the technology was introduced and after the technology was introduced.*

2. The independent variable is "Instructional Method," and the two levels are "Before technology" and "After technology."

3. The dependent variable is Student Grades.

4. Here the mean scores are very close; the grades actually went down slightly over the term. The p value of .434 shows that there is no significant difference.

The Case of Worrying about Our Neighbors

1. Here the citizens are concerned that their property values will go down. Because of that, we must state a directional "less than" research hypothesis:

   ■ *Values after annexation will be significantly lower than values before annexation.*

   The null hypothesis would read:

   ■ *There will be no significant difference in property values prior to the annexation and after the annexation.*

2. The independent variable is "Annexation"; the levels are "Before" and "After."

3. The dependent variable is Property Value.

4. We can see that the citizens may have reason for concern; the property values dropped nearly $10,000. The p value of .024 shows the difference to be significant.

The Case of SPAM

1. The entrepreneur is attempting to sell his service by advertising fewer junk e-mails; he is stating a one-tailed "less than" research hypothesis:

   ■ *There will be significantly fewer junk e-mails after customers start using his service than prior to using his service.*

   The null hypothesis would be:

   ■ *There is no significant difference in the amount of SPAM between the entrepreneur's service and that of other services.*

2. The independent variable is "Internet Provider," and there are two levels, "New Service" and "Old Service."

3. The dependent variable is the amount of SPAM received.

4. It looks as if the customers might have a valid complaint; they are receiving nearly 20 more junk e-mails per month with the new service than with the older service. The p value of .000 indicates this difference is very significant.

The Case of "We Can't Get No Satisfaction"

1. While the faculty members were fairly happy when the new dean was hired, apparently their satisfaction went down over time. That means they need to state a directional "less than" research hypothesis:

■ *Satisfaction after the dean has been employed for 3 months is significantly lower than when the dean was hired.*

The null hypothesis is:

■ *There is no significant difference in faculty satisfaction when the new dean was hired and after he had been there 3 months.*

2. We could call the independent variable "Satisfaction" with two levels, "Satisfaction Before" and "Satisfaction After."

3. The dependent variable is Faculty Satisfaction.

4. It's apparent that there is a large drop in satisfaction; the average of 60.13 before the new dean is dramatically higher than the mean of 42.27 after the new dean was hired. We can see from the two-tailed p value of .000 that this is a very significant difference.

Chapter 9 (pp. 284–289)

The Case of Degree Completion

1. The university is interested in determining whether age is a predictor of the time it takes to complete a doctoral degree; our research hypothesis is:

■ *There will be a significant difference in the time it takes to complete a doctoral degree between various age groups.*

The null hypothesis would be:

■ *There is no significant difference in the time it takes to complete a doctoral degree between various age groups.*

2. The independent variable is "Age Group."

3. The four levels are "20–29," "30–39," "40–49," and "50 and above."

4. The dependent variable is the number of years it takes to complete the degree.

5. The descriptive statistics show that the average number of years for degree completion are relatively close, ranging from 3.6 for students 40–49 to 4.8 for those 50 or older. The Levene test shows no problems with the homogeneity of variance, and the p value of .162 indicates this difference in mean scores is not significant. Because of that, the null hypothesis cannot be rejected, and the research hypothesis cannot be supported.

The Case of Seasonal Depression

1. The research hypothesis is very straightforward:

■ *There will be a significant difference in anxiety attacks during the four seasons of the year.*

This means our null hypothesis is:

- *There will be no significant difference in anxiety attacks during the four seasons of the year.*

2. The independent variable is "Season of the Year."

3. There are four levels: winter, spring, summer, and fall.

4. The dependent variable is the number of anxiety attacks.

5. The descriptive statistics show a much larger number of anxiety attacks in the winter than in the other seasons and there is no problem with the Levene test. The p value indicates that a significant difference does exist, and the Bonferroni post-hoc test supports our observation that the number of anxiety attacks in the winter is significantly higher than in the other three seasons.

The Case of Driving Away

1. Since the teachers are interested in determining if there is a difference in the number of absences between the three groups, the null hypothesis would read:

   - *There will be no significant difference in the number of absences between teens with cars, teens with access to cars, and teens without a car.*

   The research hypothesis would state:

   - *There will be no significant difference in the number of absences between teens with cars, teens with access to cars, and teens without a car.*

2. The independent variable could be something like "Transportation Status."

3. The levels are "owns car," "has access to car," and "doesn't have access to a car."

4. The dependent variable is Number of Absences.

5. According to the Levene statistics, there is no problem with the variance between the groups. It does appear, however, that the average number of absences varies from group to group. The low p value indicates there is a significant difference, and the post-hoc tests show that each group is significantly different from the other. Interestingly, the students who own their cars have the fewest absences; those with no access have the greatest number of absences. This makes sense—maybe they can't get to school!

The Case of Climbing

1. There are three null hypotheses:

   a. The interaction null hypothesis would read:

      - *There will be no significant difference in altitude climbed and the interaction between the route climbers take and whether or not they used a professional guide.*

b. The first main effect null hypothesis would read:

- *There will be no significant difference in the altitude climbed between climbers who used a professional guide and those who did not.*

c. The second main effect hypothesis would read:

- *There will be no significant difference in altitude climbed between the different routes climbers used.*

2. There are three research hypotheses:

a. The interaction research hypothesis would read:

- *There will be a significant difference in altitude climbed and the interaction between the route climbers take and whether or not they used a professional guide.*

b. The first main effect research hypothesis would read:

- *There will be a significant difference in the altitude climbed between climbers who used a professional guide and those who did not.*

c. The second main research hypothesis would read:

- *There will be a significant difference in altitude climbed between the different routes climbers used.*

3. For the independent variables we could label them "Guide Service Used" and "Route."

4. "Guide Service Used" has two levels: "yes" and "no." Route has four levels: "Ingraham Direct," "Gibraltar Rock," "Disappointment Cleaver," and "Nisqually Ice Cliff."

5. The dependent variable is the altitude the climbers reached.

6. There seem to be huge differences in some of the variables; we can see that climbers who went up the Nisqually Ice Cliff with a guide averaged over 11,000 feet, while those without a guide only managed about 8,000 feet. Before we make decisions, though, let's look further at our output. We can see the p value for the interaction between whether or not a guide was used and the route taken is insignificant ($p = .356$). Given that, we have to look at the main-effect hypotheses. We can see that the altitude differences between climbers who used a guide and those who didn't is insignificant ($p = .299$). There is also no significant difference in altitude gained between the four routes ($p = .754$). In short, climbers can sit around the fire and debate all night, but it really doesn't make a difference.

Chapter 10 (pp. 316–319)

The Case of Prerequisites and Performance

1. Since the professor is interested in determining the relationship between major and grade, he would be investigating this null hypothesis:

> ▪ *College major has no significant effect on grade earned in a graduate psychology class.*

The research hypothesis would read:

> ▪ *College major has a significant effect on the grade earned in a graduate psychology class.*

2. The independent variables are grades, "A" through "F," and whether or not the student's undergraduate degree was in psychology, "Yes" or "No."

3. The dependent variable is the frequency of each grade.

4. Since there are two independent variables and the professor wants to see if an interaction exists between the undergraduate degree and letter grade, he would use the chi-square test of independence.

5. Because of the *p* value of .027, he would reject the null hypothesis. While there isn't a post-hoc test, it appears that those students with undergraduate degrees in psychology are far more likely to make an A grade than their classmates with other majors. At the same time, if you're willing to settle for a B grade or less, it appears it is best not to have an undergraduate degree in psychology!

The Case of Getting What You Asked For

1. The null hypothesis is:

> ▪ *There will be no significant difference in the actual number of each grade assigned and the expected number of each grade.*

The research hypothesis is:

> ▪ *There will be a significant difference in the number of each grade assigned and the expected number of each grade.*

2. The independent variable is grade, and there are three levels, "A," "B," and "C."

3. The dependent variable is the frequency of each grade.

4. Because the professor is comparing the observed number of each grade to the expected number of each grade, he will use the chi-square goodness of fit.

5. While there are fewer "A" grades and more "C" grades than expected, the *p* value of .07 shows there is not a significant difference in the number of each grade that was expected and the number of each grade that was actually received.

The Case of Money Meaning Nothing

1. In this case, we are interested in looking at the relationship between drug use and socioeconomic status. Our null hypothesis would be:

> ▪ *Socioeconomic status has no significant effect on whether or not a person uses drugs.*

Our research hypothesis would be:

- *Socioeconomic status has a significant effect on whether or not a person uses drugs.*

2. The first independent variable is "Socioeconomic" status, and there are four levels: "less than $20,000," "$20,000 to $39,999," "$40,000 to $59,999," "$60,000 to $79,999," and "$80,000 and above." The second independent variable is "Drug Use," and the two levels are "Yes" and "No."

3. The dependent variable is the number of each occurrence within the cells (e.g., the number of drug users who make below $20,000).

4. Since there are two independent variables and we want to see if an interaction exists between socioeconomic status and drug use, we would use the chi-square test of independence.

5. In this case, we would fail to reject the null hypothesis given the *p* value of .421. Simply put, it doesn't make any difference how much money someone makes when it comes to whether or not they use drugs.

The Case of Equal Opportunity

1. The null hypothesis is:

- *There will be no significant difference in the actual number of each gender and the expected number of each gender.*

The research hypothesis would state that a significant difference exists.

2. The independent variable is "Gender," and the two levels are "Male" and "Female."

3. The dependent variable is the actual count of males and females.

4. Because the manager is comparing the number of each gender to the expected number of each gender, he will use the chi-square goodness of fit.

5. The *p* value of .421 indicates that the number of males and females in the company is not significantly different from the national average.

Chapter 11 (pp. 355–358)

The Case of "Like Father, Like Son"

1. Pearson's *r* says there is a strong relationship (i.e., *r* = .790) between a father's age at death and the son's age at death.

2. The coefficient of determination is .6241. This tells us that about 62.41% of change in the criterion variable (i.e., the son's age) is due to the predictor variable (i.e., the father's age).

3. Given a father's age at death of 65, the son would be expected to live about 73 years.

4. The line-of-best-fit flow would move from the bottom left to the upper right at slightly less than a 45-degree angle.

The Case of "Can't We All Just Get Along?"

1. We would use Spearman's rho because we are dealing with rank (i.e., ordinal) data.

2. The rho value of −.515 means there is a moderate negative relationship between the Democrats' rankings and the Republicans' rankings.

3. I would tell the president that I was surprised the correlation wasn't closer to −1.00! After all, aren't these two parties supposed to vote exactly the opposite of one another?

The Case of More Is Better

Following is the output of a regression procedure using data showing highest level of education attained and average salary in the workplace:

1. I would tell a group of potential high school dropouts that there is a fairly strong relationship (i.e., $r = .437$) between the number of years they go to school and how much money they can expect to make.

2. I would tell a student that the average income is nearly $15,000 annually, while a ninth-grade dropout can only expect to make about $11,600.

3. The coefficient of determination is .1909. This tells us that about 19.09% of change in the criterion variable (i.e., salary) is due to the predictor variable (i.e., the highest grade completed). While the number of years of education is a predictor, over 80% of the time, other factors are involved.

4. The line of best fit would rise, from left to right, at about a 25- to 30-degree angle.

The Case of More Is Better Still

Suppose you followed up the prior case by telling the potential dropouts that students who drop out of school earlier tend to get married earlier and have more children.

1. In this case, Pearson's r of −.259 shows a small negative correlation; people with more education tend to have a slightly smaller number of children.

2. The coefficient of determination is .0670. This tells us that about 6.7% of change in the criterion variable (i.e., number of children) is due to the predictor variable (i.e., the highest grade completed). While the number of years of education is a slight predictor, other factors are much more likely to determine the number of children.

3. I would tell the potential dropouts that students who drop out after the eighth grade have, on average, more children than the overall average.

4. The line of best fit would drop slightly from the left upper corner down to the right.

5. I would point out to the students considering leaving school that the combination of lower wages and more children makes for a difficult life. Stay in school!

Index

About the Author

Steven R. Terrell, PhD, is Professor in the Graduate School of Computer and Information Sciences at Nova Southeastern University, where he teaches quantitative and qualitative research methods. He is the author of over 100 journal articles and presentations focusing primarily on student motivation and attrition at all levels of education. Dr. Terrell is active in the American Psychological Association and the American Counseling Association, and served as president of the American Educational Research Association's Online Teaching and Learning Special Interest Group.